# Transmission Line
# Design Handbook

# Transmission Line Design Handbook

## Brian C. Wadell

Teradyne, Inc.
Boston, Massachusetts

Artech House
Boston • London

Library of Congress Cataloging-in-Publication Data
Wadell, Brian C.
  Transmission line design handbook/ Brian C. Wadell
     p.   cm.
  Includes bibliographical references and index.
  ISBN 0-89006-436-9
     1. Strip transmission lines.  2. Microwave transmission lines.
I. Title.
TK7876.W29  1991                                     91-13328
621.381'31—dc20                                      CIP


British Library Cataloguing in Publication Data
Wadell, Brian C.
  Transmission line design handbook.
  1.Transmission lines.
  1.Title
  621.3192

  ISBN 0-89006-436-9




© 1991 ARTECH HOUSE, INC.
685 Canton Street
Norwood, MA 02062

International Standard Book Number: 0-89006-436-9
Cataloging-In-Publication: 91-13328

20 19 18 17 16 15 14 13

To all my teachers—especially my father and my beautiful wife, Andrea.

# *Contents*

# *Preface*

This book is a reference to designing transmission lines and designing with transmission lines. It has evolved from my interest in and desire to understand the design process and the many exceptions to the standard design procedures. As I explored this, I found that the standard equations were not valid for the standard configurations. For example, it is very common to have embedded microstripline and off-center stripline, yet the commonly used equations have significant errors for these configurations. What some may consider a mathematical exception, in fact is common and introduces significant errors in the design.

I found that the necessary references are spread out over dozens of texts and journals over seventy years. As a result, many engineers are using equations that are nearest at hand rather than the most correct available for their purpose. My goal became to find the most accurate and authoritative articles and extract the equations one needs for design. This seemed simple; however, it became a huge task of collecting references, analyzing them, and attempting calculations to verify their completeness and accuracy. The result, I think, is one of the most complete and easy-to-use references available.

This handbook compiles these equations and references in a uniform format for easy reference. It is meant to be a guide to making the calculations involved in high-frequency design and as such does not directly cover any EM field theory. There are several good EM texts listed in the references.

The equations collected herein are so-called closed-form equations. Many of these are exact solutions of the problem or of transformed versions of the problem. A few are the result of curve-fitting exact numerical solutions of Maxwell's equations. Where only curves are found in the references, CAE equations have been fit. These equations apply only in the ranges specified and are supplied for convenience in computer designs.

I have included material for the design of certain inductors and capacitors, which are not normally considered transmission lines. This is important material to have available when designing in the 200 to 500 MHz region where transmission line lengths may be unwieldy and off-the-shelf lumped elements self-resonate.

In most cases, the major source of error or uncertainty in designing with transmission lines is the fabrication process. I have gathered tables of the available dielectrics and metals together with the expected tolerances. These can be used together with the closed-form equations to predict for yield.

Finally, I have included the special functions required for calculation of transmission lines. In the past, transmission line equations were often approximated for simple hand calculation or reduced to graphs, tables, and nomographs for common cases. However, with modern computers and calculators readily available to every engineer there is no reason to use anything other than the most accurate equations. Subroutines for complicated functions are relatively straightforward to write and libraries of these functions are readily available on many computers. Because of this, my philosophy has been to present the exact equations wherever available. The computer itself takes the place of the graphs, tables, and nomographs formerly required. This allows a great number of equations to be solved quickly to evaluate design trade-offs. The computer also does the calculation to the source accuracy without introducing any additional uncertainty due to interpolation of the table or reading of graphs. For many of the structures, the significant errors remaining will be attributable to fabrication process variations.

If you wish to write your own programs, you are now free to make any necessary trade-offs in accuracy and speed as suits your requirements and that of your computing platform. If your interest is solely in using the equations for design work, software is available to solve these equations.

As with any significant task, this book is the product of a number of people's contributions. I would like to acknowledge the assistance and encouragement of my friends and colleagues at Teradyne. In particular, Bobbi Bernard was instrumental in tracking down the references used in early versions of this book and Dan Rosenthal reviewed the draft. The anonymous reviewer and the editors and staff of Artech House, especially Mark Walsh, provided timely aid and encouragement. I would also like to thank the libraries of the Massachusetts Institute of Technology and Cornell University. My wife, Andrea C. Wadell, wrote software to check and solve the equations found in this book and to generate some of the more difficult figures. She also provided a great deal of moral support, which I greatly appreciate. Most of all, I would like to thank all of the authors referenced herein for the pleasure I have had in reading, studying, and using their articles and books.

Brian C. Wadell
Reading, MA
March 25, 1990

# *Chapter 1*

# *Introduction*

## 1.1   NOTATION USED IN THIS BOOK

An attempt has been made to be consistent with the notation throughout this book. This was done whenever multiple notations existed in the literature for consistency, clarity, and ease of comparison. In the process of doing this, the notation herein may be different than that of a referenced paper. This may seem a bit nitpicky to some, but anyone who has tried to compare several equations where the authors arbitrarily interchanged $w$ and $h$, or pulled $k$ out of the air, or referred to the relative dielectric constant as the dielectric constant (two completely different numbers!) will understand. The frequent figures are clearly labeled with these variables. In general, I have adopted the following conventions.

All dimensions are measured as labeled. Wherever possible these dimensions are specified as they would be measured on a given physical feature rather than a calculation from other dimensions or features. Most equations will work with any set of units (as long as all dimensions are the same unit). These are recognizable by the fact that they are expressed as ratios of dimensions. Where this is not true, I explicitly state the dimensions used and try to stick to SI units, which are used as summarized in Table 1.1.1 below.

| MKS Units | | | |
|---|---|---|---|
| Measure | Unit | Unit Symbol | Variables |
| length | meter | m | $l, d, D, w, t, h, b,$ etc. |
| weight | kilogram | kg | |
| time | second | s | |
| current | amp | A | $I, i$ |
| temperature | degrees Kelvin | °K | |
| frequency | hertz | Hz | $f$ |
| power | watt | W | |
| charge | coulomb | C | $q$ |
| resistance | ohm | $\Omega$ | $R$ |
| conductance | mho, siemen | $1/\Omega$, S | $G$ |
| resistivity | ohm-meter | $\Omega$-m | $\rho$ |
| conductance | mho-meter, Siemen-meter | m/$\Omega$, S-m | $\sigma$ |
| voltage | volt | V | $V$ |
| capacitance | farad | F | $C$ |
| inductance | henry | H | $L$ |
| permittivity | farad / m | F/m | $\epsilon$ |
| permeability | henry / meter | H/m | $\mu$ |

**Table 1.1.1:   MKS Units**

The three-dimensional physical structure of the lines is represented as two-dimensional views either in the x-z plane as shown in Figure 1.1.1, or in the x-y plane as shown in Figure 1.1.2. The x-y view is used mainly for planar discontinuities. Top-view figures may have equations that implicitly reference the cross-sectional figure's dimensions. For example, a y-direction length dimension, $l$, is assumed for every physical transmission line structure.

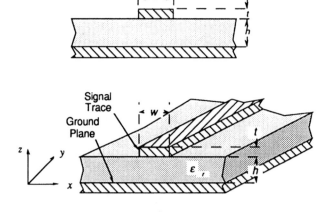

**Figure 1.1.1: Cross-Section Figure Notation**

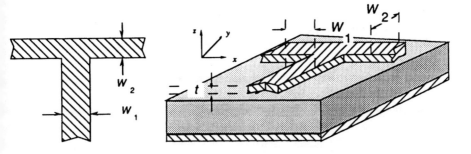

**Figure 1.1.2: Top-View Figure Notation**

The **dielectric constant** is $\varepsilon = \varepsilon_0 \varepsilon_r$. The **relative dielectric constant** of the material is $\varepsilon_r$. The relative dielectric constant ($\varepsilon_r$) of air[1] is 1; the dielectric constant of air is $\varepsilon_0$. Dielectrics are shown with a shaded (dotted) pattern. If several dielectrics are involved, each is shown with a different shading density. Air dielectric is shown without any pattern (white) and is labeled $\varepsilon_0$ or $\varepsilon_{air}$ (as a reminder). The effective relative dielectric constant, $\varepsilon_{eff}$, is used to represent the effect of a mixed dielectric for quasi-TEM lines. Wherever possible $\varepsilon_r$ and $\varepsilon_{eff}$ are explicitly shown in the equations (some authors imply a $1 / \sqrt{\varepsilon_r}$ term in their equations for $Z_0$).

Dielectric thicknesses are $h$ and $b$. The thickness of the microstrip line dielectric is $h$, and $b$ is used for striplines. A single variable $h$ could have been used for both, however, $b$ is commonly used for striplines. For circular and square dielectrics, the corresponding dimension is $D$.

Center or signal conductor width and thickness are $w$ and $t$, respectively. Conductors are drawn with a diagonal hatching pattern. For circular and square conductors, the corresponding dimension is $d$.

Coupled lines are separated by a spacing $s$. This is used strictly for distances between nongrounded conductors. A confusing situation occurs in the references for coplanar waveguide and similar structures where the width of the trace and its distance from the adjacent ground plane are both specified. In terms of fabrication, the relevant dimensions are the trace width and the spacing to the ground plane, however many references specify the trace width and the distance from the ground on one side of the trace to the ground on the other side of the trace. Some may even specify the centerline distances. In this book the latter $a$, $b$ notation (e.g., Figure 3.4.1.1) is used, although I reserve the right to come up with a more convenient notation in the future.

The odd mode characteristic impedance is $Z_{0,o}$. That is, $Z_0$ (read "zee zero") subscript $o$ for odd. Similarly $Z_{0,e}$ ("zee zero even") is the even mode characteristic

---

[1]When an air dielectric is present in a structure and you wish to calculate the impedance for a similar structure with another dielectric ($\varepsilon_{r,2}$) replacing the air, make the following equation substitutions: replace $\varepsilon_r$ with $\varepsilon_r / \varepsilon_{r,2}$; replace $\eta_0$ with $\eta_0 / \varepsilon_{r,2}$.

impedance. Admittances are $Y$s with the same conventions. In the past (partly due to lack of computer-aided typesetting), some authors got sloppy and omitted, substituted, or interchanged zeroes and $O$s. (It really wouldn't matter if only it were consistent and clear.)

Coupling, $k$, is a unitless ratio. When coupling in dB is intended, $k_{dB}$ is used for clarity. The dielectric constant will never be $k$.

The wavelength is represented by $\lambda$. The symbol $\lambda$ should <u>only</u> be used for wavelength. It is confusing to use it as a function parameter, particularly if it should end up inside a hyperbolic function. There are usually two wavelengths involved; $\lambda_0$ is the wavelength without a dielectric (free-space wavelength) and $\lambda_g$ is the guide wavelength (wavelength of the actual structure). We will avoid omitting the subscript unless the context is clear.

We have tried to use natural logarithms (logarithm base e) throughout for ease of comparison. Many equations appear in the references with the base 10 logarithms. The conversion to natural logarithms is quickly accomplished via:

$$\frac{\ln_e x}{\log_{10} x} = \ln_e 10 = 1 / \log_{10} e = 2.3026\ldots \tag{1.1}$$

The logarithms are explicitly subscripted here for clarity, but this will not be done anywhere else.

Losses that are expressed in nepers can be converted to dB with the following relations:

$$\text{Loss} = \ln x \quad \text{(nepers)} \tag{1.2}$$

$$\text{Loss} = 20 \log_{10} x \quad \text{(dB)} = \frac{20 \ln x}{\log 10} = 8.6858 \ln x = 8.6858 \times \text{nepers} \tag{1.3}$$

For other mathematical functions the notation conventions of the *Handbook of Mathematical Functions* by Abramowitz and Stegun [1] are followed. This is most convenient because it is an excellent all-in-one reference containing definitions, equations, tables of values, graphs, and approximation formulas for virtually every special function we could encounter.

The notation commonly used for expressing impedance or admittance is:

$$Z = R + j\,X \quad \Leftrightarrow \quad Impedance = Resistance + j\,Reactance \tag{1.4}$$

$$Y = G + j\,B \quad \Leftrightarrow \quad Admittance = Conductance + j\,Susceptance \tag{1.5}$$

The impedance (admittance) is often normalized (divided by) a reference impedance (admittance) which is denoted $Z_0$ ($Y_0$). Impedance is expressed in Ohms ($\Omega$), and Admittance (1 / Impedance) is expressed in Siemens (S) or mhos. When susceptances or reactances are given in a reference they can be converted to inductance and capacitance with

$$L = \frac{-1.0}{\omega \, B_L} \qquad (1.6)$$

$$L = \frac{X_L}{\omega} \qquad (1.7)$$

$$C = \frac{B_C}{\omega} \qquad (1.8)$$

$$C = \frac{-1.0}{X_C \, \omega} \qquad (1.9)$$

## 1.2 HOW TO USE THIS BOOK

As a guide to using this book, I would just like to outline briefly the expected design procedure depicted in Figure 1.2.1.

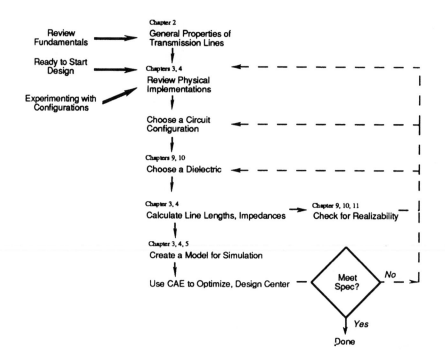

**Figure  1.2.1  Design  Flowchart**

Generally, the designer starts with a general requirement such as "design a 15 dB coupler at 3 GHz." Using the equations of Chapter 2, the design is converted to a pair of transmission lines with specified impedances and electrical lengths. We then look at *Chapter 4, Coupled Lines,* to get a feel for the physical implementations possible. Choose a substrate or dielectric (*Chapter 9, 10*) and try a few designs with the programs. Decide if the specs can be met with simple circuits or whether a cascaded or compensated design is required.

With this information, choose the type of line—microstrip line, stripline, coplanar waveguide, or other from *Chapter 4.* Synthesize the physical lines with equations from *Chapter 4.* Check for realizability of these dimensions using *Chapter 9, 10, 11* and iterate as necessary until the design is achievable. You may have to change to a different type of coupler or to a different physical realization.

Now using the dimensions and the physical layout add the discontinuities (*Chapter 5*) to your model. Enter the model into an analysis program (*e.g.,* Touchstone ™ SuperCompact ™,[2] etc.). Use the program's optimization and yield analysis features to correct for the effect of the strays and center the design. Design centering assures

---

[2]Touchstone is a registered trademark of EEsof, Inc.  SuperCompact is a registered trademark of Compact Software, Inc.

you that the values in use are the best for the expected parameter variations. Now you are ready to cut metal! At this point we may have designed dozens of transmission lines and tested them—all in software.

I have also encountered a second type of design process. In this case, the designer is familiar with the capability and limitations of a particular structure and is beginning a new design. Due to mechanical constraints of the system it is desired to make modifications to a standard structure. Electrically the requirements are relatively straightforward. This text and a transmission line program will allow the various alternatives to be tried without the delays of multiple prototypes.

Most designs will proceed in a similar fashion from spec to design, to physical realization, to layout with discontinuities, and finally to optimization and design centering. The use of computer tools allows this process to be fast and accurate and gives the designer the freedom to play "what if" with the goal of improving the design's reproducibility.

## REFERENCES

[1]    Abramowitz, Milton, and Irene A. Stegun, *ed.*, *Handbook of Mathematical Functions with Formulas, Graphs, and Mathematical Tables,* Dover Publications, 1972.

[2]    ANSI/IEEE Std 1004-1987, "IEEE Standard Definitions of Planar Transmission Lines," September 8, 1988.

# *Chapter 2*

# *Generalized Transmission*

# *Lines*

There are only a few properties necessary to characterize completely the behavior of transmission lines for RF and microwave designs. When these are understood we will have an intuitive feel for their use in transmission line or *distributed element* circuits. The goal of this chapter is to develop this base.

## 2.1 DISTRIBUTED ELEMENT LINES

A transmission line circuit can be modeled in different ways. One particularly useful way is to classify a transmission line by the number of orthogonal physical dimensions, which are large relative to a wavelength in the transmission line (See Figures 1.1.1 and 1.1.2). Thus, a coaxial line might be considered one dimensional (y, its length), a right-angle bend in stripline might be considered two dimensional (x-y), and a transition from a coaxial line to a microstrip line would be considered three dimensional. A zero-dimensional transmission line is a lumped element. Remember that this classification is based on both the wavelength of the propagating signal and the discontinuity size. Depending on the frequency range and the accuracy desired, the choice of model may be made differently in seemingly similar circumstances.

No matter how many dimensions are actually significant, we will almost always try to reduce the problem to one of a circuit made up of RLC networks, ideal transformers, and ideal transmission lines. This type of calculation is usually much faster than the available field-solution techniques and it provides us with models in terms of more familiar concepts. All of the models presented in this book were developed from solutions of Maxwell's equations by various analytic tools. Unfortunately, field-solving programs can require hours to calculate results for even simple structures and have other limitations that restrict their use to special two- and three-dimensional problems. Field solvers are also not of much use when synthesizing a design. Fortunately, closed-form equations have been developed by researchers from direct solutions and EM programs. Using these we have a highly accurate set of easily used models available for many designs. Certainly as hardware and software continues to become more powerful we will be hearing more about field-solving programs in the future.

A one-dimensional transmission line is modeled as a lumped element circuit with a very large number of elements as shown in Figure 2.1.1. The total line length, $l$, has been divided into an infinite number of sections of length $\Delta y$.

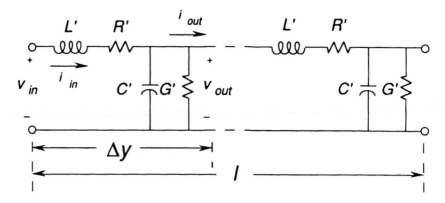

**Figure 2.1.1: Transmission Line Distributed Parameters**

Each lumped element represents an infinitesimal piece of the physical transmission line. The values $L'$, $C'$, $R'$, and $G'$ are the inductance per length, capacitance per length, resistance per length, and conductance per length, respectively. These are the line's distributed parameters. For example, in a coaxial wire, $L'$ represents the series inductance of the center conductor, $C'$ represents the center conductor's capacitance to the shield, $R'$ represents the resistance of the center conductor and shield, and $G'$ represents the dielectric's leakage resistance. Table 2.1.1 below summarizes many of the contributions to the model parameters:

| Model Parameter | Origin | Comment |
|---|---|---|
| R | Conductance | atomic property of metal used |
| | Geometry | larger cross section = lower R |
| | | longer = higher R |
| | Radiation | fields not contained are losses |
| | Skin Effect | current flows in smaller cross-section as frequency increases |
| | Proximity Affect | adjacent conductor's field forces current into smaller cross-section |
| L | Permeability | property of metal used |
| | Geometry | larger cross section = lower L |
| | | longer = higher L |
| | Skin Effect | currents at different depths in conductor have different phases, net effect is inductive |
| | Proximity | adjacent conductor's field forces current into smaller cross-section |
| | Radiation | |
| C | Geometry | larger area = higher C |
| | | closer conductors = higher C |
| | Permittivity | higher $\varepsilon_r$ = higher C |
| G | Loss Tangent | higher loss tan = higher losses in dielectric |
| | Conductance | semiconductor dielelectrics |
| | Geometry | |

Table 2.1.1: Transmission Line Model Parameters and Sources

## 2.2 WAVE PROPAGATION AND CHARACTERISTIC IMPEDANCE

In this section we show the solution of wave propagation in the distributed element line. It is very informative to understand how the simple model in Figure 2.1.1 creates the wave propagation equations and a definition of characteristic impedance; however, one can successfully undertake designs and make use of most of this book even if this derivation is skipped.

The line we will analyze is shown in Figure 2.1.1. The length of the line is the $y$ dimension. By inspection, and replacing differences with differentials (by taking the limit as $\Delta y$ approaches zero) we can write equations for the signal on our infinitesimal section of line as:

$$\frac{\partial i}{\partial y} = -v \, G' - \frac{\partial v_{out}}{\partial t} C' \qquad (2.2.1)$$

$$\frac{\partial v}{\partial y} = -i\,R' - L'\,\frac{\partial i_{in}}{\partial t}$$

(2.2.2)

That is, each section $\partial y$ of the line creates the changes in the voltage and current described above. Equations (2.2.1) and (2.2.2) are referred to as the *Telegrapher's equations*.

If we differentiate each of these equations with respect to $t$ and $y$, we obtain:

$$\frac{\partial^2 v}{\partial y^2} = -R'\,\frac{\partial i}{\partial y} - L'\,\frac{\partial^2 i}{\partial y\,\partial t}$$

(2.2.3)

$$\frac{\partial^2 v}{\partial y\,\partial t} = -R'\,\frac{\partial i}{\partial t} - L'\,\frac{\partial^2 i}{\partial t^2}$$

(2.2.4)

$$\frac{\partial^2 i}{\partial y\,\partial t} = -G'\,\frac{\partial v}{\partial t} - C'\,\frac{\partial^2 v}{\partial t^2}$$

(2.2.5)

$$\frac{\partial^2 i}{\partial y^2} = -G'\,\frac{\partial v}{\partial t} - C'\,\frac{\partial^2 v}{\partial t\,\partial y}$$

(2.2.6)

We can now combine equations to eliminate variables. Substituting (2.2.1) and (2.2.5) into (2.2.3) and (2.2.2), and (2.2.4) into (2.2.6):

$$\frac{\partial^2 v}{\partial y^2} = -R'\left(-v\,G' - \frac{\partial v}{\partial t}\,C'\right) - L'\left(-G'\,\frac{\partial v}{\partial t} - C'\,\frac{\partial^2 v}{\partial t^2}\right)$$

$$= v\,(R'\,G') + \frac{\partial v}{\partial t}\,(R'\,C' + L'\,G') + \frac{\partial^2 v}{\partial t^2}\,(L'\,C')$$

(2.2.7)

$$\frac{\partial^2 i}{\partial y^2} = -G'\left(-i\,R' - L'\,\frac{\partial i}{\partial t}\right) - C'\left(-R'\,\frac{\partial i}{\partial t} - L'\,\frac{\partial^2 i}{\partial t^2}\right)$$

$$= i\,(G'\,R') + \frac{\partial i}{\partial t}\,(L'\,G' + C'\,R') + \frac{\partial^2 i}{\partial t^2}\,(L'\,C\ )$$

(2.2.8)

Now we have two differential equations that describe our transmission line, but can we gain understanding of the wave's propagation from it? (Much of the following discussion will be for $v$, but the reasoning for $i$ is parallel.) Examining (2.2.7) we see an equation for a voltage that is a function of its time derivatives. Any function $v$ that is a solution of (2.2.7) describes a possible method of propagation on our line. There may be many such equations, so how can we proceed? An equation that is a solution and is fortunately quite general is the complex exponential function, which has derivatives [note the similarities to (2.2.7) and (2.2.8)]:

$$\frac{\partial\ e^u}{\partial x} = e^u \frac{\partial u}{\partial x} \qquad (2.2.9)$$

$$\frac{\partial^2\ e^u}{\partial x^2} = e^u \frac{\partial^2 u}{\partial x^2} + e^u \left(\frac{\partial\ u}{\partial x}\right)^2 = e^u \left[\frac{\partial^2 u}{\partial x^2} + \left(\frac{\partial\ u}{\partial x}\right)^2\right] \qquad (2.2.10)$$

We also have one other problem: $v$ is a function of $y$ and $t$. To solve this, we make the asumption that we can separate the variables in our solution. This means that we can write our function $v$ as the product of two functions $v'$ and $v''$. Similarly, we rewrite $i$ and obtain

$$v(y, t\ ) = v'(y)\ v''(t) \qquad (2.2.11)$$

$$i(y, t) = i'(y)\ i''(t) \qquad (2.2.12)$$

Now we are ready to solve (2.2.7) and (2.2.8). We choose

$$v(y, t\ ) = v'(y)\ e^{j\,\omega\,t} \qquad (2.2.13)$$

$$\frac{\partial v}{\partial t} = j\omega\ v(y, t) \qquad (2.2.14)$$

$$\frac{\partial v^2}{\partial t^2} = -\omega^2\ v(y, t) \qquad (2.2.15)$$

$$i(y, t\ ) = i'(y)\ e^{j\,\omega\,t} \qquad (2.2.16)$$

$$\frac{\partial i}{\partial t} = i(y,t)\ j\ \omega \qquad (2.2.17)$$

$$\frac{\partial^2 i}{\partial t^2} = -\omega^2 i(y,t) \qquad (2.2.18)$$

Rewriting (2.2.1) and (2.2.2) with (2.2.13) and (2.2.14)

$$\frac{\partial i}{\partial y} = v(y, t)\ \left(G' + j\ \omega\ C\ '\right) \qquad (2.2.19)$$

$$\frac{\partial v}{\partial y} = -i(y, t)\left(R' + j\,\omega\,L'\right) \tag{2.2.20}$$

and substituting into (2.2.7)

$$\frac{\partial^2 v}{\partial y^2} = v(R'G') + j\,\omega\,v(y, t)(R'C' + L'G') + -\omega^2 v(y, t)(L'C')$$

$$= v(y, t)\left[R'G' + j\,\omega\,(R'C' + L'G') - \omega^2 L'C'\right]$$

$$= v(y, t)\left[(R + j\,\omega\,L)(G + j\omega C)\right] \tag{2.2.21}$$

Again we recognize that this is an equation whose solution is of the exponential form:

$$v(y) = K_0\,e^{-\gamma y} + K_1\,e^{+\gamma y} \tag{2.2.22}$$

$$\gamma = \sqrt{\left[(R' + j\,\omega\,L')(G' + j\,\omega\,C')\right]} \tag{2.2.23}$$

It is worth examining Equation (2.2.22) further. First we see that there are two pieces of the solution for generality. Both need not exist, however, they may. Each term represents a sinusoidal wave propagating in the $y$ or $-y$ direction. This means that our lumped-element line will support propagation in either direction alone or both directions simultaneously. Each wave may have its own amplitude. This means that the voltage on the line at location $y$ will be the sum of two sine waves having the same period in distance.

We can substitute (2.2.22) and (2.2.23) into our rewritten (2.2.2)

$$-\gamma\,K_0\,e^{-\gamma y} + \gamma\,K_1\,e^{+\gamma y} = -i(y, t)\left(R' + j\,\omega\,L'\right)$$

$$i(y, t) = \frac{-\gamma\,K_0\,e^{-\gamma y} + \gamma\,K_1\,e^{+\gamma y}}{\left(R' + j\,\omega\,L'\right)} \tag{2.2.24}$$

The ratio of $v$ to $i$ or the impedance at a point $y$ on the line is

$$\frac{v(y, t)}{i(y, t)} = Z_0 = \sqrt{\frac{R' + j\,\omega\,L'}{G' + j\,\omega\,C'}} \tag{2.2.25}$$

This is known as the ***characteristic impedance*** of the transmission line. Again, notice that for our line this is not a function of the propagating waveform, the position on the line, or time. For this uniform line, it is a function of the model parameters only. In general, these parameters may be a function of $y$ which would make $Z_0$ a function of the position in the line. If $R' = G' = 0$, a lossless line, $Z_0$ is not a function of frequency either.

$$Z_0 = \sqrt{\frac{L'}{C'}} \qquad (2.2.26)$$

In the above equations we defined a parameter $\gamma$. This is the *propagation constant*

$$\gamma = \alpha + j\beta = \sqrt{j\,\omega\,\mu\,(\sigma + j\omega\varepsilon)} \qquad (2.2.27)$$

which is made up of the *attenuation constant*, $\alpha$,

$$\alpha = \omega \sqrt{\frac{\mu\varepsilon}{2} \left( \sqrt{1.0 + \frac{\sigma^2}{\omega^2\varepsilon^2}} - 1.0 \right)} \quad (\text{Np}/\text{m}) \qquad (2.2.28)$$

and the *phase constant*, $\beta$

$$\beta = \omega \sqrt{\frac{\mu\varepsilon}{2} \left( \sqrt{1.0 + \frac{\sigma^2}{\omega^2\varepsilon^2}} + 1.0 \right)} \quad (\text{rad}/\text{m}) \qquad (2.2.29)$$

The phase constant is sometimes written as $k$. We will avoid this usage to prevent confusion with a number of other $k$'s that occur in transmission line problems, *i.e.*, the modulus of the complete elliptic integral, the dielelectric constant, and coupling. The attenuation constant can be written in dB / m by multiplying by the factor, $20 / \ln 10$.

## 2.3   MAXWELL'S EQUATIONS

Although we have described our transmission line as a lumped equivalent, this solution happily corresponds to the solution of Maxwell's equations. We will not discuss Maxwell's equations in depth, however, for completeness, in differential form they are:

$$\nabla \times \mathbf{E} = -\frac{\partial \mathbf{B}}{\partial t} \qquad \mathbf{D} = \varepsilon\,\mathbf{E}$$

$$\nabla \times \mathbf{H} = \mathbf{J} + \frac{\partial \mathbf{D}}{\partial t} \qquad \mathbf{B} = \mu\,\mathbf{H}$$

$$\nabla \cdot \mathbf{D} = q_e \qquad \mathbf{J} = \sigma\,\mathbf{E}$$

$$\nabla \cdot \mathbf{B} = q_m$$

$$(2.3.1)$$

where

$$\nabla \times \mathbf{V} = \left(\frac{\partial V_z}{\partial y} - \frac{\partial V_y}{\partial z}\right)\overline{x} + \left(\frac{\partial V_x}{\partial z} - \frac{\partial V_z}{\partial x}\right)\overline{y} + \left(\frac{\partial V_y}{\partial x} - \frac{\partial V_x}{\partial y}\right)\overline{z} = \text{curl } \mathbf{V}$$

$$= \begin{bmatrix} \dfrac{\partial}{\partial x} & \dfrac{\partial}{\partial y} & \dfrac{\partial}{\partial z} \\[2mm] V_x & V_y & V_z \\[2mm] \overline{x} & \overline{y} & \overline{z} \end{bmatrix} \tag{2.3.2}$$

$$\mathbf{u} = \begin{bmatrix} \overline{x} \\ \overline{y} \\ \overline{z} \end{bmatrix} = \text{unit vector} \tag{2.3.3}$$

where **E** and **H** are the electric and magnetic field intensities, and **D** and **B** are the electric and magnetic flux densities. The current density is **J**, and $q_e$ and $q_m$ are the electric and magnetic charge densities. The magnetic current density is **M**. All bold-faced variables are functions of $x$, $y$, $z$, and $t$.

In solving Maxwell's equations for a lossless TEM line and equating them to the distributed parameter solution, we find other relations in terms of the parameters of the dielectric and the conductors. The characteristic impedance is

$$Z_0 = \frac{\sqrt{\mu_0 \, \varepsilon_0 \, \varepsilon_r}}{C} \ (\Omega) \tag{2.3.4}$$

$$Z_0 = \frac{L}{\sqrt{\mu_0 \, \varepsilon_0 \, \varepsilon_r}} \ (\Omega) \tag{2.3.5}$$

By combining (2.3.4) and (2.3.5) we define the *wave impedance*:

$$\eta = \sqrt{\frac{\mu}{\varepsilon}} \ (\Omega) \tag{2.3.6}$$

The *characteristic impedance of free space* is:

$$\eta_0 \equiv \sqrt{\frac{\mu_0}{\varepsilon_0}} = 120 \, \pi \ (\Omega) \cong 377 \ (\Omega) \tag{2.3.7}$$

The *speed of light* is the propagation velocity or phase velocity in free space (a vacuum) and is defined as:

$$c = \frac{1.0}{\sqrt{\mu_0 \varepsilon_0}} = 2.997925 \times 10^8 \quad (m / s) \tag{2.3.8}$$

From this, we define the *phase velocity* or the velocity of light in the medium:

$$v_p = \frac{1.0}{\sqrt{\mu \ \varepsilon}} = \frac{1.0}{\sqrt{\mu_0 \ \mu_r \ \varepsilon_0 \ \varepsilon_r}} = \frac{c}{\sqrt{\mu_r \ \varepsilon_r}} \tag{2.3.9}$$

In most materials $\mu_r = 1$ (nonmagnetic), so we often write (2.3.9) as:

$$v_p = \frac{c}{\sqrt{\varepsilon_r}} \tag{2.3.10}$$

The phase velocity can be used to calculate the wavelength in the medium, also known as the *guide wavelength*:

$$\lambda_g = \frac{c}{f \sqrt{\mu_r \ \varepsilon_r}} = \frac{\lambda_0}{\sqrt{\mu_r \ \varepsilon_r}} \tag{2.3.11}$$

where $f$ is the frequency of the wave and $\lambda_0$ is the wavelength in free space. Again for nonmagnetic media we can simplify:

$$\lambda_g = \frac{c}{f \sqrt{\varepsilon_r}} \tag{2.3.12}$$

and in free space:

$$\lambda_0 = \frac{c}{f} \tag{2.3.13}$$

For a nonhomogenous, nonmagnetic dielectric,

$$\lambda_g = \frac{c}{f \sqrt{\varepsilon_{eff}}} = \frac{\lambda_0}{\sqrt{\varepsilon_{eff}}} \tag{2.3.14}$$

The *electrical length* $\theta$ of a transmission line having physical length $l$ at frequency $f$ is

$$\theta = 2.0 \ \pi \frac{l}{\lambda_g} \quad (radians) \ = 360.0 \ \frac{l}{\lambda_g} \quad (degrees)$$

$$= 2.0 \ \pi \ l f \frac{\sqrt{\varepsilon_{eff}}}{c} \quad (radians) \ = 360.0 \ l f \frac{\sqrt{\varepsilon_{eff}}}{c} \quad (degrees) \tag{2.3.15}$$

## 2.3.1   Example:   Calculating Impedance from Capacitance

We are given the incremental capacitance of a coaxial line as:

$$C = \frac{2 \pi \varepsilon_0 \varepsilon_r}{\ln D / d}$$

What is the equation for $Z_0$?

**Solution:**

Since the dielectric is homogenous and the shield surrounds the center conductor, we know that $\varepsilon_{eff} = \varepsilon_r$. Using (2.3.4):

$$Z_0 = \frac{\sqrt{\mu_0 \varepsilon_0 \varepsilon_r} \ln D / d}{2 \pi \varepsilon_0 \varepsilon_r} = \frac{\sqrt{\mu_0} \ln D / d}{2 \pi \sqrt{\varepsilon_0 \varepsilon_r}} = \boxed{\frac{\eta_0 \ln D / d}{2 \pi \sqrt{\varepsilon_r}}}$$

## 2.4    MATERIAL PROPERTIES AND LUMPED MODEL

As we saw in the previous section, there are several properties of the material which are important to signal propagation. The properties of the medium in which we are propagating are expressed in terms of the **permeability**:

$$\mu = \mu_0 \mu_r \tag{2.4.1}$$

the **permittivity**,

$$\varepsilon = \varepsilon_0 \varepsilon_r \tag{2.4.2}$$

and the **conductivity**.

$$\sigma = \frac{1.0}{\rho} \tag{2.4.3}$$

Conductivity is also encountered as its reciprocal, **resistivity**, $\rho$. The permeability and permittivity of free space have the values:

$$\mu_0 \equiv 4.0 \; \pi \times 10^{-7} \; (H/m) \tag{2.4.4}$$

$$\varepsilon_0 \equiv 8.854 \times 10^{-12} \; (F/m) \cong \frac{1.0}{36\pi \times 10^{-9}} \; (F/m) \tag{2.4.5}$$

The use of $\mu_0$ and $\varepsilon_0$ allows us to express all of our equations in terms of the unitless relative permeability and permittivities, $\mu_r$ and $\varepsilon_r$, for convenience. Resistivity is often similarly normalized to the resistivity of copper or gold.

The specific values of $\varepsilon_r$, $\mu_r$, and $\sigma$ are determined by interactions at the atomic level. For example, the conductivity is a function of the mobility and the number of free charge carriers (electrons and holes). In nonconductive materials the charges are bound or locked in position. They cannot move freely when a field is applied. The permittivity[1] is a measure of the ability to store energy by shifting the center of the bound charges against the atomic forces (*polarizing* the material) to cancel the applied field. There are three mechanisms for this: dipole, ionic, and electronic. Each of these can create resonance-like behavior in the values of $\varepsilon_r$ and $\tan \delta$ at high frequencies. Similarly, the permeability is a measure of the ability of the material to store magnetic energy. An impressed magnetic field stores energy in the material by displacing the magnetic moments of the material. Again, there are several atomic mechanisms that can be involved: electric loops, spin, and nuclear spin.

---

[1]The *index of refraction* is the square root of the permittivity.

Values of relative permeability range from slightly below one to as high as a million (supermalloy). It is a measure of the ability of the medium to store energy in the magnetic field. It is frequency dependent and has both a real and imaginary component. Permeabilities of some metals are tabulated in Chapter 9.

Relative permittivity is 1.0 for free space and can be as high as 6000. It is also a function of frequency with both real and imaginary components. Permittivities are tabulated in Chapters 7 and 9.

Resistivity varies over a very wide range ($10^{25}$). It is a function of frequency. Resistivities are tabulated for metals in Chapter 9.

## 2.4.1   Complex Permeability and Permittivity

Permeability and permittivity are complex numbers so we write

$$\varepsilon_r^* = \varepsilon_r{}' + j\varepsilon_r{}'' \tag{2.4.1.1}$$

$$\mu_r^* = \mu_r{}' + j\mu_r{}'' \tag{2.4.1.2}$$

The real part represents the ability of the medium to store electric and magnetic energy while the imaginary part represents losses. Usually $\varepsilon_r{}''$, and $\mu_r{}''$ are small so we drop the primed and starred notations when referring to the real part.

The meaning of the complex values of permittivity and permeability can be seen by comparison with the more common power factor. The power factor is defined by circuit theory to be the ratio of the average power dissipated to the voltage × current product. This is the cosine of the angle between the actual or real power dissipated to the instantaneous VI product as shown in Figure 2.4.1.

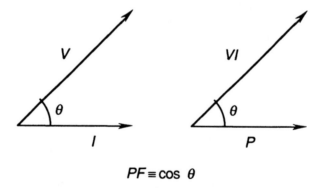

$$PF \equiv \cos \theta$$

**Figure 2.4.1: Power Factor**

For dielectric and magnetic losses, we define a similar and related function, the *loss tangent*. The loss tangent is depicted in Figure 2.4.2.

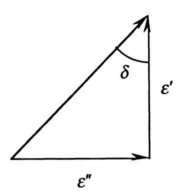

**Figure 2.4.2: Loss Tangent**

In both Figures 2.4.1 and 2.4.2 the horizontal axis represents lost energy and the vertical axis represents stored energy. The power factor of the dielectric and the loss tangent are related ways of expressing the material's losses. For small power factors (less than 0.15, which is generally the case for dielectrics we use), the loss tangent approximates the power factor within 1%. The relationships are:

$$PF = \text{power factor} = \sin\left(\frac{\text{energy loss per cycle}}{\text{energy stored per cycle}}\right) = \sin\left[\tan^{-1} D\right] \qquad (2.4.8)$$

$$D = \text{dissipation factor} = \frac{1.0}{Q} = \frac{\varepsilon''}{\varepsilon'} + \frac{\sigma}{\omega\,\varepsilon'} \qquad (2.4.9)$$

For a dielectric, the last term is much less than one so

$$D \cong \tan\delta \qquad (2.4.10)$$

and for small $D$

$$PF = \sin\left[\tan^{-1} D\right] \cong \sin\left[\sin^{-1} D\right] = D \cong \tan\delta \qquad (2.4.11)$$

When we use $\varepsilon_r$ we will mean $\varepsilon_r'$, and similarly we replace $\mu_r'$ with the notation $\mu_r$. The imaginary parts are then represented by the dielectric and magnetic loss tangents

$$\varepsilon_r'' \cong \varepsilon_r'\,\tan\delta_d \qquad (2.4.12)$$

$$\mu_r'' \cong \mu_r'\,\tan\delta_m \qquad (2.4.13)$$

These are referred to as dielectric or magnetic "loss tan" or "tan delta." You will most often see tan $\delta$ used without any subscript where the dielectric loss tan is implicit since tan $\delta_m$ is less often encountered in actual design.

## 2.5   MATERIAL CLASSIFICATIONS

The parameters we have defined create a number of natural classifications. These classes allow us to simplify our analysis of transmission line characteristics. Although we often assume that the materials making up our transmission lines have constant parameters, this is not true in all cases.

If $[\sigma / (\omega \varepsilon)]^2 \gg 1.0$ we have a *conductor*. When $[\sigma / (\omega \varepsilon)]^2 \ll 1.0$ the material is a *dielectric*. Semiconductors lie in between. *Superconductors* are conductors for which $\sigma = \infty$.

The permittivity, conductivity, and permeability are all functions of frequency and direction in the medium. When these parameters vary with frequency, the phase velocity also varies with frequency and the line is said to have *dispersion*. A dispersionless line has constant signal delay over frequency. Such a line propagates wideband signals (*e.g.*, pulses) without distortion.

Woven dielectrics are *anisotropic*; they will have slightly different $\varepsilon_r$ values in different directions within the material. We also encounter *nonhomogeneous* dielectrics in semiconductor dielectrics, multilayer printed circuit boards, and in microstrip line. A nonhomogenous dielectric has an $\varepsilon_r$ that is a function of position ($x$, $y$, $z$) in the line.

A material may also be *nonlinear*—that is, $\varepsilon_r$ or $\mu_r$ is a function of any applied fields. Permeability is commonly a function of the previous magnetization of the material.

## 2.6   QUASI-TEM TRANSMISSION LINE PARAMETERS

Inhomogeneous dielectrics occur when we have several dielectrics or when the $\varepsilon_r$, $\mu_r$ varies with position in the dielectric. In this case, the wave is not propagating in a strictly TEM fashion. However, it is convenient and useful to assume that for low frequencies we can substitute an *effective* parameter calculated from the geometry and the parameters of the individual regions making up the medium . These parameters—the effective permeability, $\mu_{eff}$, and effective permittivity, $\varepsilon_{eff}$—are relative numbers, like $\mu_r$ and $\varepsilon_r$.

For structures with mixed dielectrics (*e.g.*, air and substrate as in microstrip line) we define the *effective relative dielectric constant*:

$$\varepsilon_{eff} = \frac{C_{actual}}{C_{air}} \tag{2.6.1}$$

which is the ratio of the actual structure's capacitance (Figure 2.6.1a) to the capacitance when the dielectric is replaced with air (Figure 2.6.1b). This effective dielectric constant of a homogeneous structure (Figure 2.6.1c) is used wherever $\varepsilon_r$ would be used in a homogenous dielectric calculation. This definition allows us to treat propagation as propagation in an equivalent or *quasi-TEM* line.

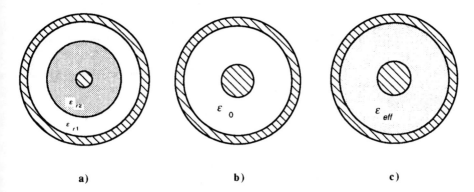

a)                              b)                              c)

**Figure 2.6.1: Effective Dielectric Constant**

Another commonly used number is the *filling factor*, $q$:

$$q = \frac{\varepsilon_{eff} - 1}{\varepsilon_r - 1} = \begin{cases} 0 \text{ if all air} \\ 1 \text{ if all dielectric} \end{cases} \qquad (2.6.2)$$

or rearranging:

$$\varepsilon_{eff} = q\,(\varepsilon_r - 1) + 1 \qquad (2.6.3)$$

## 2.6.1   Example:   Time Delay, Electrical Length and $\varepsilon_{eff}$

What is the time delay of a 3 cm length of microstrip line having an $\varepsilon_{eff}$ of 3.1? What is the phase delay of this line at the fundamental and first two harmonics of a 500 MHz square wave?

**Solution:**        We can calculate the phase velocity in the medium with (2.3.10):

$$v_p = \frac{3.0 \times 10^8 \text{ m / s}}{\sqrt{3.1}} = 1.7039 \times 10^8 \text{ m / s}$$

so the time delay between input and output is

$$\boxed{t = \frac{3 \text{ cm} \times 1 \text{ m / } 100 \text{ cm}}{1.7039 \times 10^8 \text{ m / s}} = 176 \text{ ps}}$$

and we calculate the guide wavelengths of the fundamental and harmonics as (2.3.12):

$$\lambda_{g1} = \frac{3.0 \times 10^8 \text{ m / s}}{500 \times 10^6 \sqrt{3.1}} = 0.3407 \text{ m}$$

$$\lambda_{g2} = \frac{3.0 \times 10^8 \text{ m / s}}{1000 \times 10^6 \sqrt{3.1}} = 0.1704 \text{ m}$$

$$\lambda_{g3} = \frac{3.0 \times 10^8 \text{ m / s}}{1500 \times 10^6 \sqrt{3.1}} = 0.1136 \text{ m}$$

Thus, the 3 cm represents a phase delay $\phi_n$ at the $n$th harmonic of

$$\boxed{\phi_1 = 360° \times .03 \text{ m} / .3407 \text{ m} = 31.6994°}$$

$$\boxed{\phi_2 = 360° \times .03 \text{ m} / .1704 \text{ m} = 63.3803°}$$

$$\boxed{\phi_3 = 360° \times .03 \text{ m} / .1136 \text{ m} = 95.0704°}$$

## 2.7    TRANSMISSION LINE LOSSES

The total losses in a transmission line are described by:

$$\alpha = \alpha_c + \alpha_d + \alpha_r + \alpha_l \tag{2.7.1}$$

where

$\alpha_c$ = conductor losses $\tag{2.7.2}$

$\alpha_d$ = dielectric losses $\tag{2.7.3}$

$\alpha_r$ = radiation losses $\tag{2.7.4}$

$\alpha_l$ = leakage losses $\tag{2.7.5}$

Losses are converted to and from series resistances with the equation

$$\alpha \cong \frac{R_{series}}{2.0 \ Z_0} \tag{2.7.6}$$

which is valid for small $R_{series}$.

## 2.7.1 Conductor Losses

A number of variables and concepts commonly occur when calculating conductor losses. For a complete derivation of the following equations see, e.g., Ramo et al. [17]. The dc resistance of a conductor is calculated with

$$R_{dc} = \frac{\rho \, l}{A} \quad (\Omega) \tag{2.7.1.1}$$

where $\rho$ is the resistivity, $l$ is the conductor length, and $A$ is the cross-sectional area.

The internal impedance of the conductor is a function of frequency

$$Z_s = R_s + j \, \omega \, L_i \tag{2.7.1.2}$$

where

$$R_s = \frac{1.0}{\sigma \, \delta} \tag{2.7.1.3}$$

$$= \sqrt{\frac{\pi f \mu}{\sigma}} = \text{\textit{surface resisitivity}} \text{ or } \text{\textit{sheet resistivity}} \quad (\Omega \, / \, \text{square})$$

$$L_i = R_s / \omega \tag{2.7.1.4}$$

and

$$\delta = \sqrt{\frac{2.0}{\omega \mu \sigma}} = \frac{1.0}{\sqrt{\pi f \mu \sigma}} = \text{skin depth} \tag{2.7.1.5}$$

A normalized frequency is defined as

$$P = \sqrt{2.0 \, \mu \, \sigma f A} \tag{2.7.1.6}$$

The variable $R_s$ is the skin effect resistance. As frequency increases our conductor losses also increase. The skin depth is the depth at which the current density is reduced to $1/e$. It can also be thought of as the equivalent thickness of a conductor at dc having the same resistance. The skin depth for several conductors is shown in Figure 2.7.1.1.

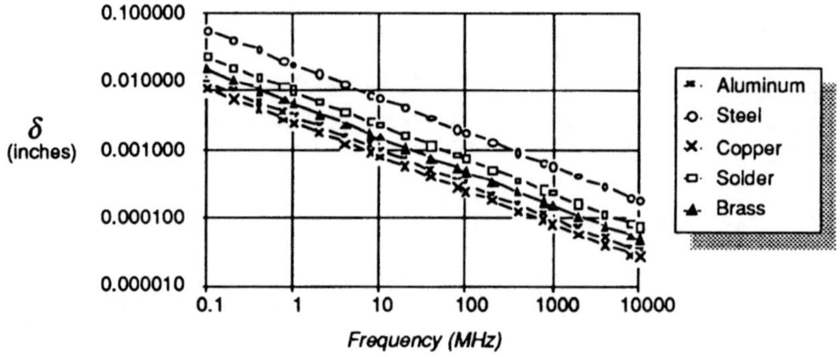

**Figure 2.7.1.1:    Skin Depth in Conductors**

The magnitude of the real and imaginary parts of the internal impedance are equal at all frequencies and the phase angle of $Z_s$ is therefore a constant 45°. This fact is the origin of the *incremental inductance* rule used to calculate conductor losses. We can use the incremental inductance rule to calculate the conductor losses of any transmission line when

- losses are low
- losses are due to skin effect
- internal inductance is small
- conductor thickness >> skin depth
- conductor radii >> skin depth

Using the equivalence of (2.7.7) we can calculate the loss resistance as

$$R = \omega \left[ L \text{ (nominal dimensions)} - L \text{ (conductors reduced by } \delta / 2) \right]$$

(2.7.1.7)

The meaning of the inductances is shown in Figure 2.7.1.2.

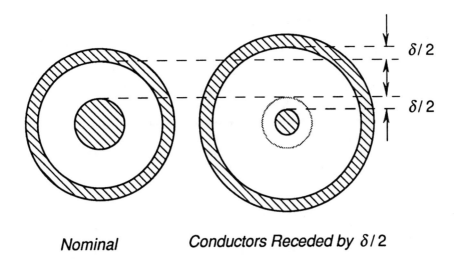

| Nominal | Conductors Receded by $\delta/2$ |

**Figure 2.7.1.2:**     Incremental Inductance Defined

The presence of the dielectric has no effect on the inductance. Because we can use this relation for virtually any transmission line for which we have an expression for the inductance or impedance, it will be of great value in loss calculations.

## 2.7.2 Dielectric Losses

Dielectric losses are created by the atomic mechanisms described earlier. Inhomogeneous (in the x-z plane of Figure 1.1.1 and 1.1.2) dielectric losses are reduced by the presence of the (lossless) air in the cross-section. The loss of these structures is modified by a dielectric loss filling factor [10]

$$q_{\tan \delta} = \frac{\varepsilon_{eff} - 1.0}{\varepsilon_{eff} - \varepsilon_{eff}/\varepsilon_r} \tag{2.7.2.1}$$

and

$$\alpha_d = \frac{\pi f \, q_{\tan \delta} \tan \delta \sqrt{\varepsilon_{eff}}}{c_0} \tag{2.7.2.2}$$

## 2.7.3 Radiation Losses and Leakage Losses

Signal power which is radiated into space or to another conductor is not available in the load. This loss is heavily dependent on the physical structure. Discontinuities tend to increase the radiation losses. Usually the design keeps these losses small and they are neglected. Some structures (*e.g.*, open ends) have been characterized for calculation purposes.

Leakage losses are present in transmission lines having dielectrics with finite resisitivity. An example of this is transmission lines fabricated in MMICs which have semiconductor material as the dielectric. The propagation is complicated in this situation; however, Hoffman [10] gives an approximation useful in some cases.

When the electric field lines lie along (or approximately along) the dielectric boundary the following may be used

$$\alpha_l = 1.64 \times 10^3 \; \sigma \frac{\sqrt{\varepsilon_{eff}} - 1 / \sqrt{\varepsilon_{eff}}}{\varepsilon_r - 1.0} \quad (dB / cm) \qquad (2.7.3.1)$$

where conductivity is in $(\Omega\text{-cm})^{-1}$.

## 2.8   TRANSMISSION LINE CALCULATIONS

The solution of Telegrapher's equations found above is only one possible form. Another useful form is expressed in terms of hyperbolic functions as shown in the following sections. A number of calculations making use of this form will be illustrated in this section.

### 2.8.1   Line Terminated with an Arbitrary Load

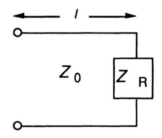

Figure  2.8.1.1:     Line Terminated With $Z_R$

The impedance seen looking into a generalized transmission line terminated in $Z_R$ (as shown in Figure 2.8.1.1) is:

$$Z_{in} = Z_0 \left( \frac{Z_R \cosh \gamma l + Z_0 \sinh \gamma l}{Z_0 \cosh \gamma l + Z_R \sinh \gamma l} \right) \quad (\Omega) \qquad (2.8.1.1)$$

This equation comes from the solution for the RLCG line (2.2.25) where we have chosen hyperbolic trigonometric functions rather than $e^x$ when simplifying. For a lossless line, the above reduces to:

$$Z_{in} = Z_0 \left( \frac{Z_R \cos \beta l + j Z_0 \sin \beta l}{Z_0 \cos \beta l + j Z_R \sin \beta l} \right) \qquad (2.8.1.2)$$

where

$$\alpha = 0$$

$$\beta = \frac{2\pi}{\lambda}$$

$$\gamma = j\beta = \frac{2\pi j}{\lambda}$$

Note that in both (2.8.1.1) and (2.8.1.2) the impedance seen is a function of the load impedance, $Z_R$, the line parameters, $\alpha$ and $\beta$, and the distance from the load, $l$. Note that the form of the equations for $\alpha$ and $\beta$ is in units per wavelength. That is, the loss per wavelength is a constant. Note that $\beta l$ is the electrical length of a line with length $l$ in radians.

### 2.8.1.1    Example:  Quarter-Wave Line with Open Circuit Load

What is the impedance looking into a lossless line one-quarter of a wavelength long terminated in an open circuit?

**Solution:**      We have been given that

$$l = \lambda / 4$$

$$Z_R = \infty \ \Omega$$

So from Equation (2.8.1.1):

$$Z_{in} = Z_0 \left( \frac{Z_R \cos \beta l + j Z_0 \sin \beta l}{Z_0 \cos \beta l + j Z_R \sin \beta l} \right) = Z_0 \left( \frac{\infty \cos \frac{\pi}{2} + j Z_0 \sin \frac{\pi}{2}}{Z_0 \cos \frac{\pi}{2} + j \infty \sin \frac{\pi}{2}} \right)$$

$$\boxed{Z_{in} = \frac{Z_0}{\infty} = 0 \ \Omega}$$

The quarter-wavelength line has transformed the output open circuit into a short circuit at the input of the line.

### 2.8.1.2    Example:  Quarter-Wave Line Terminated with a Generalized Load

What happens when we put a general impedance $Z_L$ at the end of a quarter-wavelength line?

**Solution:**    We have been given that

$$l = \lambda / 4$$

$$Z_R = Z_L$$

So again using (2.8.1.1):

$$Z_{in} = Z_0 \left( \frac{Z_R \cos \beta l + j\, Z_0 \sin \beta l}{Z_0 \cos \beta l + j\, Z_R \sin \beta l} \right) = Z_0 \left( \frac{j\, Z_0}{j\, Z_L} \right) = \frac{Z_0^2}{Z_L}$$

This is often expressed as the quarter-wave line impedance, $Z_0$ required to match a desired $Z_{in}$ to a specified $Z_L$ as:

$$Z_0 = \sqrt{Z_{in}\, Z_L} \qquad\qquad (2.8.1.2.1)$$

For example, if our source impedance is 50 $\Omega$ and we wish to match it to a 68 $\Omega$ load, we could use a quarter-wave line having impedance:

$$\boxed{Z_0 = \sqrt{50 \times 68} = \sqrt{3400} = 58.31\ \Omega}$$

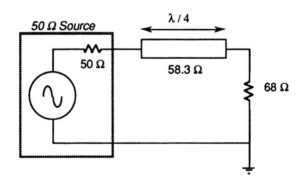

**Figure  2.8.1.2.1:**    Quarter-Wave Transformer

The quarter-wavelength required would be calculated from the frequency of the source and the $\varepsilon_{eff}$ of the transmission line. As shown in Figure 2.8.1.2.2, the impedance is matched exactly at the design frequency (in this case 1 GHz). The impedance looking from the source into the line is 50 $\Omega$, and the impedance looking back from the load is 68 $\Omega$ at 1 GHz. The match is good to within some tolerance over a wider bandwidth and the response repeats.

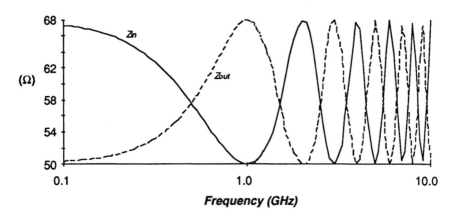

**Figure 2.8.1.2.2:** **Input and Output Impedance of Quarter-Wavelength Transformer**

This property is the basis for multiple-step transformers and tapered lines used for matching unequal impedances.

## 2.8.2 Line Terminated with $Z_0$

If we let the arbitrary load in Equation (2.8.1.1), $Z_R$, equal the characteristic impedance of the line, we get:

$$Z_{in} = Z_0 \left( \frac{Z_0 \cosh \gamma l + Z_0 \sinh \gamma l}{Z_0 \cosh \gamma l + Z_0 \sinh \gamma l} \right) = Z_0 \qquad (2.8.2.1)$$

which is no longer a function of $l$, the distance from the load. This condition is known as *matched*.

## 2.8.3   Line Terminated with Short Circuit

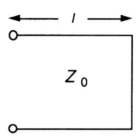

**Figure  2.8.3.1:**    Short-Circuited Line

If we terminate the line in a short circuit ($Z_R = 0$), equation (2.8.1.1) gives:

$$Z_{in} = Z_{sc} = Z_0 \left( \frac{Z_0 \, \sinh \, \gamma l}{Z_0 \, \cosh \, \gamma l} \right) = Z_0 \tanh \gamma l \tag{2.8.3.1}$$

For lossless lines, Equation (2.8.1.1) reduces to:

$$Z_{sc} = j \, Z_0 \tan \beta l = j \, Z_0 \tan \left( \frac{2\pi l}{\lambda} \right) \tag{2.8.3.2}$$

If $\tan x$ is now approximated with $x$ (true for small $x$):

$$Z_{sc} \cong j \, Z_0 \, \beta l = j \, Z_0 \left( \frac{2\pi l}{\lambda} \right) = j \, \omega \, \frac{(Z_0 \, l)}{c} \tag{2.8.3.3}$$

This equation looks a great deal like the impedance of an inductor ($Z_{ind} = j \, \omega L$) with inductance $Z_0 l$. In fact, physically short, short-circuited lines are used in this fashion as inductors in frequency ranges where lumped inductors become unusable. We can increase the apparent inductance by either increasing the line length or by increasing its characteristic impedance, $Z_0$.[2]

---

[2]In designing transmission line filters the second part of (2.8.3.2) is known as *Richards' variable*, $\Omega$

$$\Omega = \tan \left( \frac{2\pi l}{\lambda} \right) = \tan \left( \frac{\pi f_c}{2.0 \, f_0} \right) \tag{2.8.3.4}$$

For example, a low pass filter could be designed by using $\lambda / 4$ lines (at $f_0$ in the reject band). The cutoff frequency, $f_c$ is then used with (2.8.3.2), (2.8.3.4), and the filter $g_k$s to calculate the required line lengths.

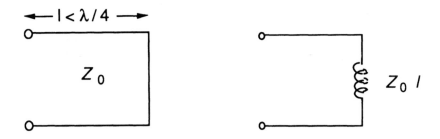

**Figure 2.8.3.2:** **Short-Circuited Line Used as an Inductor**

Notice that the tangent function in (2.8.3.2) is periodic in its parameter and $Z_{sc}$ will therefore also repeat. Also, the goodness of our lumped (2.8.3.3) approximation to the tangent function degrades as the line's length relative to a wavelength increases, which effectively makes the line's equivalent inductance a function of frequency.

Similarly, for:

$$\pi/2 < \left(\frac{2\pi l}{\lambda}\right) < \pi$$

the tangent function is negative and $Z_{sc}$ is approximately:

$$Z_{sc} \cong -j\,Z_0\,\beta l = -j\,Z_0\left(\frac{2\pi l}{\lambda}\right) = -j\,\omega\,\frac{Z_0\,l}{c} = \frac{1.0}{j\left(\frac{c}{\omega\,Z_0\,l}\right)} \qquad (2.8.3.5)$$

which is the impedance of a capacitor of value

$$C = \frac{1.0}{\omega^2 Z_0\,l} \qquad (2.8.3.6)$$

Finally, for

$$2\,\pi l/\lambda = \pi/2 \qquad (2.8.3.7)$$

or equivalently,

$$l = \lambda/4$$

the tangent function of (2.8.3.2) and the line's impedance is infinite. This property is used to create series resonant circuits.

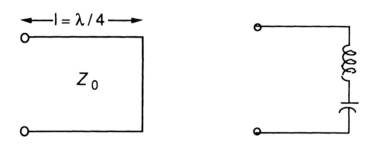

**Figure 2.8.3.3:**      Short-Circuited Line Used as a Resonant Circuit

Summarizing our results for the short-circuited line:

$$\text{"inductive"}\begin{cases} l < \dfrac{\lambda}{4} \\[2mm] \dfrac{\lambda}{2} < l < \dfrac{3\lambda}{4} \end{cases}$$

$$\text{"capacitive"}\begin{cases} \dfrac{\lambda}{4} < l < \dfrac{\lambda}{2} \\[2mm] \dfrac{3\lambda}{4} < l < \lambda \end{cases}$$

$$\text{"series LC resonator"}\left\{ l = \frac{n\,\lambda}{4} \text{ where } n \in I \text{ (the integers)} \right.$$

### 2.8.3.1    Example:    Resonant Line

Design a series resonator from a shorted 50 Ω coaxial cable having a PTFE dielectric at 402 MHz.

**Solution:**        We have been given:

$Z_0 = 50\ \Omega$

$\varepsilon_r = 2.10$

$Z_R = 0\ \Omega$

We calculate the wavelength of 402 MHz in PTFE from Equation (2.3.12):

$$\lambda_g = \frac{c}{f\sqrt{\varepsilon_r}} = \frac{3 \times 10^8\ \text{m/s}}{402 \times 10^6\ /\text{s}\ \sqrt{2.10}} = 0.5149\ \text{m} = 20.27\ \text{in}$$

$$\boxed{\lambda_g / 4 = 0.5149 \text{ m} / 4 = 0.1287 \text{ m} = 5.07 \text{ in}}$$

### 2.8.3.2 Example: Line Used as a Capacitor

Design a capacitive reactance from a shorted 50 Ω coaxial cable with an air dielectric equivalent to a 1 pF capacitor at 1.5 GHz.

**Solution:** We have been given:

$$Z_0 = 50 \text{ Ω}$$

$$\varepsilon_r = 1$$

$$Z_{in} = \frac{1}{j \, \omega \, C} = \frac{1}{j \, 2 \, \pi \times 1.5 \times 10^9 \, /s \times 1.0 \times 10^{-12} \text{ F}} = -j \, 106.103 \text{ Ω}$$

$$Z_R = 0 \text{ Ω}$$

We assume:

$$\frac{\lambda}{4} < l < \frac{\lambda}{2}$$

We calculate the required length with (2.8.3.2):

$$Z_{sc} = j \, Z_0 \tan \beta l = j \, Z_0 \tan \left( \frac{2\pi l}{\lambda} \right) = -j \, 106.103 \text{ Ω}$$

and

$$\lambda_g = \frac{c}{f \sqrt{\varepsilon_r}} = \frac{3 \times 10^8 \text{ m/s}}{1.5 \times 10^9 \, /s} = 0.2 \text{ m} = 7.874 \text{ in}$$

$$\lambda_g / 4 = 0.2 \text{ m} / 4 = 0.05 \text{ m} = 1.969 \text{ in (our answer will be longer than this)}$$

Solving,

$$\tan^{-1} \left( \frac{-50 \text{ Ω}}{106.103 \text{ Ω}} \right) = -0.440 \text{ radians} = -25.216°$$

The calculator's inverse tangent function gives us a value in the fourth quadrant but we know we want an answer in the second quadrant so we add the above to $\pi$:

$$\pi + (-0.440) = 2.701 \text{ radians} = 154.768°$$

$$l = \frac{2.701 \text{ radians } \lambda_g}{2 \pi}$$

$$\boxed{l = 0.086 \text{ m} = 3.385 \text{ in}}$$

## 2.8.4  Line Terminated with Open Circuit

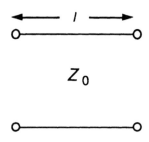

**Figure 2.8.4.1:**    Open-Circuited Line

As we did for the short-circuited line in the last section, we simplify (2.8.1.1). The equation for the input impedance of an open circuited line is:

$$Z_{oc} = Z_0 \coth \gamma l \qquad (2.8.4.1)$$

for a lossless line:

$$Z_{oc} = -j \, Z_0 \cot \beta l = -j \, Z_0 \cot \left( \frac{2 \pi l}{\lambda} \right) \qquad (2.8.4.2)$$

and we can similarly derive that for an open circuited line:

$$\text{"capacitive"} \begin{cases} l < \dfrac{\lambda}{4} \\[2mm] \dfrac{\lambda}{2} < l < \dfrac{3\lambda}{4} \end{cases}$$

$$\text{"inductive"} \begin{cases} \dfrac{\lambda}{4} < l < \dfrac{\lambda}{2} \\[2mm] \dfrac{3\lambda}{4} < l < \lambda \end{cases}$$

"parallel LC resonator" $\left\{ l = \dfrac{n\,\lambda}{4} \right.$ where $n \in I$ (the integers)

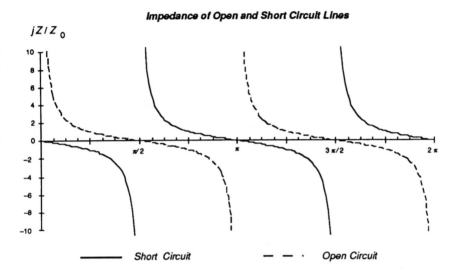

**Impedance of Open and Short Circuit Lines**

$jZ/Z_0$

Short Circuit       — — · Open Circuit

**Figure 2.8.4.2:**     Impedance of Open- and Short-Circuited Lines

## 2.8.4.1   Example:   Chebyschev LPF

From the specification it is determined that an $n = 5$, 0.5 dB Chebyschev LPF is required. The required corner frequency is 4.0 GHz and the reject band is 8.0 to 10.0 GHz. Design the lumped filters. Using the relations of the previous two sections, convert this to a distributed equivalent.

**Solution:**       Using the usual filter tables or equations we find the $g_k$s required

$$g_1 = g_5 = 1.81$$

$$g_2 = g_4 = 1.3$$

$$g_3 = 2.69$$

The lumped element values are then

$$\boxed{C_2 = C_4 = \frac{g_k}{2.0\,\pi f_c\,Z_0} = 3.60 \text{ nH}}$$

$$L_1 = L_5 = \frac{gk\,Z_0}{2.0\,\pi\,f_c} = 1.03 \text{ pF}$$

$$L_3 = 5.35 \text{ nH}$$

The distributed element filter is calculated by using lines which are $\lambda/4$ at the center of the reject band or 9.0 GHz. The inductors of the lumped element design will be replaced with short-circuited lines and the capacitors with open-circuited lines. For physical reasons, it is most convenient to use series short-circuited lines and shunt open-circuited lines.

Richards' variable at the corner frequency is

$$\Omega_c = \tan\left(\frac{\pi\,4.0 \text{ GHz}}{2.0 \times 9.0 \text{ GHz}}\right) = 0.8391$$

We calculate the line impedances to be

$$Z_1 = Z_5 = \frac{gk\,Z_0}{\Omega_c} = 107.9\ \Omega \qquad \text{series short-circuit line}$$

$$Z_2 = Z_4 = \frac{gk}{Z_0\Omega_c} = 32.3\ \Omega \qquad \text{shunt open-circuit line}$$

$$Z_3 = 160.3\ \Omega \qquad \text{series short-circuit line}$$

The line dimensions (lengths and widths) are calculated with the equations of Chapter 3 for the desired method of construction. In an actual design we would then add step junction discontinuities to our model, and correct the open-circuited line lengths for fringing effects.

The response of the two filters constructed with ideal elements is shown in Figure 2.8.4.3 below. The limited number of simulation points creates the finite depth of the nulls and the roughness in the distributed element plot. As expected, the distributed filter response repeats limiting its useful bandwidth (a lumped element filter constructed with real-world components would also show returns due to strays, etc.). This design is an approximation to the lumped design because there is no broadband equality between lumped and distributed elements.

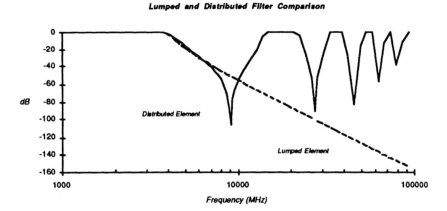

**Figure 2.8.4.3:** Comparison of Lumped and Distributed Filters

## 2.8.5 Line Terminated with a Lumped Element

In addition to open circuits and short circuits we can place lumped elements on the end of our transmission lines. This effectively changes the length of the line as we shall see in the next example.

### 2.8.5.1 Example: Varactor Tuned Line Length

Calculate the value of a tunable capacitor (varactor) required to vary an open-circuited 50 Ω PTFE dielectric coaxial line's apparent length from 0.1 to 0.2 λ at 2.5 GHz.

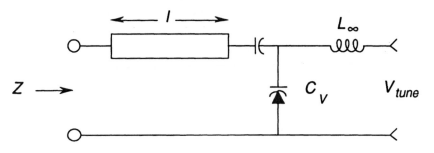

**Figure 2.8.5.1.1:** Varactor Tuned Line

**Solution:**                    We have been given

$\varepsilon_r = 2.10$

$f = 2.5 \times 10^9$ /s

$Z_0 = 50 \ \Omega$

$Z_R = \dfrac{1}{j \ \omega C}$

$$\lambda_g = \frac{c}{f \sqrt{\varepsilon_r}} = \frac{3 \times 10^8 \text{ m/s}}{2.5 \times 10^9 \text{ /s} \times \sqrt{2.10}} = 0.083 \text{ m} = 3.260 \text{ in}$$

We know that we wish to tune the capacitance to increase the length. As the length of an open-circuited line ($l < \lambda / 4$) is increased, the impedance decreases and the effective capacitance therefore is increased. So, we know that as the capacitor is tuned from its minimum to its maximum capacitance, the line's apparent length will increase. We now calculate the effective capacitance of an open-circuited line at the two requested line lengths:

$$Z_{oc} \ (0.1 \ \lambda) = -j \ Z_0 \cot \left( \frac{2 \ \pi \ l}{\lambda} \right) = -j \ 68.819 \ \Omega$$

$$Z_{oc} \ (0.2 \ \lambda) = -j \ Z_0 \cot \left( \frac{2 \ \pi \ l}{\lambda} \right) = -j \ 16.246 \ \Omega$$

We will assume that we can obtain a capacitor that can be set to 0 pF for this example, although in real life we would need to shorten the line to compensate for the adjustable capacitor's minimum capacitance. In our case, the line's physical length will be 0.1 $\lambda$.

We now use the general form equation for the impedance of an arbitrarily terminated line (2.8.1.1):

$l = 0.1 \ \lambda$

$$Z_{in} = Z_0 \left( \frac{Z_R \cos 0.2 \ \pi + j \ Z_0 \sin 0.2 \ \pi}{Z_0 \cos 0.2 \ \pi + j \ Z_R \sin 0.2 \ \pi} \right) = -j \ 16.246 \ \Omega$$

Solving,

$$-j\,0.325 = \left(\frac{Z_R\cos 0.2\,\pi + j\,Z_0\sin 0.2\,\pi}{Z_0\cos 0.2\,\pi + j\,Z_R\sin 0.2\,\pi}\right)$$

$$-j\,0.325 = \left(\frac{Z_R\,0.8090 + j\,29.3893}{40.4508 + j\,Z_R\,0.5878}\right)$$

$$(-j\,0.325)(40.4508 + j\,Z_R\,0.5878) = Z_R\,0.8090 + j\,29.3893$$

$$-13.1465\,j + 0.1910\,Z_R = Z_R\,0.8090 + j29.3893$$

$$Z_R = \frac{42.5358\,j}{-0.6180} = -j\,68.8282\ \Omega = \frac{-j}{\omega C}$$

$$\boxed{C = 0.9249\ \text{pF}}$$

A similar analysis could have been used to tune the line's $\lambda/4$ resonant frequency, giving us a tunable resonator.

## 2.9   ABCD MATRIX OF A GENERALIZED TRANSMISSION LINE

Use of the ABCD matrix is described in Chapter 12. The ABCD matrix for a series transmission line is:

$$\begin{bmatrix} \cosh\gamma l & Z_0\sinh\gamma l \\ \dfrac{\sinh\gamma l}{Z_0} & \cosh\gamma l \end{bmatrix} \quad \text{general case}$$

$$\begin{bmatrix} \cos\beta l & j\,Z_0\sin\beta l \\ \dfrac{j\sin\beta l}{Z_0} & \cos\beta l \end{bmatrix} \quad \text{lossless case}$$

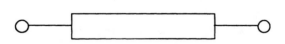

Figure 2.9.1: Series Transmission Line

The ABCD matrix for a shunt open-circuited transmission line is:

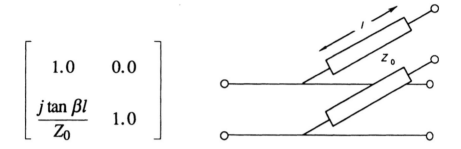

$$\begin{bmatrix} 1.0 & 0.0 \\ \dfrac{j \tan \beta l}{Z_0} & 1.0 \end{bmatrix}$$

**Figure 2.9.2: Shunt Open-Circuited Transmission Line**

The ABCD matrix for a shunt short-circuited transmission line is:

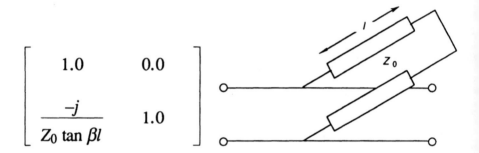

$$\begin{bmatrix} 1.0 & 0.0 \\ \dfrac{-j}{Z_0 \tan \beta l} & 1.0 \end{bmatrix}$$

**Figure 2.9.3: Shunt Short-Circuited Transmission Line**

## 2.10 SUMMARY

In this chapter, we have seen, in a general way, how transmission line calculations are made and how transmission lines can be used as circuit elements. We have seen that they can be used to emulate lumped elements and resonators. We have learned that the impedance of a transmission line element does not vary with frequency the way that a lumped element does. In fact, it is clear that a good definition of a *lumped* element is a section of transmission line that is electrically short. As we saw, when the assumption of short electrical length was made in the above examples, the characteristics of lines reduce to that of lumped elements.

In Chapter 3 we will see how to translate these general line equations into actual physical realizations.

REFERENCES

[1]     *Ultrashort Electromagnetic Waves*, American Institute of Electrical Engineers, New York Section, 1943.

[2]     Balanis, Constantine A., *Advanced Engineering Electromagnetics*, John Wiley & Sons, New York, 1989. (A modern EM text that covers everything; very good.)

[3]     Bianco, Bruno, *et al.*, "Comments on 'Microstrip Characteristic Impedance'," *IEEE Transactions on Microwave Theory and Techniques*, Vol. MTT-28, No. 2, February 1980, p. 152.

[4]     Bianco, Bruno, *et al.*, "Some Considerations About the Frequency Dependence of the Characteristic Impedance of Uniform Microstrips," *IEEE Transactions on Microwave Theory and Techniques*, Vol. MTT-26, No. 3, March 1978, pp. 182–185.

[5]     Cheng, David K., *Field and Wave Electromagnetics*, Addison-Wesley Publishing Co., Reading, MA, 1989.

[6]     Collin, Robert E., and Robert Plonsey, *Principles and Applications of Electromagnetic Fields,* McGraw-Hill Book Co., New York, 1961.

[7]     Collin, Robert E., *Field Theory of Guided Waves*, IEEE Press, 1991.

[8]     Durney, Carl H., and Curtis C. Johnson, *Introduction to Modern Electromagnetics*, McGraw-Hill Book Co., New York, 1969.

[9]     Getsinger, William J., "Microstrip Characteristic Impedance," *IEEE Transactions on Microwave Theory and Techniques*, Vol. MTT-27, No. 4, April 1979, p. 293.

[10]    Hoffman, Reinmut, *Handbook of Microwave Integrated Circuits,* Artech House, Norwood, MA, 1987.

[11]    Jansen, Rolf H., and Norbert H.L. Koster, "New Aspects Concerning the Definition of Microstrip Characteristic Impedance as a Function of Frequency," *1982 IEEE MTT-S International Microwave Symposium Digest*, Dallas, Texas, June 15–17, 1982, pp. 305–307.

[12]    Johnk, Carl T.A., *Engineering Electromagnetic Fields and Waves,* John Wiley & Sons, New York, 1975.

[13]    Jordan, Edward C., and Keith G. Balman, *Electromagnetic Waves and Radiating Systems,* Prentice-Hall, New York, 1968.

[14]    Lerner, C.M., *Problems and Solutions in Electromagnetic Theory*, John Wiley & Sons, New York, 1985. (A rare source of worked EM example problems.)

[15]    Liboff, Richard L., and G. Conrad Dalman, *Transmission Lines, Waveguides, and Smith Charts*, Macmillan Publishing, New York, 1985.

[16]    Metzger, Georges, *et al.*, *Transmission Lines with Pulse Excitation*, Academic
       Press, New York, 1969. (Transmission line propagation from the digital
       engineer's perspective.)

[17]    Ramo, Simon, John R. Whinnery, and Theodore Van Duzer, *Fields and Waves in
       Communication Electronics*, John Wiley & Sons, New York, 1984. (A classic text
       that has been revised several times since the first version, *Fields and Waves in
       Modern Radio* in 1944. Good example problems.)

[18]    Sander, K.F., and G.A.L. Reed, *Transmission and Propagation of Electromagnetic
       Waves*, Cambridge University Press, Cambridge, 1986.

[19]    Steele, Charles W., *Numerical Computation of Electric and Magnetic Fields*, Van
       Nostrand Reinhold Company, New York, 1987.

[20]    Waldow, Peter, and Ingo Wolff, "The Skin-Effect at High Frequencies," *IEEE
       Transactions on Microwave Theory and Techniques*, Vol. MTT-33, No. 10,
       October 1985, pp. 1076–1082.

[21]    Young, Leo, ed., *Advances in Microwaves*, Volume 1, "Directional Couplers," by
       R. Levy, Academic Press, New York, 1966, pp. 115–209. (An excellent
       discussion.)

[22]    Young, Leo, ed., *Parallel Coupled Lines and Directional Couplers*, Artech House,
       Norwood, MA, 1972. (A very good collection of papers relating to the
       understanding and design of coupled lines of various types.)

# Chapter 3

# Physical Transmission Lines

## 3.1 INTRODUCTION

In this chapter we transform our knowledge about the generalized transmission line calculations into actual physical realizations for a wide variety of configurations. Structures are grouped into major categories: coaxial, paired lines, coplanar, microstrip line, stripline, waveguide, and others. Within each category, a number of nonstandard configurations are included to illustrate the effects of variations from the standard structure.

ABCD matrices for the lines in this chapter are found in Chapter 2. The equations of this chapter allow calculation of $Z_0$ in the matrices from the line's physical dimensions. The electrical length is calculated from the operating frequency, $f$, and the physical length, $l$, using $\varepsilon_{eff}$.

$$\theta = 2.0 \; \pi \, l f \; \frac{\sqrt{\varepsilon_{eff}}}{c} \; \text{(radians)} \; = 360.0 \, l f \; \frac{\sqrt{\varepsilon_{eff}}}{c} \; \text{(degrees)} \tag{3.1.1}$$

A number of the older and commonly referenced articles are available as part of reprint collections listed below. Two of these, [2] and [4] are out of print but they may be found in libraries. Consulting these can save considerable time.

### REFERENCES

[1]     Itoh, T., ed., *Planar Transmission Line Structures,* IEEE Press, NY, 1987.

[2]     Frey, J., ed., *Microwave Integrated Circuits*, Artech, Norwood, MA, 1975.

[3]     Frey, J., and Kul Bhasin, eds., *Microwave Integrated Circuits*, 2nd Ed., Artech House, Norwood, MA, 1985. (The second edition of [2] with a new group of papers.)

[4]    Young, L., ed., *Parallel Coupled Lines and Directional Couplers,* Artech, Norwood, MA, 1972.

## 3.2    COAXIAL STRUCTURES

### 3.2.1    Round Coaxial Cable

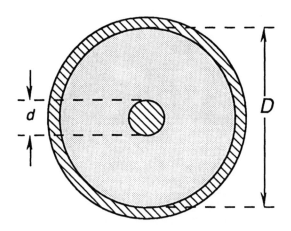

**Figure 3.2.1.1:**    Round Coaxial Cable

The equations for this transmission line can be derived easily and are exact for the case of an infinitely long line (no fringing fields):

$$Z_0 = \frac{\eta_0}{2.0 \; \pi \; \sqrt{\varepsilon_r}} \ln\left(\frac{D}{d}\right) \quad (\Omega) \tag{3.2.1.1}$$

$$\alpha_c = \left(\frac{0.014272 \; \sqrt{f}}{Z_0}\right)\left(\frac{1}{d} + \frac{1}{D}\right) \quad (\text{dB} / \text{m}) \tag{3.2.1.2}$$

$$\alpha_d = 0.091207 \, f \sqrt{\varepsilon_r} \tan \delta \quad (\text{dB} / \text{m}) \tag{3.2.1.3}$$

For inner and outer conductors having equal conductivity, the optimal $D / d$ for breakdown voltage is 2.718; for optimal power transfer, 1.65; and for minimum attenuation, 3.59 [7, 13]. See [7] for information on optimizing these parameters.

Coaxial connectors have mechanical tolerances of ±0.0025 mm and ±0.005 mm for laboratory and general precision, respectively. For a 50 Ω connector, this corresponds to about 0.5% and 0.8% tolerance on the impedance.

To reduce losses, we might increase $d$ and $D$, keeping their ratio constant to lower $\alpha_c$. However, the coaxial wire of Figure 3.2.1.1 will propagate higher order modes.

The cutoff frequency of these modes decreases with increasing $d$, and $D$ placing a limitation on the cable size for a desired useful frequency range.  Marcuvitz [11] gives the relations for the higher order modes, which begin at approximately 1.5 times the $TE_{11}$ mode.  The first higher order mode begins to propagate when

$$\lambda_{C_{11}} = \frac{2.0 \ \pi}{\beta_{z_1}}$$

(3.2.1.4)

where $\beta_{zm}$ is the first $(m = 1)$ solution of

$$\frac{J_1{}'(\beta_{zm} R)}{N_1{}'(\beta_{zm} R)} = \frac{J_1{}'(\beta_{zm} r)}{N_1{}'(\beta_{zm} r)}$$

(3.2.1.5)

where

$$r = d / 2.0$$

(3.2.1.6)

$$R = D / 2.0$$

(3.2.1.7)

Dimitrios [5] rewrites the above by replacing derivative forms of Bessel functions with nonderivative forms.  This form is more accurate when used with available tables of Bessel functions:

$$\frac{J_0{}'(\beta_{zm} R) - \dfrac{1}{\beta_{zm} R} J_1{}'(\beta_{zm} R)}{N_0{}'(\beta_{zm} R) - \dfrac{1}{\beta_{zm} R} N_1{}'(\beta_{zm} R)} = \frac{J_0{}'(\beta_{zm} r) - \dfrac{1}{\beta_{zm} r} J_1{}'(\beta_{zm} r)}{N_0{}'(\beta_{zm} r) - \dfrac{1}{\beta_{zm} r} N_1{}'(\beta_{zm} r)}$$

(3.2.1.8)

Green [8] derives an expression for the wavelength at which this occurs ($TE_{11}$ mode begins to propagate):

$$\lambda_c = 2 \ \pi \, r_m \left[ 1.0 - \frac{1.0}{6.0} \left( \frac{t}{2.0 \ r_m} \right)^2 - \frac{7.0}{120.0} \left( \frac{t}{2.0 \ r_m} \right)^4 - \ ... \right]$$

(3.2.1.9)

where

$$r_m = \frac{d + D}{2.0}$$

(3.2.1.10)

$$t = D - d$$

(3.2.1.11)

The stated agreement with Marcuvitz's Bessel function solution is better than 1% for $D/d$ < 2.5, and better than 4% for $D/d$ < 4.0. The number of terms used to reach this agreement is not stated.

A simple approximation to the above is:

$$\lambda_c \cong \frac{\pi (D + d)}{2.0} \text{ (units of } D, d) \tag{3.2.1.12}$$

which is the circumference of a circle with radius midway between the inner and outer conductors. Dimitrios [5] compared the accuracy of the equation above to the exact solution and found that it is generally accurate to 3% for 50 $\Omega$ lines with various dielectrics.

A lesser known fact about real-world coaxial cables is that they can introduce nonlinear distortion at a low level. Amin [3] presents data at the L, S, and C (390 MHz to 10.9 GHz) bands for various commercial cables showing that braided cables can have distortion products down 90 to 115 dB from the carrier depending on construction. Solid shields had no distortion. Nickel-plated, stainless steel, and Al alloy braids had distortion down 90 to 95 dB from the carrier. Distortion increased with power, frequency, and cable length.

### 3.2.1.1    Example:    Coaxial Pogo Pin

We wish to design a 50 $\Omega$ controlled impedance path that launches from a microstrip line and makes use of a standard size (0.040" o.d.) pogo pin (a spring-loaded pin used in ATE equipment) as the center conductor. This configuration is of use in test equipment where a controlled Z path must be blind-mated repetitively, as in Figure 3.2.1.2. Calculate the required shield dimensions for both air and PTFE dielectrics.

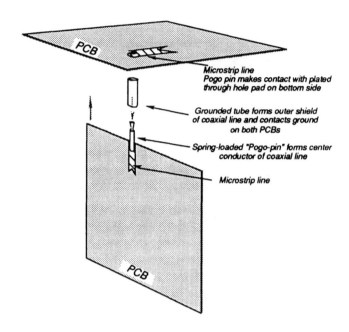

*Microstrip line*
*Pogo pin makes contact with plated*
*through hole pad on bottom side*

*Grounded tube forms outer shield*
*of coaxial line and contacts ground*
*on both PCBs*

*Spring-loaded "Pogo-pin" forms center*
*conductor of coaxial line*

*Microstrip line*

**Figure 3.2.1.2:** **Coaxial Pogo-Pin Example**

**Solution:**

$$Z_0 = 50 \ \Omega = \frac{\eta_0}{2.0 \ \pi \ \sqrt{\varepsilon_r}} \ln\left(\frac{D}{d}\right) \Omega$$

For air:

$$50 \ \Omega = \frac{377}{2 \ \pi \sqrt{1}} \ln\left(\frac{D}{0.040}\right) \Omega$$

$$\boxed{D = 0.092"}$$

For PTFE:

$$50 \ \Omega = \frac{377}{2 \ \pi \sqrt{2.1}} \ln\left(\frac{D}{0.040}\right) \Omega$$

$$\boxed{D = 0.134"}$$

A number of companies have attempted to combine the 360° shielding benefits of the coaxial structure with the fabrication advantages of printed-circuit boards (PCBs). This is increasingly an issue as the density and speeds of digital circuitry increase. See also [15] for a planar structure constructed with additive techniques. See Section 3.2.5 for more information.

Swengel et al. [18] describe a technique for machine wiring coaxial lines on PCB. The technique point-to-point wires 3.14 mil wires coated with a 10 mil thick layer of PTFE. The wire ends are then plated together. The PTFE is plated, creating the coaxial shield. A final selective etch separates the wire ends as required. Because of the solid ground plane, wave-soldering is difficult. The technique is also sensitive to handling and vibration.

A similar technique, described in [21], uses a semirigid cable (PTFE, $D = 9.5$ mil, $d = 3.14$ mil), which is also machine routed point-to-point. The wires are laid down on an adhesive layer during routing and then the shields are plated together. The wire is stripped at its ends, coated with epoxy and the center conductors are plated to plated-through holes. This structure is easily used as part of a multilayer PCB.

REFERENCES

[1]    Adair, Robert T., and Eleanor M. Livingston, "Coaxial Intrinsic Impedance Standards," U.S. Department of Commerce, NIST Technical Note 1333, October 1989.

[2]    Alford, Andrew, "Higher Modes in Insulating Beads," Microwave Journal, March 1990, pp. 146–156.

[3]    Amin, M.B., and I.A. Benson, "Nonlinear Effects in Coaxial Cables at Microwave Frequencies," Electronics Letters, December 8, 1977, Vol. 13, No. 25, pp. 768–770.

[4]    Daywitt, William C., "First-Order Symmetric Modes for a Slightly Loss Coaxial Transmission Line," IEEE Transactions on Microwave Theory and Techniques, Vol. 38, No. 11, November 1990, pp. 1644–1650. (Maxwell's equations are solved for a slightly lossy line resulting in new, more accurate results. A significant correction is made to calculations of the distributed resistance.)

[5]    Dimitrios, James, "Exact Cutoff Frequencies of Precision Coax," Microwaves, June 1965, pp. 28–31.

[6]    Franke, Ernie, "Minimum Attenuation Geometry for Coaxial Transmission Line," RF Design, May 1989, pp. 58–62.

[7]    Green, E.I., et al., "The Proportioning of Shielded Circuits for Minimum High-Frequency Attenuation," The Bell System Technical Journal, 1936, pp. 248–283.

[8]    Green, Harry E., "Determination of the Cutoff of the First Higher Order Mode in a Coaxial Line by the Transverse Resonance Technique," IEEE Transactions on Microwave Theory and Techniques, Vol. 37, No. 10, October 1989, pp. 1652–1653.

[9]     Hubbard, George, "Technique Links SMA Connectors to Flexible Coax Cables," *Microwaves & RF*, June 1990, pp. 156–158, 163.

[10]    MacKenzie, T.E. and A.E. Sanderson, "Some Fundamental Design Principles for the Development of Precision Coaxial Standards and Components," *IEEE Transactions on Microwave Theory and Techniques*, MTT-14, No. 1, January 1966, pp. 29–39.

[11]    Marcuvitz, N., *Waveguide Handbook*, McGraw-Hill, 1955. Reprint with errata and preface by N. Marcuvitz. Peter Peregrinus Ltd., 1986.

[12]    Maury, Mario A., Jr., "Microwave Coaxial Connector Technology: A Continuing Evolution," *Microwave Journal*, 1990 State of the Art Reference, pp. 39–59. (A good overview and history of coaxial connectors.)

[13]    Moreno, T., ed., *Microwave Transmission Design Data*, Sperry Gyroscope Company, Great Neck, New York, Publication No. 23-80, 1944. (Reprint available from Artech House, Norwood, MA, 1989.)

[14]    Neubauer, H., and F.R. Huber, "Higher Modes in Coaxial RF Lines," *The Microwave Journal*, June 1969, pp. 57–66.

[15]    Salvage, Seward T., *et al.*, "Design and Construction of a Thick-Film Shielded Strip Transmission Line," *1981 IEEE Southeastcon Conference Proceedings*, pp. 7–11.

[16]    Schelkunoff, S.A., "The Electromagnetic Theory of Coaxial Transmission Lines and Cylindrical Shields," *The Bell System Technical Journal*, October 1934, Vol. XIII, No. 4, October 1934, pp. 532–579. (Probably the best single reference for the coaxial line.)

[17]    Spaderna, Conan H., "A New Formula for Attenuation in Coaxial Cables," *IEEE Transactions on Microwave Theory and Techniques*, Vol. MTT-12, No. 5, May 1964, pp. 363–364.

[18]    Swengel, Sr., Robert C., *et al.*, "A Coaxial Interconnection System for High Speed Digital Processors," *IEEE Transactions on Parts, Hybrids, and Packaging*, Vol. PHP-10, No. 3, September 1974, pp. 181–187.

[19]    Weinschel, Bruno O., "Errors in Coaxial Air Line Standards Due to Skin Effect," *Microwave Journal*, November 1990, pp. 131–143.

[20]    Wheeler, Harold A., "Transmission-Line Properties of a Round Wire in a Polygon Shield," *IEEE Transactions on Microwave Theory and Techniques*, Vol. MTT-27, No. 8, August 1979, pp. 717–721.

[21]    Wong, Kenneth H., "Using Precision Coaxial Air Dielectric Transmission Lines as Calibration and Verification Standards," *Microwave Journal*, December 1988, pp. 83–92. (Discusses mechanical precision requirements for coaxial standards.)

[22]   "Embedded Coaxial Wires Speed Printed-Circuit Board," *Electronics*, January 27, 1986, pp. 56–57.

[23]   Young, Leo, ed., *Advances in Microwaves*, Volume 6, "Precision Coaxial Connectors," by Robert C. Powell, Academic Press, New York, 1971.

### 3.2.2   Partially Filled Round Coaxial Cable

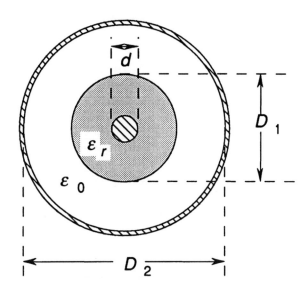

**Figure  3.2.2.1:**      **Partially  Filled  Round  Coax**

This case is encountered where a low-loss line is being constructed and the dielectric losses need to be reduced by making part of the dielectric air.  It also occurs when the dielectric diameter is reduced over a portion of its length for assembly.

$$Z_0 = \frac{\eta_0}{2\,\pi} \ln\left(\frac{D_2}{d}\right) \sqrt{\frac{\varepsilon_r \ln\left(\frac{D_2}{d}\right) + \ln\left(\frac{D_1}{d}\right)}{\varepsilon_r \ln\left(\frac{D_2}{d}\right)}} \quad (\Omega) \tag{3.2.2}$$

Ragan [3] also discusses the matching of such structures by proper choice of dimensions.  Hatsuda [1] analyzes a round coaxial structure with wedges of dielectric removed.

### 3.2.2.1      *Example:  Calculating q for a Partially Filled Coaxial Line*

Calculate $q$ for the partially filled round coaxial structure of 3.2.2.1.

**Solution:**        The ratio of the structure's actual capacitance to its capacitance with solid air as the dielectric is defined as $q$. We can easily get $C_{air}$ from the solid dielectric structure:

$$C_{air} = \frac{2\,\pi\,\varepsilon_0}{\ln\left(\frac{D_2}{d}\right)}$$

and using the relationships in Chapter 1 we can write the capacitance in terms of the known $Z_0$:

$$C_{actual} = \frac{\sqrt{\mu_0\,\varepsilon}}{Z_0}$$

$$q = \frac{C_{actual}}{C_{air}} = \frac{\dfrac{\sqrt{\mu_0\,\varepsilon}}{Z_0}}{\dfrac{2\,\pi\,\varepsilon_0}{\ln\left(\frac{D_2}{d}\right)}} = \frac{\dfrac{\eta_0}{2\,\pi}\ln\left(\frac{D_2}{d}\right)\sqrt{\dfrac{\varepsilon_r\ln\left(\frac{D_2}{d}\right)+\ln\left(\frac{D_1}{d}\right)}{\varepsilon_r\ln\left(\frac{D_2}{d}\right)}}}{\dfrac{2\,\pi\,\varepsilon_0}{\ln\left(\frac{D_2}{d}\right)}}$$

$$q = \frac{\varepsilon_r}{\sqrt{\dfrac{\varepsilon_r\ln\left(\frac{D_2}{d}\right)+\ln\left(\frac{D_1}{d}\right)}{\varepsilon_r\ln\left(\frac{D_2}{d}\right)}}}$$

REFERENCES

[1]    Hatsuda, Takeshi, "Computation of Impedance of Partially Filled and Slotted Coaxial Line," *IEEE Transactions on Microwave Theory and Techniques*, MTT-15, No. 11, November 1967, pp. 643–644.

[2]    Kraus, J.D., *Electromagnetics*, McGraw-Hill, New York, 1953.

[3]    Ragan, George L., ed., *Microwave Transmission Circuits*, MIT Radiation Laboratory Series: Volume 9, McGraw-Hill, New York, 1948.

[4]    Sullivan, D.J., and D.A. Parkes, "Stepped Transformers for Partially Filled Transmission Lines," *IEEE Transactions on Microwave Theory and Techniques*, MTT-8, No. 2, March 1960, pp. 212–217.

### 3.2.3    Eccentric Round Coaxial Cable

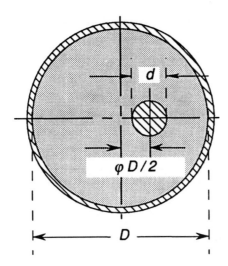

**Figure  3.2.3.1:**        **Eccentric  Round  Coaxial  Cable**

The eccentric coax equations enable us to analyze the effects of tolerances in the manufacture of the cable. This structure has also been used as a continuously adjustable $\lambda / 4$ line. The center conductor is moved within the outer shield by a mechanical probe resulting in smooth variation of the characteristic impedance.

For center conductors off center in one direction, [1] gives:

$$Z_0 = \frac{\eta_0}{2.0 \ \pi \ \sqrt{\varepsilon_r}} \cosh^{-1} \left[ \frac{D}{2.0 \ d} \left( 1.0 - \varphi^2 \right) + \frac{d}{2.0 \ D} \right] (\Omega) \qquad (3.2.3.1)$$

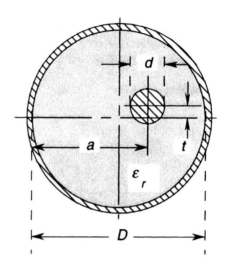

**Figure 3.2.3.2: Eccentric Round Coaxial Line**

An alternative relation for this structure dimensioned as in Figure 3.2.3.2 [2]:

$$Z_0 = \left[ \frac{\eta_0}{2\pi\sqrt{\varepsilon_r}} \ln\left(\frac{D}{d}\right) \right] \left[ 1.0 - \frac{e^2 k^2}{(k^2 - 1.0) \log k} \right] \quad (\Omega) \tag{3.2.3.2}$$

where

$$b = \frac{d}{2.0} \tag{3.2.3.3}$$

$$c = t \tag{3.2.3.4}$$

$$e = \frac{c}{a} \tag{3.2.3.5}$$

$$k = \frac{a}{b} \tag{3.2.3.6}$$

The conductor losses of this line are

$$\alpha_c = \alpha_{c,centered} \left[ 1.0 + \frac{2.0\, e^2}{k} + \frac{e^2 k^2}{(k^2 - 1.0) \log k} \right] \quad (dB\,/\,m) \tag{3.2.3.7}$$

$\alpha_{c,centered}$ = conductor losses of the centered line calculated with (3.2.1.2).

REFERENCES

[1]    Moreno, T., ed., *Microwave Transmission Design Data*, Sperry Gyroscope Company, Great Neck, New York, Publication No. 23-80, 1944. (Reprint by Artech House, Norwood, MA, 1989.)

[2]    Schelkunoff, S.A., "The Electromagnetic Theory of Coaxial Transmission Lines and Cylindrical Shields," *The Bell System Technical Journal*, October 1934, Vol. XIII, No. 4, pp. 532–579. (Probably the best single reference for the coaxial line.)

### 3.2.4  Square Coaxial Cable

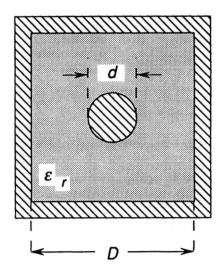

**Figure 3.2.4.1:**    Square Coaxial With Circular Center Conductor

The square coaxial wire configurations are useful at very high frequencies as smaller replacements for waveguide. The square cross section makes them easier to fabricate than a round coax.

For the round center conductor as shown in Figure 3.2.4.1:

$$Z_0 = \frac{\eta_0}{2\pi\sqrt{\varepsilon_r}} \ln\left[\frac{1.0787\,D}{d}\right] \quad (\Omega) \tag{3.2.4.1}$$

and for $Z_0 \leq 2.0\ \Omega$

$$Z_0 = 21.2\ \sqrt{D/d - 1.0} \quad (\Omega) \tag{3.2.4.2}$$

which was derived by S. Frankel. Error of the first equation is less than 1.5% above 17 $\Omega$ and very small above 30 $\Omega$ [2]. The second equation is accurate to 0.5 $\Omega$.

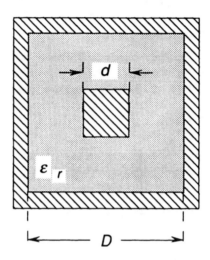

**Figure 3.2.4.2:** **Square Coax with Square Center Conductor**

For the square center conductor configuration shown in Figure 3.2.4.2 a conformal mapping solution is available [6]:

$$Z_0 = 5.0 \; \pi \; v_p \times 10^{-8} \frac{K(k')}{K(k)} = \frac{5.0 \; \pi \; c}{\sqrt{\varepsilon_r}} \frac{K(k')}{K(k)} \times 10^{-8} \quad (\Omega) \qquad (3.2.4.3)$$

$$k = \frac{(\Lambda' - \Lambda)^2}{(\Lambda' + \Lambda)^2} \qquad (3.2.4.4)$$

$$\Lambda' = \sqrt{1.0 - \Lambda^2} \qquad (3.2.4.5)$$

where $\Lambda$ is the solution to:

$$\frac{K(\Lambda')}{K(\Lambda)} = \frac{D - d}{D + d} \qquad (3.2.4.6)$$

This can be solved iteratively with an equation solver or using the relations of Chapter 12. Observe that the above is always $\geq 0$ and $\leq 1$.

An approximation [6] is:

$$Z_0 \cong \frac{1.0}{4.0 \, v_p \, (C_{pp} + C_c)} = \frac{\eta_0}{\sqrt{\varepsilon_r}} \left[ \frac{1.0}{4.0 \left( \dfrac{2.0 \, d}{D - d} + 0.558 \right)} \right] (\Omega) \qquad (3.2.4.7)$$

which can be seen to be the equation for four parallel plate capacitors ($C_{pp}$) added to the four corner capacitances ($C_c$). The accuracy of the above equation is better than 1% for

$$D/d \leq 4.0$$

### REFERENCES

[1]  Allessandri, F., and R. Sorrentio, "Analysis of T-Junction in Square Coaxial Cable," *18th European Microwave Conference Proceedings*, 1988, pp. 162–167.

[2]  Cohn, Seymour B., "Beating a Problem to Death," *Microwave Journal*, Vol. 12, No. 11, November 1969, pp. 22–24.

[2a]  Green, H.E., and J.D. Cashman, "Higher Order Mode Cutoff in Polygonal Transmission Lines," *IEEE Transactions on Microwave Theory and Techniques*, Vol. MTT-33, No. 1, January 1985, pp. 67–69.

[3]  Hillberg, Wolfgang, "From Approximations to Exact Relations for Characteristic Impedances," *IEEE Transactions on Microwave Theory and Techniques*, Vol. MTT-17, No. 5, May 1969, pp. 259–265.

[4]  Liao, Samuel Y., *Microwave Circuit Analysis and Amplifier Design*, Prentice-Hall, Englewood Cliffs, NJ, 1987.

[5]  Wheeler, Harold A., "Transmission-Line Properties of a Round Wire in a Polygon Shield," *IEEE Transactions on Microwave Theory and Techniques*, Vol. MTT-27, No. 8, August 1979, pp. 717–721.

[6]  Young, Leo, ed., *Advances in Microwaves*, Volume 2, "The Numerical Solution of Transmission Line Problems," by Harry E. Green, Academic Press, New York, 1967.

### 3.2.5 Rectangular Coaxial Line

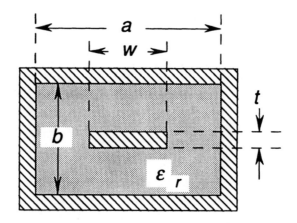

**Figure 3.2.5.1:**    Rectangular Coaxial Line

This structure is a transitional structure between round coax and stripline, microstrip line or other planar lines. Many attempts have been made to utilize this structure for photolithographic construction of fully shielded controlled impedance lines. Rotating the center conductor relative to the shield also allows this structure to have a continuously variable characteristic impedance [4].

As reported in [9] rectangular coax can be used in PCBs constructed with additive techniques to obtain a completely shielded structure. The process by Augat Microtec builds the structure using photolithographic techniques from polyimide and copper.

A related structure [14], Figure 3.2.5.2, was constructed with thick-film techniques. The main limitation appears to have been the thickness, b, achievable without excessive layered printings. Without a sufficiently thick dielectric, center conductor to shield shorts were common. Low dielectric constant pastes were recommended.

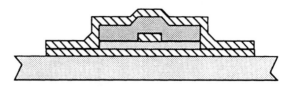

**Figure 3.2.5.2:**    Thick-Film Coaxial Line

$$Z_0 = \frac{\eta_0}{4.0 \sqrt{\varepsilon_r}} \left[ \frac{1.0}{\frac{w/b}{1.0 - t/b} + \frac{2.0}{\pi} \ln \left( \frac{1.0}{1.0 - t/b} + \coth \frac{\pi a}{2.0 b} \right)} \right] (\Omega)$$

(3.2.5.1)

for a square shield and center conductor, $a = b$ and $w = t$:

$$Z_0 = \frac{\eta_0}{4.0 \sqrt{\varepsilon_r}} \left[ \frac{1.0}{\frac{1.0}{b/t - 1.0} + \frac{2.0}{\pi} \ln \left( \frac{1.0}{1.0 - t/b} + \coth \frac{\pi}{2.0} \right)} \right] (\Omega)$$   (3.2.5.2)

As stated earlier, rectangular coaxial structures are also useful at extremely high frequencies as a smaller replacement for waveguide that is more easily fabricated than round coax.

Rotation of the center conductor varies the characteristic impedance of the line. The equations are in Cruz and Brooke, [4]:

$$Z(\theta) = \frac{Z(0°) (1.0 + \cos 2.0\theta) + Z(90°) (1.0 - \cos 2.0\theta)}{2.0} \quad (\Omega)$$   (3.2.5.3)

where $Z(0°)$ and $Z(90°)$ are calculated with any of the equations for a rectangular line above. Although the calculations of $Z(0°)$ and $Z(90°)$ were out of range for the [2] and [1] equations, agreement with experimental data to better than 1.4% was achieved.

REFERENCES

[1]   Chen, Tsung-Shan, "Determination of the Capacitance, Inductance, and Characteristic Impedance of Rectangular Lines," *IRE Transactions on Microwave Theory and Techniques*, Vol. MTT-8, No. 9, September 1960, pp. 510–519.

[2]   Cohn, Seymour B., "Characteristic Impedance of the Shielded-Strip Transmission Line," *Transactions of the IRE*, Vol. MTT-2, July 1954, pp. 52–57.

[3]   Conning, S.W., "The Characteristic Impedance of Square Coaxial Line," *IEEE Transactions on Microwave Theory and Techniques*, Vol. 12, No. 4, April 1964, pp. 468–469.

[4]   Cruz, J.E., and R.L. Brooke, "A Variable Characteristic Impedance Coaxial Line," *IEEE Transactions on Microwave Theory and Techniques*, MTT-13, No. 4, April 1965, pp. 477–478.

[5]   Cruzan, O.R., and R.V. Garver, "Characteristic Impedance of Rectangular Coaxial Transmission Lines," *IEEE Transactions on Microwave Theory and Techniques*, Vol. MTT-12, No. 9, September 1964, pp. 488–495 and "Correction to

"Characteristic Impedance of Rectangular Coaxial Transmission Lines,"" Vol. MTT-32, No. 2, February 1984, p. 219.

[6]     Garver, Robert V., "$Z_0$ of Rectangular Coax," *IEEE Transactions on Microwave Theory and Techniques*, Vol. MTT-9, No. 3, May 1961, pp. 262–263.

[7]     Green, Harry E., "The Characteristic Impedance of Square Coaxial Line," *IEEE Transactions on Microwave Theory and Techniques*, Vol. 11, No. 11, November 1963, pp. 554–555.

[8]     Gruner, L., "Higher Order Modes in Rectangular Coaxial Waveguides," *IEEE Transactions on Microwave Theory and Techniques*, Vol. MTT-15, No. 8, August 1967, pp. 483–485.

[9]     Landis, Richard C., "Buried Coaxial Conductors for High-Speed Interconnections," *IEEE Transactions on Components, Hybrids, and Manufacturing Technology*, Vol. CHMT-10, No. 2, June 1987, pp. 204–208.

[10]    Lau, K.H., "Loss Calculations for Rectangular Coaxial Lines," *IEE Proceedings*, Vol. 135, Pt. H, No. 3, June 1988, pp. 207–209.

[11]    Riblet, Henry J., "An Expansion for the Fringing Capacitance," *IEEE Transactions on Microwave Theory and Techniques*, Vol. MTT-28, No. 3, March 1980, pp. 265–267.

[12]    Riblet, Henry J., "Upper Limits on the Error of an Improved Approximation for the Characteristic Impedance of Rectangular Coaxial Line," *IEEE Transactions on Microwave Theory and Techniques*, Vol. MTT-28, No. 6, June 1980, pp. 666–667.

[13]    Salvage, Seward T., *et al.*, "Design and Construction of a Thick-Film Shielded Strip Transmission Line," *1981 IEEE Southeastcon Conference Proceedings*, pp. 7–11.

[14]    Terakado, Ryuiti, "The Characteristic Impedance of Rectangular Coaxial Line with Ratio 2:1 of Outer-to-Inner Conductor Side Length," *IEEE Transactions on Microwave Theory and Techniques*, Vol. 24, No. 2, February 1976, pp. 124–125.

[15]    Tippet, John C., and David C. Chang, "A New Approximation for the Capacitance of a Rectangular Coaxial-Strip Transmission Line," *IEEE Transactions on Microwave Theory and Techniques*, Vol. MTT-24, No. 9, September 1976, pp. 602–604.

### 3.2.6 Trough Line or Channel Line

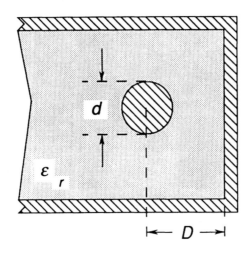

Figure 3.2.6: Trough Line

Wheeler [2] gives:

$$Z_0 = \frac{\eta_0}{4\pi\sqrt{\varepsilon_r}} \ln\left[ 1.0 + \left\{ \frac{1}{2}\left( \frac{4}{\pi}\tanh\frac{\pi}{2} \right)^2 \left[ (D/d)^2 - 1.0 \right] \right\} \right.$$

$$\left. + \sqrt{ \left\{ \frac{1}{2}\left( \frac{4}{\pi}\tanh\frac{\pi}{2} \right)^2 \left[ (D/d)^2 - 1.0 \right] \right\}^2 + \frac{4}{9}\left[ (D/d)^2 - 1.0 \right] } \right]$$

$$(\Omega) \qquad (3.2.6)$$

with a stated accuracy of about 1%.

### REFERENCES

[1]  Liao, Samuel Y., *Microwave Circuit Analysis and Amplifier Design*, Prentice-Hall, Englewood Cliffs, NJ, 1987.

[2]  Wheeler, Harold A., "Transmission-Line Properties of a Round Wire in a Polygon Shield," *IEEE Transactions on Microwave Theory and Techniques*, Vol. MTT-27, No. 8, August 1979, pp. 717–721.

### 3.2.7   Strip-Centered Coaxial Line

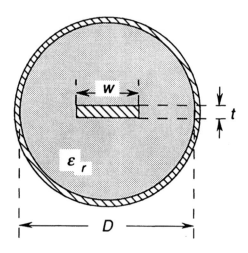

**Figure  3.2.7.1:**      **Strip-Centered Coaxial Cable**

The configuration above is used to obtain lower loss by building a coaxial cable with a larger center conductor surface area.  Bongianni [1] reports cables made with extremely fine dimensions—the center conductor was 150 μm × 12.5 μm.  For $Z_0 \leq 30.0 \; \pi / \sqrt{\varepsilon_r}$ (Ω):

$$Z_0 = \frac{15.0 \; \pi^2}{\sqrt{\varepsilon_r}} \; \frac{1.0}{\ln \left[ \frac{2.0(D + w)}{D - w} \right]} \quad (\Omega) \tag{3.2.7.1}$$

For $Z_0 \geq 30.0 \; \pi / \sqrt{\varepsilon_r}$ (Ω):

$$Z_0 = \frac{60.0}{\sqrt{\varepsilon_r}} \ln \left( \frac{2.0 \; D}{w} \right) \quad (\Omega) \tag{3.2.7.2}$$

There is a bit of a chicken-and-the-egg problem here because it is necessary to know $Z_0$ in order to choose the equation for $Z_0$; however, the two equations pass quite close and it is okay to use either to make the choice.

REFERENCES

[1]    Bongianni, Wayne L., *Proceedings of the IEEE*, Vol. 72, No. 12, December 1984, pp. 1810–1811.

[2]  Hilberg, Wolfgang, *Electrical Characteristics of Transmission Lines*, Artech House, Norwood, MA, 1979. (This is an excellent conformal mapping reference with many worked examples of the technique.)

## 3.3   PAIRED LINES

### 3.3.1   Parallel Wires

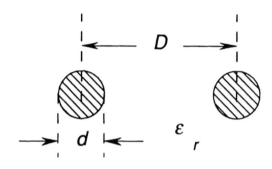

**Figure  3.3.1.1:**        **Parallel  Wires**

$$Z_0 = \frac{\eta_0}{\pi\sqrt{\varepsilon_r}} \cosh^{-1}\left(\frac{D}{d}\right) \ (\Omega) \tag{3.3.1.1}$$

or

$$Z_0 = \frac{\eta_0}{2.0 \ \pi\sqrt{\varepsilon_r}} \cosh^{-1}\left(\frac{2.0 \ D^2 - d^2}{d^2}\right) \ (\Omega) \tag{3.3.1.2}$$

valid for

$$d/D \ll 1.0$$

The first reference is found in [4] and [5]. The second was derived with conformal mapping techniques in [2] with a stated accuracy of better than 0.24%.   Green, *et al.* [1] gives the conductor losses as:

$$\alpha_c = \frac{P}{2.0 \ d} \frac{\sqrt{\dfrac{f \varepsilon}{\sigma}}}{\cosh^{-1} v} \quad \text{(nepers / cm)} \tag{3.3.1.3}$$

where

$$P = \frac{v}{\sqrt{v^2 - 1.0}} \qquad\qquad (3.3.1.4)$$

$$v = \frac{D}{d} \qquad\qquad (3.3.1.5)$$

and $P$ is a proximity factor; the equation is valid for high frequencies.

REFERENCES

[1]    Green, E.I., *et al.*, "The Proportioning of Shielded Circuits for Minimum High-Frequency Attenuation," *The Bell System Technical Journal*, 1936, pp. 248–283.

[2]    Hilberg, Wolfgang, *Electrical Characteristics of Transmission Lines*, Artech House, Norwood, MA, 1979.

[3]    Liboff, Richard L., and G. Conrad Dalman, *Transmission Lines, Waveguides, and Smith Charts*, Macmillan Publishing Co., New York, 1985.

[4]    Ramo, Simon, and John R. Whinnery, *Fields and Waves in Modern Radio*, Wiley, London, 1944.

[5]    *Reference Data For Radio Engineers*, Howard W. Sams, Indianapolis, IN, 1982.

### 3.3.2   Unequal Size Parallel Wires

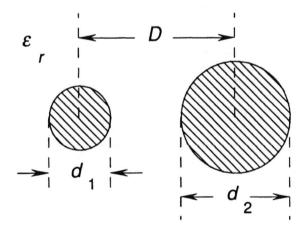

**Figure 3.3.2.1:**     **Unequal Size Parallel Wires**

$$Z_0 = \frac{\eta_0}{2.0\,\pi\sqrt{\varepsilon_r}}\cosh^{-1}\left(\frac{4.0\,D^2 - d_1^2 - d_2^2}{2.0\,d_1 d_2}\right)\ (\Omega) \qquad\qquad (3.3.2.1)$$

The stated accuracy [1] is better than 0.24%.

REFERENCES

[1]    Hilberg, Wolfgang, *Electrical Characteristics of Transmission Lines*, Artech House, Norwood, MA, 1979.

### 3.3.3   Twisted Pair

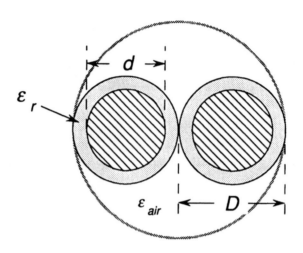

**Figure   3.3.3.1:**        Twisted   Pair

The twisted pair provides good low frequency shielding.  Undesired signals tend to be coupled equally into each line of the pair.  A differential receiver will therefore completely cancel the interference.  Lefferson [3] gives design equations for this configuration:

$$Z_0 = \frac{\eta_0}{\pi \sqrt{\varepsilon_{eff}}} \cosh^{-1}\left(\frac{D}{d}\right) \ (\Omega) \qquad\qquad (3.3.3.1)$$

$$\varepsilon_{eff} = 1.0 + q\,(\varepsilon_r - 1.0) \qquad\qquad (3.3.3.2)$$

$$q = 0.25 + 0.0004\ \theta^2 \qquad\qquad (3.3.3.3)$$

$$T = \frac{\tan \theta}{\pi\,D} = \text{twists per length} \qquad\qquad (3.3.3.4)$$

or

$$\theta = \tan^{-1}\left(T\,\pi\,D\right) \tag{3.3.3.5}$$

where $T$ and $D$ have the same length units and $\theta$ is the pitch angle of the twist; the angle between the twisted pair's center line and the twist. It was found to be optimal for $\theta$ to be between 20 and 45°. Smaller angles make the twist loose and create problems in maintaining tolerances. Angles above approximately 50.5° break the wire. The value of $q$ was determined by fitting a line to measurements of the effective dielectic constant.

For the softer insulation PTFE, a different equation should be used for $q$

$$q = 0.25 + 0.001\,\theta^2 \tag{3.3.3.6}$$

An equation for the wire's total length before twisting in terms of number of turns, $N$, is:

$$l = N\,\pi D\,\sqrt{1.0 + \frac{1.0}{\tan^2\theta}} \tag{3.3.3.7}$$

Lefferson found these equations sufficiently accurate to design transmission lines with VSWR $\leq 1.1{:}1$.

In [2] a new analysis and experimental data are presented. The derived equations are

$$Z_0 = \sqrt{\frac{L}{C}}\ (\Omega) \tag{3.3.3.8}$$

$$\varepsilon_{eff} = \frac{C}{C_{air}}$$

where

$$L = \left(\frac{\mu_0}{\pi}\right)\cosh^{-1}\left(\frac{D}{d}\right) \tag{3.3.3.9}$$

$$C = C_1 + C_2 - C_3 \tag{3.3.3.10}$$

$$C_1 = \int_a^b \frac{\varepsilon_0\,dx}{D + (1.0\,/\,\varepsilon_r - 1.0)\sqrt{D^2 - x^2} - \dfrac{\sqrt{d^2 - x^2}}{\varepsilon_r}} \tag{3.3.3.11}$$

$$C_2 = \frac{\pi\,\varepsilon_0}{\cosh^{-1}\left(\dfrac{D}{d}\right)} \tag{3.3.3.12}$$

$$C_3 = \int_a^b \frac{\varepsilon_0 \, dx}{D - \sqrt{d^2 - x^2}} \tag{3.3.3.13}$$

Calculate $C_{air}$ by replacing $\varepsilon_r$ with $\varepsilon_0$ in the equations for $C$. The integrals are evaluated by the Romberg numerical technique. Accuracy was tested by comparison with measurements of the line impedance. The technique was found to be accurate within the tolerances of the measurement and the film thickness variations.

REFERENCES

[1]    Blood, Jr., William R., *Motorola MECL System Design Handbook*, 4th Ed.

[2]    Broxon, John H., and Douglas K. Linkhart, "Twisted-Wire Transmission Lines," *RF Design*, June 1990, pp. 73–76.

[3]    Lefferson, Peter, "Twisted Magnet Wire Transmission Line," *IEEE Transactions on Parts, Hybrids, and Packaging*, Vol. PHP-7, No. 4, December 1971, pp. 148–154, and errata.

### 3.3.4    Five-Wire Line

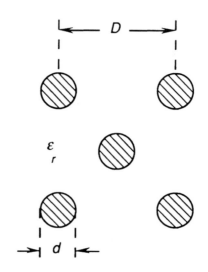

Figure 3.3.4.1:        Five-Wire Line

$$Z_0 = \frac{2.5044 \, \eta_0}{\pi \sqrt{\varepsilon_r}} \ln\left[\frac{D}{0.933 \, d}\right] \quad (\Omega) \tag{3.3.4}$$

where

$$d / D \ll 1.0$$

Of course, five conductors means lots of modes, so although the reference doesn't state it an assumption is that $(D - d)$ is small relative to a wavelength.

REFERENCES

[1]    *Reference Data For Radio Engineers*, Howard W. Sams, Indianapolis, IN, 1982.

### 3.3.5  Paired Strips

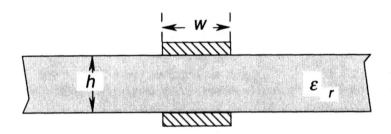

**Figure  3.3.5.1:**     **Paired Strips**

For wide strips $(a / b > 1)$:

$$Z_0 = \frac{\eta_0}{\sqrt{\varepsilon_r}} \left\{ \frac{a}{b} + \frac{1.0}{\pi} \ln 4 + \frac{\varepsilon_r + 1.0}{2 \pi \varepsilon_r} \ln \left[ \frac{\pi e \, (a / b + 0.94)}{2.0} \right] \right.$$

$$\left. + \frac{\varepsilon_r - 1.0}{2 \pi \varepsilon_r^2} \ln \frac{e \, \pi^2}{16.0} \right\}^{-1} \; (\Omega) \tag{3.3.5.1}$$

Stated error is less than 1% for wide strips.

For narrow strips $(a / b < 1)$:

$$Z_0 = \frac{\eta_0}{\pi \sqrt{\varepsilon_r}} \left[ \ln \frac{4.0 \, b}{a} + \frac{1.0}{8.0} \left( \frac{a}{b} \right)^2 - \frac{\varepsilon_r - 1.0}{2.0 \, (\varepsilon_r + 1.0)} \left( \ln \frac{\pi}{2.0} + \frac{\ln \frac{4.0}{\pi}}{\varepsilon_r} \right) \right] \; (\Omega)$$

$$\tag{3.3.5.2}$$

where

$$a = w / 2.0 \tag{3.3.5.3}$$

$$b = h / 2.0 \tag{3.3.5.4}$$

Stated error is less than 1% for narrow strips.

Equations are valid for $2b$ much smaller than half a wavelength in the dielectric ($\lambda_g$).

REFERENCES

[1]     Crampagne, Raymond, and Gratia Khoo, "Comments on 'Approximation for the Symmetrical Parallel-Strip Transmission Line,'" *IEEE Transactions on Microwave Theory and Techniques*, August 1976, Vol. MTT-24, No. 8, pp. 532–534. (Comments on inaccuracies in approximations of [2]. Makes suggestions for increased accuracy)

[2]     Rochelle, J.M., "Approximations for the Symmetrical Parallel-Strip Transmission Line," *IEEE Transactions on Microwave Theory and Techniques*, MTT-23, No. 8, August 1975, pp. 712–714.

[3]     Wheeler, Harold A., "Transmission-Line Properties of Parallel Wide Strips by a Conformal-Mapping Approximation," *IEEE Transactions on Microwave Theory and Techniques*, Vol. MTT-12, May 1964, pp. 280–289.

[4]     Wheeler, Harold A., "Transmission-Line Properties of Parallel Strips Separated by a Dielectric Sheet," *IEEE Transactions on Microwave Theory and Techniques*, MTT-13, No. 2, March 1965, pp. 172–185.

## 3.4   COPLANAR STRUCTURES

### 3.4.1   Coplanar Waveguide

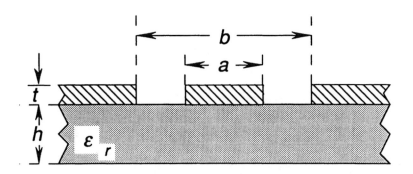

Figure 3.4.1.1:     Coplanar Waveguide

The coplanar waveguide (CPW) configuration's advantages stem from its single-sided nature. Grounding components does not require plated through-holes to a plane on the other side of the substrate. This makes it ideal for use with surface mounted components. Another feature of coplanar waveguide is that we can narrow traces to match component lead widths while keeping $Z_0$ constant. The ground plane should extend greater than $5b$ on each side of the gap or the coplanar strip analysis should be used. The ground planes on either side of the center conductor may need to be connected periodically with wire jumpers depending on the frequency. The enclosure's cover can be used to jumper the two ground planes if it is kept away by at least $(b - a)$. To prevent propagation of higher modes, $b$ should be less than $\lambda / 2$.

The design equations for coplanar waveguide are:

$$Z_0 = \frac{30.0 \, \pi}{\sqrt{\varepsilon_{eff,t}}} \frac{K(k_t')}{K(k_t)} \tag{3.4.1.1}$$

$$\varepsilon_{eff,t} = \varepsilon_{eff} - \frac{\varepsilon_{eff} - 1.0}{\dfrac{(b - a) / 2.0}{0.7 \, t} \dfrac{K(k)}{K'(k)} + 1.0} \tag{3.4.1.2}$$

$$\varepsilon_{eff} = 1.0 + \frac{\varepsilon_r - 1.0}{2.0} \frac{K(k')K(k_1)}{K(k)K(k_1')} \tag{3.4.1.3}$$

$$k_t = \frac{a_t}{b_t} \qquad\qquad k = \frac{a}{b} \qquad\qquad (3.4.1.4)$$

$$k_t' = \sqrt{1.0 - k_t^2} \qquad k' = \sqrt{1.0 - k^2} \qquad\qquad (3.4.1.5)$$

$$k_1 = \frac{\sinh\left(\dfrac{\pi\,a_t}{4.0\,h}\right)}{\sinh\left(\dfrac{\pi\,b_t}{4.0\,h}\right)} \qquad\qquad (3.4.1.6)$$

$$k_1' = \sqrt{1.0 - k_1^2} \qquad\qquad (3.4.1.7)$$

$$a_t = a + \frac{1.25\,t}{\pi}\left[1.0 + \ln\left(\frac{4.0\,\pi\,a}{t}\right)\right] \qquad\qquad (3.4.1.8)$$

$$b_t = b - \frac{1.25\,t}{\pi}\left[1.0 + \ln\left(\frac{4.0\,\pi\,a}{t}\right)\right] \qquad\qquad (3.4.1.9)$$

These equations are corrected for thickness which is important for accurate results. The dielectric losses are described in [6, 11] by:

$$\alpha_d = \frac{q\,\varepsilon_r\,\tan\delta}{\varepsilon_{eff}\,\lambda_g} \quad (\mathrm{Np\,/\,m}) \qquad\qquad (3.4.1.10)$$

where $q$ and $\lambda_g$ have been previously defined.

The conductor losses are solved numerically and plotted in [6]. Jackson [9] reports that in some cases coplanar waveguide can have lower losses and dispersion than microstrip line for impedances near 50 $\Omega$.

An equation for conductor losses [5] is:

$$\alpha_c = \frac{R_s\,\sqrt{\varepsilon_{eff}}\left[\Phi(a) + \Phi(b)\right]}{480.0\,\pi\,K(k)\,K(k')\,k'} \quad (\mathrm{Np\,/\,m}) \qquad\qquad (3.4.1.11)$$

where

$$R_s = \text{surface resistivity} = \sqrt{\pi f \mu_0 / \sigma} \quad (\Omega\,/\,\square) \qquad\qquad (3.4.1.12)$$

$$\Phi(x) = \frac{\pi}{x} + \frac{\ln\left[\dfrac{8.0\,\pi\,x\,(1.0 - k)}{t\,(1.0 + k)}\right]}{x} \qquad\qquad (3.4.1.13)$$

These equations are valid for

$$t \ll a$$

$$t \gg b - a$$

Radiation losses are

$$\alpha_r = \frac{\pi}{Q_{rad}\,\lambda_g} \tag{3.4.1.14}$$

$$Q_{rad} = \frac{K(k)\,K(k')}{\Psi(\varepsilon_r, h, k_0, b)\,\psi_{sc}\,\psi_{oc}} \tag{3.4.1.15}$$

$$\Psi(\varepsilon_r, h, k_0, b) = (\varepsilon_r - 1.0)\left[1.0 + 0.5\,(\varepsilon_r - 1.0)^2\,(k_0\,h)^2\right]$$

$$\times (k_0\,h)\,(k_0\,b)^2\left[1.0 + 0.25\,(\varepsilon_r - 1.0)^2\,(k_0\,h)^2\right]$$

$$\times \left\{ \frac{1.0}{\sqrt{\varepsilon_{eff}}\,\left[1.0 + (\varepsilon_r - 1.0)^2\,(k_0\,h)^2\right]} \right\} \tag{3.4.1.16}$$

$$\psi_{sc} = \frac{5.0\,\pi}{8}\,\frac{1.0 + \left(1.0 - \frac{\pi^2}{8.0}\right)\left[1.0 + (\varepsilon_r - 1.0)^2\,(k_0\,h)^2\,/\,4.0\right]}{3.0\,\varepsilon_{eff}}$$

$$\times \frac{\left[1.0 + (\varepsilon_r - 1.0)^2\,(k_0\,h)^2\,/\,4.0\right]}{\varepsilon_{eff}} \tag{3.4.1.17}$$

$$\psi_{oc} = \frac{\pi}{8}\,\frac{3.0 + \left(1.0 - \frac{\pi^2}{8.0}\right)\left[1.0 + (\varepsilon_r - 1.0)^2\,(k_0\,h)^2\,/\,4.0\right]}{\varepsilon_{eff}} \tag{3.4.1.18}$$

These are valid for thin dielectric.

In Houdart [8, Figure 2], the effect of ground plane width ($c$ in Figure 3.4.8.1) on the CPW is analyzed. For $c\,/\,b > 5.0$ the impedance is affected by less than about 3%.

### 3.4.1.1    Example:    Hybrid IC Probe in Coplanar Waveguide

What are the required dimensions for a 50 $\Omega$ coplanar waveguide probe to a 10 mil $\times$ 10 mil bonding pad on a thick-film hybrid IC? The dielectric is 25.2 mil thick 96% alumina.

**Solution:**        Assuming the probe is the same material as the hybrid IC substrate and using the above equations with

$h$ = 0.0252 in = 0.0640 cm

$a$ = 0.010 in = 0.0254 cm

$t$ = 0.001 in = 0.003 cm

$\varepsilon_r$ = 10.0

The program iteratively adjusts $b$ until the goal impedance 50 $\Omega$ is achieved. The final dimension of the probe is:

$b$ = 0.0221 in. = 0.0561 cm
$\varepsilon_{eff}$ = 4.45

REFERENCES

[1]     Bachert, Peter S., "A Coplanar Waveguide Primer," *RF Design*, July 1988, pp. 52–57, and errata.

[2]     Bahl, Inder, and Prakash Bhartia, *Microwave Solid State Circuit Design*, Wiley, New York, 1988.

[3]     Bellantoni, J.V., and R.C. Compton, "A New Coplanar Waveguide Vector Network Analyzer for On Wafer Measurements," *Proceedings of Cornell/IEEE Conference on Advanced Concepts in High Speed Semiconductor Development and Circuits*, Ithaca, NY, 1989, pp. 201–207.

[4]     Davis, M.E., *et al.*, "Finite-Boundary Corrections to the Coplanar Waveguide Analysis," *IEEE Transactions on Microwave Theory and Techniques*, Vol. 21, No. 9, September 1973, pp. 594–596.

[5]     Ghione, Giovanni, *et al.*, "Q-factor evaluation for coplanar resonators," *Alta Frequenza*, Vol. LII, No. 3, May-June 1983, pp. 191–193.

[6]     Gopinath, A., "Losses in Coplanar Waveguides," *IEEE Transactions on Microwave Theory and Techniques*, Vol. MTT-30, No. 7, July 1982, pp. 1101–1104.

[7]     Gupta, K. C., Ramesh Garg, and Rakesh Chadha, *Computer-Aided Design of Microwave Circuits*, Artech House, Norwood, MA, 1981.

[8]     Houdart, M., "Coplanar Lines: Application to Broadband Microwave Integrated Circuits," *6th European Microwave Conference Proceedings*, 1976, pp. 49–53.

[9]     Jackson, R.W., "Coplanar Waveguide vs. Microstrip for Millimeter Wave Integrated Circuits," *1986 IEEE MTT-S International Microwave Symposium Digest*, pp. 699–702.

[10]  Kitazawa, T., *et al.*, "A Coplanar Waveguide with Thick Metal-Coating," *IEEE Transactions on Microwave Theory and Techniques*, Vol. 24, No. 9, September 1976, pp. 604–608.

[11]  Kitazawa, T., and Y. Hayashi, "Quasistatic Characteristics of a Coplanar Waveguide with Thick Metal Coating," *IEE Proceedings*, Vol. 133, Pt. H, No. 1, February 1986, pp. 18–20.

[12]  Koshiji, Kohji, *et al.*, "Simplified Computation of Coplanar Waveguide with Finite Conductor Thickness," *IEE Proceedings*, Vol. 130, Pt. H, No. 5, August 1983, pp. 315–321.

[13]  Riaziat, Majid, *et al.*, "Propagation Modes and Dispersion Characteristics of Coplanar Waveguides," *IEEE Transactions on Microwave Theory and Techniques*, Vol. MTT-38, No. 3, March 1990, pp. 245–251.

[14]  Shibata, Tsugumichi, "Characterization of MIS Structure Coplanar Transmission Lines for Investigation of Signal Propagation in Integrated Circuits," *IEEE Transactions on Microwave Theory and Techniques*, Vol. MTT-38, No. 7, July 1990, pp. 881–890.

[15]  Veyres, C., and V. Fouad Hanna, "Extension of the Application of Conformal Mapping Techniques to Coplanar Lines with Finite Dimensions," *International Journal of Electronics*, Vol. 48, No. 1, 1980, pp. 47–56.

[16]  Wen, Cheng P., "Coplanar Waveguide: A Surface Strip Transmission Line Suitable for Nonreciprocal Gyromagnetic Device Applications," *IEEE Transactions on Microwave Theory and Techniques*, MTT-17, No. 12, December 1969, pp. 1087–1090.

### 3.4.2   Micro-Coplanar Stripline

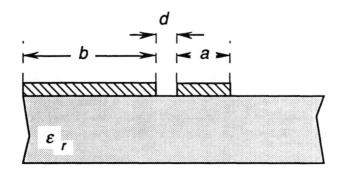

**Figure  3.4.2.1:**      **Micro-Coplanar  Stripline**

This structure has been called nonsymmetrical coplanar waveguide; however, to avoid confusion we prefer to reserve this term for coplanar waveguide having unequal gaps.

Kneppo and Gotzman [2] proposed this structure as an improvement over CPW for the connection of shunt elements. Their equations were derived with conformal transformation:

$$Z_0 = \frac{\eta_0}{\left(\sqrt{\varepsilon_{r_1}} + \sqrt{\varepsilon_{r_2}}\right)} \frac{K(k)}{K'(k)} \quad (\Omega) \qquad (3.4.2.1)$$

$$\varepsilon_{eff} = \frac{\left(\sqrt{\varepsilon_{r_1}} + \sqrt{\varepsilon_{r_2}}\right)^2}{4.0} \qquad (3.4.2.2)$$

where

$$k = \sqrt{\frac{1.0 + a/d + b/d}{(1.0 + b/d)(1.0 + a/d)}} \qquad (3.4.2.3)$$

Polynomial equations for the line width, $\varepsilon_{eff}$, and conductor losses are given in Quian and Yamashita [3] for dielectrics with $\varepsilon_r = 2.22$, 9.7, 10.1, and 12.9 and in Yamashita *et al.* [4] for $\varepsilon_r = 12.7$.

Losses may be calculated with the incremental inductance rule.

REFERENCES:

[1]     Fang, J., *et. al*, "Dispersion Characteristics of Microstrip Lines in the Vicinity of a Coplanar Ground," *Electronics Letters*, Vol. 23, No. 21, October 8, 1987, pp. 1142–1143.

[2]     Kneppo, Ivan and Jozef Gotzman, "Basic Parameters of Nonsymmetrical Coplanar Line," *IEEE Transactions on Microwave Theory and Techniques*, Vol. MTT-25, No. 8, August 1977, p. 718.

[3]     Quian, Yongxi, and Eikichi Yamashita, "Additional Approximate Formulas and Experimental Data on Micro-Coplanar Striplines," *IEEE Transactions on Microwave Theory and Techniques*, Vol. MTT-38, No. 4, April 1990, pp. 443–445.

[4]     Yamashita, Eikichi, *et al.*, "Characterization Method and Design Formulas of MCS Lines Proposed for MMIC's," *1987 IEEE MTT-S Symposium Digest*, pp. 685–688.

### 3.4.3 Coplanar Waveguide with Ground

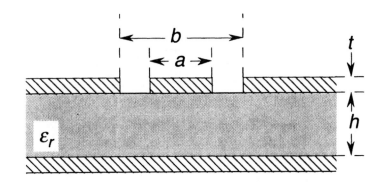

**Figure 3.4.3.1:** Coplanar Waveguide with Ground

The equations of this section can be used to analyze coplanar waveguide with ground or for microstrip lines with signal side ground plane.

$$Z_0 = \frac{\eta_0}{2.0 \sqrt{\varepsilon_{eff}}} \frac{1.0}{\frac{K(k)}{K(k')} + \frac{K(k_1)}{K(k_1')}} \tag{3.4.3.1}$$

$$k = a / b \tag{3.4.3.2}$$

$$k' = \sqrt{1.0 - k^2} \tag{3.4.3.3}$$

$$k_1' = \sqrt{1.0 - k_1^2} \tag{3.4.3.4}$$

$$k_1 = \frac{\tanh\left(\frac{\pi a}{4.0 \, h}\right)}{\tanh\left(\frac{\pi b}{4.0 \, h}\right)} \tag{3.4.3.5}$$

$$\varepsilon_{eff} = \frac{1.0 + \varepsilon_r \frac{K(k')}{K(k)} \frac{K(k_1)}{K(k_1')}}{1.0 + \frac{K(k')}{K(k)} \frac{K(k_1)}{K(k_1')}} \tag{3.4.3.6}$$

The notation varies between references. Some use the $a$, $b$ notation of Figure 3.4.2.1 or a $2a$ and $2b$ version while others use $s$ (spacing to the adjacent ground) and $w$ (trace width).

These equations "show good agreement" with spectral-domain variational calculation techniques.

The structure may propagate in three different modes: microstrip, coplanar waveguide, and coupled slotlines. To prevent slotline modes, jumpers connecting the two halves of the component side ground plane can be used. To avoid microstrip line modes, it is recommended [2] that $h \gg b$ and that the component side ground extend away from the trace on each side more than $b$.

REFERENCES:

[1]    Ghione, G., and C. Naldi, "Parameters of Coplanar Waveguide with Lower Ground Plane," *Electronics Letters*, Vol. 19, No. 18, September 1, 1983, pp. 734-735. (Error in equation for $k_1$', p. 735, corrected above.)

[2]    Riaziat, M., et. al, "Single-Mode Operation of Coplanar Waveguides," *Electronics Letters*, Vol. 23, No. 24, November 19, 1987, pp. 1281–1283.

[3]    Singh, Donald R., and Keith S. Champlin, "Coplanar Schottky Waveguides for Microwave Phase Shifting, Attenuation and Harmonic Generation," *Circuits and Systems Symposium Digest*, 1990, pp. 3073–3076.

### 3.4.4   Shielded  Coplanar  Waveguide

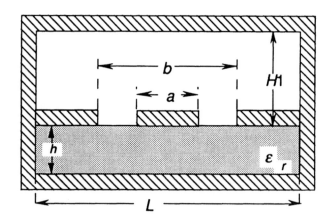

**Figure  3.4.4.1:**    Shielded  Coplanar  Waveguide

The formulas below allow calculation of coplanar waveguide including the shield's effects [1, 2]. Many times we are simply looking for guidelines on how not to be affected by the shield's presence. For $L/b \geq 1.75$, and $H_1/a \geq 2.50$, $Z_0$ is affected less than 1.5% by the shield's presence.

The equations of [2] assume $H_1$ is infinite and are:

$$Z_0 = \cfrac{1.0}{\cfrac{5.0\, q}{Z_m\,(1.0 + 5.0\, q)} + \cfrac{1.0}{Z_c(1.0 + q)}} \quad (\Omega) \tag{3.4.4.1}$$

$$q = \frac{a}{h}\left(\frac{b}{a} - 1.0\right)\left[3.6 - 2\, e^{-(\varepsilon_r + 1.0)\,/\,4.0}\right] \tag{3.4.4.2}$$

$$Z_c = \text{coplanar waveguide impedance} \tag{3.4.4.3}$$

$$Z_m = \text{microstrip impedance} \tag{3.4.4.4}$$

Use the equations given in the coplanar waveguide and microstrip line sections together with the relevant dimensions to calculate $Z_c$ and $Z_m$. Accuracy is within 2% of numerical results.

## REFERENCES

[1]   Leong, M.S., P.S. Kooi, and A.L. Satya Prakash, "Effect of a Conducting Enclosure on the Characteristic Impedance of Coplanar Waveguides," *Microwave Journal*, August 1986, pp. 105–108.

[2]   Rowe, David A., and Binneg Y. Lao, "Numerical Analysis of Shielded Coplanar Waveguides," *IEEE Transactions on Microwave Theory and Techniques*, Vol. MTT-31, No. 11, November 1983, pp. 911–915.

## 3.4.5   Asymmetric Coplanar Waveguide

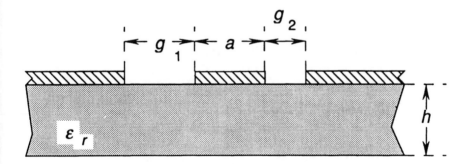

Figure  3.4.5.1:       Asymmetric  Coplanar  Waveguide

The asymmetric coplanar waveguide structure can be used to reduce the line impedance of symmetric coplanar waveguide or to analyze the effects of fabrication tolerances.

The equations of Hanna and Thebault [1, 2] derived with conformal mapping are

$$Z_0 = \frac{30.0 \, \pi}{\sqrt{\varepsilon_{eff}}} \frac{K'(k_1)}{K(k_1)} \quad (\Omega)$$

(3.4.5.1)

$$\varepsilon_{eff} = 1.0 + \frac{\varepsilon_r - 1.0}{2.0} \frac{K(k_2)}{K'(k_2)} \frac{K'(k_1)}{K(k_1)}$$

(3.4.5.2)

where

$$k_1 = \frac{0.5 \, b \left[ 1.0 + \alpha \, (0.5 \, b + d_1) \right]}{0.5 \, b + d_1 \, \alpha \, \sqrt{0.5 \, b}}$$

(3.4.5.3)

$$k_2 = \frac{w_A \, (1.0 + \alpha_1 \, w_B)}{w_B + \alpha_1 \, w_A^2}$$

(3.4.5.4)

and

$$w_A = \sinh \left( \frac{\pi \, a}{4.0 \, h} \right)$$

(3.4.5.5)

$$w_B = \sinh \left[ \frac{\pi \, (a \, / \, 2.0 + g_1)}{2.0 \, h} \right]$$

(3.4.5.6)

$$w_E = -\sinh \left[ \frac{\pi \, (a \, / \, 2.0 + g_2)}{2.0 \, h} \right]$$

(3.4.5.7)

$$\alpha = \frac{d_1 \, d_2 + 0.5 \, b \, (d_1 + d_2) \pm \sqrt{d_1 \, d_2 \, (b + d_1) \, (b + d_2)}}{\sqrt{0.5 \, b} \, (d_1 - d_2)}$$

(3.4.5.8)

$$\alpha_1 = \left( \frac{1.0}{w_B + w_E} \right) \left[ -1.0 - \frac{w_B w_E}{w_A^2} \pm \sqrt{\left( \frac{w_B^2}{w_A^2} - 1.0 \right) \left( \frac{w_E^2}{w_A^2} - 1.0 \right)} \right]$$

(3.4.5.9)

The equations assume that the traces have negligible thickness. Comparison of the equations to experimental data showed agreement to within 4% for $Z_0$. Choose the '−' sign solution of Equations (3.4.5.8) and (3.4.5.9).

Graphs of the dispersion of this structure are available in [3].

REFERENCES

[1] Hanna, Victor Fouad, and Dominique Thebault, "Theoretical and Experimental Investigation of Asymmetric Coplanar Waveguides," *1984 IEEE MTT-S Symposium Digest*, pp. 469–471.

[2] Hanna, V. Fouad, and D. Thebault, "Analysis of Asymmetrical Coplanar Waveguides," *International Journal of Electronics*, Vol. 50, No. 3, 1981, pp. 221–224.

[2] Kitlinski, M., and B. Janiczak, "Dispersion Characteristics of Asymmetric Coupled Slot Lines on Dielectric Substrates," *Electronics Letters*, Vol. 19, No. 3, February 1983, pp. 91–92.

## 3.4.6 Coplanar Strips

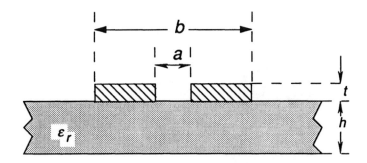

**Figure 3.4.6.1:** Coplanar Strips

The coplanar strips structure is similar to paired wire transmission line structures. Note also that coplanar strips are the complementary structure to coplanar waveguide.

$$Z_0 = \frac{\eta_0}{\sqrt{\varepsilon_{eff}}} \frac{K(k)}{K(k')} \tag{3.4.6.1}$$

$$\varepsilon_{eff} = 1 + \frac{\varepsilon_r - 1}{2} \frac{K(k')\, K(k_1)}{K(k)\, K(k_1')} \tag{3.4.6.2}$$

$$k = \frac{a}{b} \tag{3.4.6.3}$$

$$k' = \sqrt{1.0 - k^2} \tag{3.4.6.4}$$

$$k_1' = \sqrt{1.0 - k_1^2} \tag{3.4.6.5}$$

$$k_1 = \frac{\sinh\left(\dfrac{\pi\, a}{4\, h}\right)}{\sinh\left(\dfrac{\pi\, b}{4\, h}\right)} \tag{3.4.6.6}$$

Losses for this structure are [4]

$$\alpha_c = 17.34 \frac{R_s}{Z_0} \frac{P'}{\pi\, a}\left(1.0 + \frac{b-a}{2.0\, a}\right)$$

$$\times \left(\frac{\dfrac{1.25}{\pi}\ln\dfrac{4.0\,\pi\,(b-a)}{2.0\,t} + 1.0 + \dfrac{2.5\, t}{\pi\,(b-a)}}{\left\{1.0 + \dfrac{b-a}{a} + \dfrac{1.25\, t}{\pi\, a}\left[1.0 + \ln\dfrac{2.0\,\pi\,(b-a)}{t}\right]\right\}^2}\right) \quad \text{(dB / m)} \tag{3.4.6.7}$$

$$\alpha_d = \frac{20\,\pi}{\ln(10)} \frac{\varepsilon_r}{\sqrt{\varepsilon_{\mathit{eff}}}} \frac{q \tan\delta}{\lambda_0} \quad \text{(dB / unit length)} \tag{3.4.6.8}$$

for $0 \le k \le 0.707$

$$P' = \frac{k}{k'^{3/2}\,(1.0 - k')}\left[\frac{K(k)}{K'(k)}\right] \tag{3.4.6.9}$$

for $0.707 \le k \le 1.0$

$$P' = \frac{1.0}{\sqrt{k}\,(1.0 - k)} \tag{3.4.6.10}$$

Pintzos [7] gives plots of the dispersion characteristics of this line.

REFERENCES

[1]    Bahl, Inder, and Prakash Bhartia, *Microwave Solid State Circuit Design*, John Wiley and Sons, New York, 1988. (A typo in the denominator of $k_1$ was corrected above.)

[2]    Ghione, G., and C. Naldi, "Analytical Formulas for Coplanar Lines in Hybrid and Monolithic MICs," *Electronics Letters*, Vol. 20, No. 4, February 16, 1984, pp. 179-181. (Compares accuracy of various formulas, pointing out that some are incorrect.)

[3] Ghione, Giovanni, *et al.*, "Q-factor evaluation for coplanar resonators," *Alta Frequenza*, Vol. LII, No. 3, May-June 1983, pp. 191–193.

[4] Gupta, K. C., Ramesh Garg, and Rakesh Chadha, *Computer-Aided Design of Microwave Circuits*, Artech House, Norwood, MA, 1981.

[5] Hanna, V. Fouad, "Finite Boundary Corrections to Coplanar Stripline Analysis," *Electronics Letters*, July 17, 1980, Vol. 16, No. 15, pp. 604–605.

[6] Knorr, Jeffrey B., and Klaus-Dieter Kuchler, "Analysis of Coupled Slots and Coplanar Strips on Dielectric Substrate," *IEEE Transactions on Microwave Theory and Techniques*, Vol. MTT-23, No. 7, July 1975, pp. 541–548.

[7] Pintzos, Sotirios G., "Full-Wave Spectral-Domain Analysis of Coplanar Strips," *IEEE Transactions on Microwave Theory and Techniques*, Vol. MTT-39, No. 2, February 1991, pp 239–246.

### 3.4.7 Asymmetrical Coplanar Strips

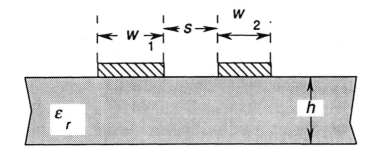

**Figure 3.4.7.1:**     **Asymmetrical Coplanar Strips**

Hoffman [2] solves the equations for this structure from those of the infinitely thick dielectric structure. His equations are

$$Z_0 = \frac{\eta_0}{2.0 \sqrt{\varepsilon_{eff}}} \frac{K(k)}{K(k')} \quad (\Omega) \tag{3.4.7.1}$$

$$\varepsilon_{eff} = 1.0 - \frac{(\varepsilon_r - 1.0) \, K(k_1) \, K(k)}{2.0 \, K(k_1') \, K(k')} \tag{3.4.7.2}$$

where

$$k = \sqrt{\frac{s}{b} \left(1.0 + \frac{b}{d} - \frac{s}{d}\right)} \tag{3.4.7.3}$$

$$k_1 = \sqrt{\frac{(t_1 - t_2)(t_3 - t_2)}{(t_1 + t_2)(t_3 + t_2)}}$$ (3.4.7.4)

$$t_n = \frac{e^{\lambda_n} - 1.0}{e^{\lambda_n} + 1.0} \quad , \quad n = 1, 2, 3$$ (3.4.7.5)

$$\lambda_1 = \frac{\pi}{2.0} \left( \frac{2.0\, w_2}{h} + \frac{s}{h} \right)$$ (3.4.7.6)

$$\lambda_2 = \frac{\pi\, s}{2.0\, h}$$ (3.4.7.7)

$$\lambda_3 = \frac{\pi}{2.0} \left( \frac{2.0\, w_1}{h} + \frac{s}{h} \right)$$ (3.4.7.8)

$$k_1' = \sqrt{1.0 - k_1^2}$$ (3.4.7.9)

$$k' = \sqrt{1.0 - k^2}$$ (3.4.7.10)

$$b = w_2 + s$$ (3.4.7.11)

$$d = w_1 + s$$ (3.4.7.12)

## REFERENCES

[1]    Gevorgian, S.S., and I.G. Mironenko, "Asymmetric Coplanar-Strip Transmission Lines for MMIC and Integrated Optic Applications," *Electronic Letters*, Vol. 26, No. 22, October 25, 1990, pp. 1916–1918.

[2]    Hoffman, Reinmut, *Handbook of Microwave Integrated Circuits,* Artech House, Norwood, MA, 1987.

### 3.4.8  Three Coplanar Strips

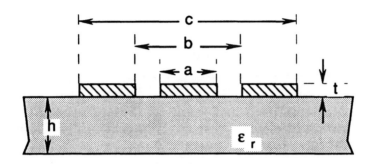

**Figure 3.4.8.1:** **Three Coplanar Strips**

This configuration was derived in [4] to analyze and design a tape automated bonding (TAB) IC package with alternate traces grounded.

$$Z_0 = \frac{\eta_0}{4.0 \sqrt{\varepsilon_{eff}}} \frac{1.0}{\dfrac{K(k_1')}{K(k_1)} + \dfrac{t}{b-a}} \quad (\Omega) \tag{3.4.8.1}$$

where

$$\varepsilon_{eff} = 1.0 + \frac{(\varepsilon_r - 1.0)}{2.0} \frac{\dfrac{K(k_2')}{K(k_2)}}{\dfrac{K(k_1')}{K(k_1)} + \dfrac{t}{b-a}} \tag{3.4.8.2}$$

$$k_1 = \frac{c}{b} \sqrt{\frac{b^2 - a^2}{c^2 - a^2}} \tag{3.4.8.3}$$

$$k_2 = \frac{\sinh\left(\dfrac{\pi\ c}{4.0\ h}\right)}{\sinh\left(\dfrac{\pi\ b}{4.0\ h}\right)} \sqrt{\frac{\sinh^2\left(\dfrac{\pi\ b}{4.0\ h}\right) - \sinh^2\left(\dfrac{\pi\ a}{4.0\ h}\right)}{\sinh^2\left(\dfrac{\pi\ c}{4.0\ h}\right) - \sinh^2\left(\dfrac{\pi\ a}{4.0\ h}\right)}} \tag{3.4.8.4}$$

$$k_n' = \sqrt{1 - k_n^2}, \quad n = 1, 2 \tag{3.4.8.5}$$

In Houdart [3, Figure 2], the effect of ground plane width ($c$ in Figure 3.4.8.1) on the CPW is analyzed. For $c / b > 5.0$ the impedance is affected by less than about 3%.

REFERENCES

[1]     Ghione, Giovanni, and Carlo U. Naldi, "Coplanar Waveguides for MMIC Applications: Effect of Upper Shielding Conductor Backing, Finite-Extent Ground Planes, and Line-to-Line Coupling," *IEEE Transactions on Microwave Theory and Techniques*, Vol. MTT-35, No. 3, March 1987, pp. 260–267.

[2]     Herrel, Dennis, and David Carey, "High-Frequency Performance of TAB," *IEEE Transactions on Components, Hybrids, and Manufacturing Technology*, Vol. CHMT-10, No. 2, June 1987, pp. 199–203.

[3]     Houdart, M., "Coplanar Lines: Application to Broadband Microwave Integrated Circuits," *6th European Microwave Conference Proceedings 1976*, Rome, Italy, pp. 49–53.

[4]     Mueller, E., "Measurement of the Effective Relative Permittivity of Unshielded Coplanar Waveguides," *Electronics Letters*, Vol. 13, No. 24, November 24, 1977, pp. 729–730.

[5]     Wentworth, Stuart M., *et al.*, "The High-Frequency Characteristics of Tape Automated Bonding (TAB) Interconnects," *IEEE Transactions on Components, Hybrids, and Manufacturing Technology*, Vol. CHMT-122, No. 3, September 1989, pp. 340–347.

### 3.4.9   Three Coplanar Strips with Ground

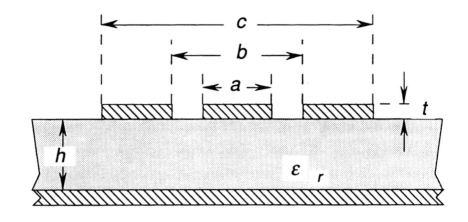

Figure   3.4.9.1:          Three Coplanar Strips With Ground

This configuration was derived in [1] to analyze and design a tape automated bonding structure where alternate traces are grounded and are connected to a ground layer with vias.

$$Z_0 = \frac{\eta_0}{2.0 \sqrt{\varepsilon_{eff}}} \frac{1.0}{\frac{K(k_1)}{K(k_1')} + \frac{K(k_2)}{K(k_2')} + \frac{2.0 \, t}{(b-a)}} \tag{3.4.9.1}$$

where

$$\varepsilon_{eff} = \frac{\frac{K(k_1)}{K(k_1')} + \frac{\varepsilon_r \, K(k_2)}{K(k_2')} + \frac{2.0 \, t}{b-a}}{\frac{K(k_1)}{K(k_1')} + \frac{K(k_2)}{K(k_2')} + \frac{2.0 \, t}{(b-a)}} \tag{3.4.9.2}$$

$$k_1 = \frac{c}{b} \sqrt{\frac{b^2 - a^2}{c^2 - a^2}} \tag{3.4.9.3}$$

$$k_2 = \frac{\tanh\left(\frac{\pi \, a}{4.0 \, h}\right)}{\tanh\left(\frac{\pi \, b}{4.0 \, h}\right)} \tag{3.4.9.4}$$

$$k_n' = \sqrt{1.0 - k_n^2} \, , \, n = 1, \, 2 \tag{3.4.9.5}$$

## REFERENCES

[1]    Wentworth, Stuart M., et al., "The High-Frequency Characteristics of Tape Automated Bonding (TAB) Interconnects," *IEEE Transactions on Components, Hybrids, and Manufacturing Technology*, Vol. CHMT-122, No. 3, September 1989, pp. 340–347.

## 3.4.10 Covered Coplanar Waveguide with Ground

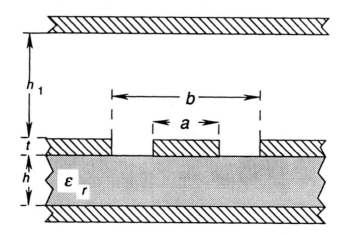

**Figure 3.4.10:**        Covered Coplanar Waveguide with Ground

The equations describing this structure are

$$Z_0 = \frac{\eta_0}{2.0 \sqrt{\varepsilon_{eff}}} \frac{1.0}{\dfrac{K(k_3)}{K(k'_3)} + \dfrac{K(k_4)}{K(k'_4)}} \tag{3.4.10.1}$$

$$\varepsilon_{eff} = 1.0 + \frac{\dfrac{K(k_3)}{K(k'_3)}}{\dfrac{K(k_3)}{K(k'_3)} + \dfrac{K(k_4)}{K(k'_4)}} (\varepsilon_r - 1.0) \tag{3.4.10.2}$$

where

$$k_3 = \frac{\tanh\left(\dfrac{\pi\,a}{h}\right)}{\tanh\left(\dfrac{\pi\,b}{h}\right)} \tag{3.4.10.3}$$

$$k_4 = \frac{\tanh\left(\dfrac{\pi\,a}{h_1}\right)}{\tanh\left(\dfrac{\pi\,b}{h_1}\right)} \tag{3.4.10.4}$$

which are strictly valid for

$$h = h_1$$

and were found to give good results elsewhere.

REFERENCES

[1]    Ghione, Giovanni, and Carlo U. Naldi, "Coplanar Waveguides for MMIC Applications: Effect of Upper Shielding Conductor Backing, Finite-Extent Ground Planes, and Line-to-Line Coupling," *IEEE Transactions on Microwave Theory and Techniques*, Vol. MTT-35, No. 3, March 1987, pp. 260–267.

### 3.4.11 Covered Coplanar Waveguide

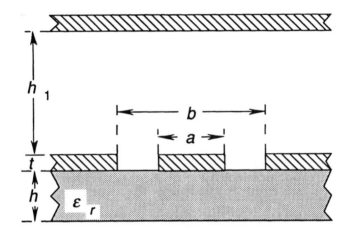

Figure 3.4.11.1:    Covered Coplanar Waveguide

The equations of Ghione and Naldi [1] are

$$Z_0 = \frac{\eta_0}{2.0 \sqrt{\varepsilon_{eff}}} \frac{1.0}{\dfrac{K(k_2)}{K(k'_2)} + \dfrac{K(k)}{K(k')}} \qquad (3.4.11.1)$$

$$\varepsilon_{eff} = 1.0 + \frac{\dfrac{K(k_1)}{K(k'_1)}}{\dfrac{K(k_2)}{K(k'_2)} + \dfrac{K(k)}{K(k')}} (\varepsilon_r - 1.0) \qquad (3.4.11.2)$$

where

$$k = a / b \qquad (3.4.11.3)$$

$$k_1 = \frac{\sinh\left(\dfrac{\pi a}{h}\right)}{\sinh\left(\dfrac{\pi b}{h}\right)} \qquad k_1' = \sqrt{1.0 - k_1} \qquad\qquad (3.4.11.4)$$

$$k_2 = \frac{\tanh\left(\dfrac{\pi a}{h_1}\right)}{\tanh\left(\dfrac{\pi b}{h}\right)} \qquad k_2' = \sqrt{1.0 - k_2} \qquad\qquad (3.4.11.5)$$

These equations are valid to $X$-band for usual dimensions of PCBs, hybrids, and ICs.

### REFERENCES:

[1]    Ghione, Giovanni, and Carlo U. Naldi, "Coplanar Waveguides for MMIC Applications: Effect of Upper Shielding Conductor Backing, Finite-Extent Ground Planes, and Line-to-Line Coupling," *IEEE Transactions on Microwave Theory and Techniques*, Vol. MTT-35, No. 3, March 1987, pp. 260–267.

## 3.5 MICROSTRIP LINE STRUCTURES

Microstrip line may well be the most popular transmission line structure. Ease of fabrication by photolithographic techniques and a good range of impedances and couplings allow it to be used for a wide variety of circuit components. Harold Wheeler developed a planar transmission line (two coplanar strips) which could be rolled up in 1936 and a stripline-like structure in 1942. Flat coaxial transmission line was used by V.H. Rumsey and H.W. Jamieson in WWII. Coaxial cable was first adapted to a flat configuration using printed circuit techniques by Barrett. This then evolved into stripline after WWII. The first use of the microstrip line configuration was reported by engineers at the Federal Telecommunications Research Laboratories (a division of ITT) sometime after 1949 [2].

REFERENCES

[1]    Ayer, D.R., and C.A. Wheeler, "The Evolution of Strip Transmission Line," *Microwave Journal*, Vol. 12, No. 5, May 1969, pp. 31–40.

[2]    Barrett, Robert M., "Microwave Printed Circuits—A Historical Survey," *IRE Transactions on Microwave Theory and Techniques*, Vol. MTT-3, No. 2, March 1955, pp. 1–9.

[3]    Howe, Harlan, Jr., "Microwave Integrated Circuits—An Historical Perspective," *IEEE Transactions on Microwave Theory and Techniques*, Vol. MTT-32, No. 9, September 1984, pp. 991–996.

### 3.5.1  Microstrip Line

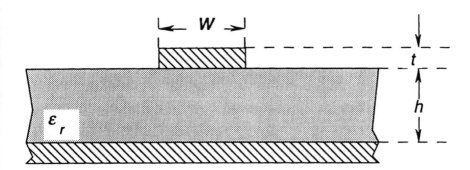

**Figure 3.5.1.1:     Microstrip Line**

This is probably the most common and most analyzed transmission line structure. It is easy to use and has a good range of practical impedances. Equations have been calculated with an incredible variety of techniques both analytic and computational. A FORTRAN program, MSTRIP2, is commonly encountered in references as a tool for checking the results of new analysis and synthesis equations. It is a numerical analysis program that assumes quasistatic conditions, zero thickness strips, and perfect conductivity. It is also assumed that the dielectric thickness and trace widths are thin relative to a wavelength.

Bogatin [10] experimentally compared various calculation techniques for this structure and recommends using the Wheeler equations with Schneider's $\varepsilon_{eff}$.

$$Z_0 = \frac{\eta_0}{2.0 \sqrt{2.0} \, \pi \sqrt{\varepsilon_r + 1.0}} \ln \left\{ 1.0 + \frac{4.0 \, h}{w'} \left[ \frac{14.0 + 8.0 \, /\varepsilon_r}{11.0} \frac{4.0 \, h}{w'} \right. \right.$$

$$\left. \left. + \sqrt{\left(\frac{14.0 + 8.0 \, /\varepsilon_r}{11.0}\right)^2 \left(\frac{4.0 \, h}{w'}\right)^2 + \frac{1.0 + 1.0 \, /\varepsilon_r}{2.0} \pi^2} \right] \right\} \quad (\Omega) \quad (3.5.1.1)$$

Improvements in Schneider's $\varepsilon_{eff}$ made by Hammerstad and Bekkadal [22] are given here. For $w \, / \, h \le 1.0$:

$$\varepsilon_{eff} = \frac{\varepsilon_r + 1}{2} + \frac{\varepsilon_r - 1}{2} \left[ \left(1 + \frac{12 \, h}{w}\right)^{-0.5} + 0.04 \left(1.0 - \frac{w}{h}\right)^2 \right] \quad (3.5.1.2)$$

and for $w \, / \, h \ge 1$:

$$\varepsilon_{eff} = \frac{\varepsilon_r + 1.0}{2.0} + \frac{\varepsilon_r - 1.0}{2.0} \left(1 + \frac{12.0 \, h}{w}\right)^{-0.5} \quad (3.5.1.3)$$

The equations for $\varepsilon_{eff}$ are accurate to within 1% for:

$$\varepsilon_r \le 16 \quad (< 2\% \text{ error } \varepsilon_r > 16)$$

$$0.05 \le \frac{w}{h} \le 20.0 \quad (< 2\% \text{ error } \frac{w}{h} < 0.05)$$

The thickness of the trace can be corrected for by relating it to an equivalent change in the width. Owens and Potok [40] examined a number of formulas for this correction and show that Wheeler's is the most accurate:

$$\frac{\Delta w}{t} = \frac{1.0}{\pi} \ln \left[ \frac{4 \, e}{\sqrt{(t \, / \, h)^2 + \left(\frac{1/\pi}{w \, / \, t + 1.1}\right)^2}} \right] \quad (3.5.1.4)$$

$$w' = w + \Delta w' \quad (3.5.1.5)$$

$$\Delta w' = \Delta w \left( \frac{1.0 + 1.0 / \varepsilon_r}{2.0} \right) \qquad (3.5.1.6)$$

Error in $Z_0$ is less than 2% for any $\varepsilon_r$, $w$.

### 3.5.1.1  Frequency Dependencies of Microstrip Line

The frequency dependency (dispersion) of the microstrip line's $\varepsilon_{eff}$ can be calculated with ([28], see also [35]:)

$$\varepsilon_{eff}(f) = \varepsilon_r - \frac{\varepsilon_r - \varepsilon_{eff}(f = 0)}{1.0 + P(f)} \qquad (3.5.1.7)$$

$$P(f) = P_1 P_2 [(0.1844 + P_3 P_4) \times (10.0 f h)]^{1.5763} \qquad (3.5.1.7a)$$

$$P_1 = 0.27488 + \left[ 0.6315 + \frac{0.525}{(1.0 + 0.157 f h)^{20}} \right] \frac{w}{h} - 0.065683 \, e^{-8.7513 w / h}$$

$$(3.5.1.8)$$

$$P_2 = 0.33622 \left[ 1.0 - e^{-.03442 \, \varepsilon_r} \right] \qquad (3.5.1.9)$$

$$P_3 = 0.0363 \, e^{-4.6 w / h} \left[ 1.0 - e^{-(f h / 3.87)^{4.97}} \right] \qquad (3.5.1.10)$$

$$P_4 = 1.0 + 2.751 \left[ 1.0 - e^{-(\varepsilon_r / 15.916)^8} \right] \qquad (3.5.1.11)$$

where $f$ is in GHz and $h$ is in cm. The accuracy of this correction is better than 0.6% for

$$0.1 \leq w/h \leq 100.0$$

$$1.0 \leq \varepsilon_r \leq 20.0$$

$$0 \leq h / \lambda \leq 0.13$$

Atwater [2] compares several corrections for $\varepsilon_{eff}(f)$ with actual measurement data from eight published papers and found the above correction to be the best (by an admittedly small margin). His version of the equation is a rearranged equivalent.

### 3.5.1.2  Conductor Losses

Conductor losses are a result of several contributing factors: the actual conductance of the material, the frequency-dependent skin effect losses, and the surface roughness losses caused by the lengthened path at the surface. Conductor losses due to skin effect and the metal's conductivity are given in [53]. For $w/h \leq 1$:

$$\alpha_c = \frac{10.0\, R_s}{\pi \ln 10.0} \frac{\left(\frac{8.0\, h}{w} - \frac{w}{4\, h}\right)\left(1.0 + \frac{h}{w} + \frac{h}{w}\frac{\partial w}{\partial t}\right)}{h\, Z_0\, e^{Z_0/60}} \quad \text{(dB/ unit length)}$$

(3.5.1.12)

For $w/h \geq 1$:

$$\alpha_c = \frac{Z_0\, R_s}{720.0\, \pi^2\, h \ln 10.0}\left[1.0 + \left(\frac{0.44\, h^2}{w^2}\right) + \frac{6.0\, h^2}{w^2}\left(1.0 - \frac{h}{w}\right)^5\right]$$

$$\times \left(1.0 + \frac{w}{h} + \frac{\partial w}{\partial t}\right) \quad \text{(dB/ unit length)}$$

(3.5.1.13)

where for $\dfrac{w}{h} \leq \dfrac{1}{2\,\pi}$

$$\frac{\partial w}{\partial t} = \left(\frac{1.0}{\pi}\right)\ln\frac{4.0\,\pi\,w}{t}$$

(3.5.1.14)

for $\dfrac{w}{h} \geq \dfrac{1}{2\,\pi}$

$$\frac{\partial w}{\partial t} = \left(\frac{1.0}{\pi}\right)\ln\frac{2.0\, h}{t}$$

(3.5.1.15)

These equations are valid for ground and strip conductivities the same and

$t \ll h$

$t < h/2.0$

$\partial w/\partial t > 1.0$

Pucel *et al.* [43] also give equations for the conduction losses.

### 3.5.1.2.1  *Effect of Ground Plane on Conductor Losses*

The ground plane's resistance can be a significant factor as frequency increases. At low frequencies the current in the ground plane is spread over the full width of the ground plane and the current is evenly distributed in the center conductor. Because of this, the ground plane contributes little resistance relative to the center conductor. As frequency increases the current in the line begins to concentrate in the sides and bottom of the trace. Simultaneously, the ground current begins to concentrate in the area directly under the strip, increasing the effective resistance. Faraji-Dana and Chow [16] give a closed-form equation suitable for design calculations. The resistance of the ground plane is $R_g$

$$R_g = 0.55\, R_{dc}\, \sqrt{\frac{\pi}{2.0}}\, \sqrt{\frac{t}{w}}\, \left(1.0 - e^{-w'\,/\,1.2\,\pi}\right) P \qquad (3.5.1.16)$$

where

$$w' = w\,/\,h \qquad (3.5.1.17)$$

$$P = \sqrt{2.0\, \mu\, \sigma\, f\, w\, t} \qquad (3.5.1.18)$$

$R_{dc}$ = the dc resistance of the signal conductor

The total ac resistance for use in the microstrip line model is $R_t$.

$$R_t = R_s + R_g \qquad (3.5.1.19)$$

### 3.5.1.2.2   Effect of Conductor Surface Roughness

The conductor surface analyzed in the formulas above are perfectly smooth. The physical processes which make real world conductors create scratches, bumps, and random bumps. See Tanaka and Okada [62] for some scanning microscope photographs of typical conductor surfaces.

Conductor surface roughness losses are discussed thoroughly in Sanderson [53]. Sanderson also reports on a technique used for calculating the roughness effect on the losses, $Z_0$, and $v_p$. Higher than expected losses were experienced in stainless steel conductors that can be modeled by a lumped resistance. Good agreement between the predicted performance and the experimental measurements was found from 200 MHz to 10 GHz in the other cases. Saad [50] gives a plot from Lending [32] for losses as a function of surface roughness. In [20], an approximate formula is given for the conductor losses due to roughness:

$$\alpha_c = \alpha_{c,0}\left\{ 1.0 + \frac{2.0}{\pi} \tan^{-1}\left[ 1.40 \left(\frac{\Delta}{\delta}\right)^2 \right] \right\} \qquad (3.5.1.20)$$

where

$\alpha_{c,0}$ = the conductor losses calculated for a perfectly smooth conductor; equations (3.5.1.12) and (3.5.1.13)

$\Delta$ = the rms surface roughness

$\delta$ = the skin depth

The results of a number of references [20, 32, 39, 60, 66] are summarized in Figure 3.5.1.2 below. This figure shows good agreement between Morgan's data and the Goldfarb closed form equation.

**Increase in Conductor Resistivity Due to Surface Roughness (%)**

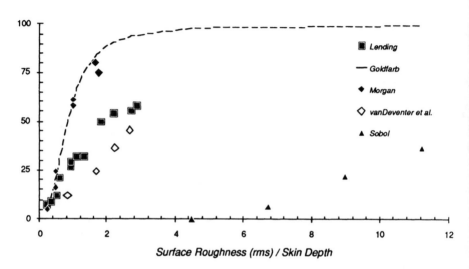

Figure 3.5.1.2.2.1:  Microstrip Line Losses vs. Surface Roughness

### 3.5.1.3    Dielectric Losses

Dielectric losses are given in Schneider [50]:

$$\alpha_d = \frac{20.0\ \pi}{\ln\ (10.0)}\ \frac{q\ \tan\ \delta}{\lambda}\ \text{(dB / unit length)} \qquad (3.5.1.21)$$

where $q$ is the filling factor and $\lambda_g$ is the wavelength in microstrip line both previously defined.

### 3.5.1.4    Radiation Losses

Radiative losses were discussed in [1]. The equations are:

$$\alpha_r = 60 \left( \frac{2 \pi h}{\lambda_0} \right)^2 F(\varepsilon_{eff}) \tag{3.5.1.22}$$

where $F(\varepsilon_{eff})$ is:

$$F(\varepsilon_{eff}) = \frac{\varepsilon_{eff} + 1.0}{\varepsilon_{eff}} - \frac{(\varepsilon_{eff} - 1.0)^2}{2.0 \, \varepsilon_{eff}^{3/2}} \log \left( \frac{\sqrt{\varepsilon_{eff}} + 1.0}{\sqrt{\varepsilon_{eff}} - 1.0} \right) \quad \text{, for an open-circuited line}$$

$$\tag{3.5.1.23}$$

$$F(\varepsilon_{eff}) = 1.0 - \frac{(\varepsilon_{eff} - 1.0)}{2.0 \sqrt{\varepsilon_{eff}}} \log \left( \frac{\sqrt{\varepsilon_{eff}} + 1.0}{\sqrt{\varepsilon_{eff}} - 1.0} \right) \quad \text{, for a matched line}$$

$$\tag{3.5.1.24}$$

### 3.5.1.5    Higher Order Modes in Microstrip Line

The microstrip line structure is not strictly TEM; other modes may propagate. The lowest order $TE_1$ mode frequency is [69]:

$$f_c = \frac{c}{4.0 \, h \sqrt{\varepsilon_r - 1.0}} \tag{3.5.1.25}$$

These non-TEM modes are not to be confused with waveguide modes which can be propagated when a microstrip line is placed in a metallic enclosure for shielding. These modes are launched or stimulated by radiation or discontinuities in the transmission line. These modes can be suppressed by decreasing the enclosure dimensions, adding shorting posts, lossy films, or by damping materials. See [71, 72] for more information.

### 3.5.1.6    Magnetic Substrates

Magnetic substrates are useful for building voltage programmable transmission lines. This is useful, for example, in constructing programmable phase shifters. Magnetic substrates are analyzed in [42, 43] which propose the following modifications to the above:

$$Z_0 = Z_0(\varepsilon_r = 1, \mu_r = 1, h, w, t) \sqrt{\frac{\mu_{eff}}{\varepsilon_{eff}}} \quad (\Omega) \tag{3.5.1.26}$$

$$\mu_{eff}(w / h, \mu) = \frac{1.0}{\varepsilon_{eff}(w / h, 1.0 / \mu)} \tag{3.5.1.27}$$

where

$$Z_0(\varepsilon_r = 1, \mu_r = 1, h, w, t) \text{ is given by (3.5.1.1)} \tag{3.5.1.28}$$

Equation (3.5.1.27) can be rewritten using a simpler version of (3.5.1.2) and (3.5.1.3)

$$\mu_{eff}(w / h, \mu) = \frac{1.0}{\dfrac{1.0 / \mu + 1}{2.0} + \dfrac{1.0 / \mu - 1.0}{2} \left(1.0 + \dfrac{10.0\, h}{w}\right)^{-0.5}}$$

$$= \frac{2.0\, \mu}{(1.0 + \mu) + (1.0 - \mu) \left(1.0 + \dfrac{10.0\, h}{w}\right)^{-0.5}} \tag{3.5.1.29}$$

The dielectric is magnetic, so we must use the complete equation for the guide wavelength, (2.3.11):

$$\lambda_g = \frac{\lambda_0}{\sqrt{\varepsilon_{eff}\, \mu_{eff}}} \tag{3.5.1.30}$$

The dielectric losses are modified in the presence of a magnetic substrate to

$$\alpha_d = \frac{27.3}{\lambda_g} \left(\tan \delta_{d,eff} + \tan \delta_{m,eff}\right) \quad \text{(dB)} \tag{3.5.1.31}$$

where

$$\tan \delta_{d,eff} = \left(\frac{1.0 - 1.0 / \varepsilon_{eff}}{1.0 - 1.0 / \varepsilon_r}\right) \tan \delta_d = \left(\frac{\varepsilon_{eff}\, \varepsilon_r - \varepsilon_r}{\varepsilon_{eff}\, \varepsilon_r - \varepsilon_{eff}}\right) \tan \delta \tag{3.5.1.32}$$

$$\tan \delta_{m,eff} = \left(\frac{1.0 - \mu_{eff}}{1.0 - \mu_r}\right) \tan \delta_m \tag{3.5.1.33}$$

The magnetic loss tangent can be calculated [20]:

$$\tan \delta_m = \frac{\omega_m / T}{\omega^2} \tag{3.5.1.34}$$

where

$$T = \frac{2.0}{\gamma\, \Delta H} = \text{relaxation time} \tag{3.5.1.35}$$

$$\gamma = \frac{g}{m} \frac{e}{c} \qquad\qquad (3.5.1.36)$$

$\Delta H$ = resonance-line width

$M_s$ = saturation magnetization    (Wb / m$^2$)

$g$ = spectroscopic splitting factor

$e$ = electron charge

$m$ = electron mass

$\omega_m = \gamma M_s$ = resonance frequency of the ferrite

$\omega$ = operating frequency

### 3.5.1.7    Conductors of Trapezoidal Cross-section

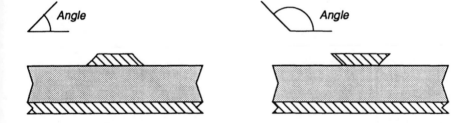

**Figure 3.5.1.7.1:**    **Trapezoidal Microstrip Line Traces**

Conventionally it is assumed that the microstrip line conductor has a rectangular cross-section, however this is only an approximation to the physical reality. The additive and subtractive processes used to create conductors for PCBs, thin-, and thick-film hybrids produce a conductor of approximately trapezoidal cross-section. For wide strips (large $w / h$) the thickness and edge effects will be small since most of the capacitance is parallel plate rather than fringing and we can ignore the cross-section. For narrow strips which are common in smaller structures this condition is violated and the thickness and edge shape become important.

Information on this structure is found in references [7, 36, 47, 48, 49, 56]. Barsotti, *et al.* [7] develop a modified incremental inductance rule which can be used for calculating the loss of any shape conductor once a number of parameters are determined. It has been found in Rizzoli [49] that the trapezoidal cross-section results in a characteristic impedance lying somewhere between that of a rectangular cross-section (having width equal to the widest side of the trapezoid) and a zero thickness conductor.

The program CALIF[1] was used to calculate the effect of trapezoidal cross-section on a microstrip line. The width of the wider side was kept constant and the angle is as shown in Figure 3.5.1.7.1 above. The case examined was:

$w = 18.02$ mil  $(w / h = 1.82)$

$t = 1.4$ mil  $(t / h = .14, t / w = .078)$

$h = 9.9$ mil

$\varepsilon_r = 4.30$

The limiting cases define the bounds of the impedance effects. The rectangular trace having width $w$ is the lowest impedance we can have (51.58 $\Omega$). The zero-thickness trace is the highest impedance we can achieve with etching of the type shown in the leftmost figure of Figure 3.5.1.7.1 (53.30 $\Omega$). The highest impedance we can achieve with etching similar to the right-hand figure of Figure 3.5.1.7.1 is that of a zero-thickness trace suspended by $t$ over the dielectric substrate (70.20 $\Omega$). The actual impedance variation between these cases is shown in Figure 3.5.1.7.2.

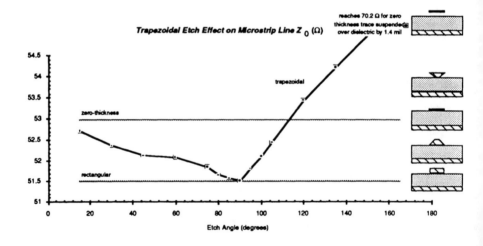

**Figure 3.5.1.7.2:**    **Trapezoidal Trace Effect on $Z_0$ ($t / h$ = 0.14)**

[1]CALIF is a registered trademark of is a product of Integrity Engineering. The program (version 2.0) uses the boundary element method to solve a number of transmission line structures.

It can be expected that the increase of $t/h$ (more square trace) will move the two lines labeled *rectangular* and *zero-thickness* relatively further apart on the plot above. In fact, it is seen in the plot below that the asymptotic lines are indeed further apart.

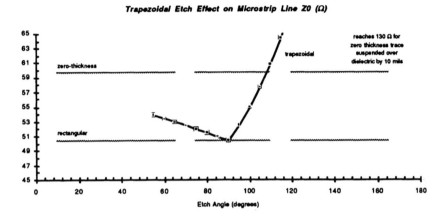

**Figure 3.5.1.7.3:** **Trapezoidal Trace Effect on $Z_0$ ($t/h = 1.01$)**

Based on the two extreme cases shown in Figure 3.5.1.7.2 and 3.5.1.7.3, it appears that calculating the effect of the trapezoidal trace (wide side near dielectric) on $Z_0$ by linearly interpolating the zero-thickness and rectangular impedances with a line over the 0 to 90° range. This is valid up until the point that the trace becomes an equilateral triangle. At this point, the thickness of the limiting rectangular trace can be decreased and the impedance interpolated in the same fashion. The equation is

$$Z_{0,trapezoid} = Z_0 (t = 0) - \frac{\theta}{90°} [Z_0 (t = 0) - Z_0 (t)] \qquad (3.5.1.7.1)$$

which is valid for angles, $\theta$

$$\tan^{-1}\left(\frac{2.0\ t}{w}\right) \le \theta \le 90.0° \qquad (3.5.1.7.2)$$

For the trapezoidal trace with the wide side further from the dielectric the relation is not linear. The impedance reaches the zero-thickness impedance by approximately 110°. Again, a linear interpolation will provide reasonable results over the 90 to 110° range.

The effective dielectric constant is plotted for the above two cases in figures 3.5.1.7.4 and 3.5.1.7.5 below.

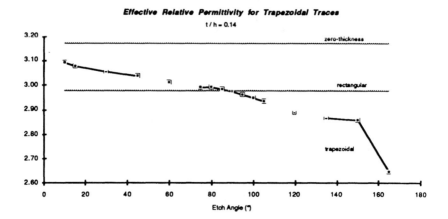

**Figure   3.5.1.7.4:**    $\varepsilon_r$  for  Trapezoidal  Trace  ($t / h = 0.14$)

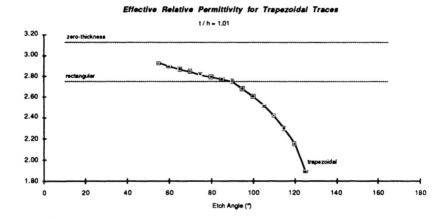

**Figure   3.5.1.7.5:**    $\varepsilon_r$  for  Trapezoidal  Trace  ($t / h = 1.01$)

## 3.5.1.8   *Superconducting  Microstrip  Line  Losses*

A relation for conductor losses in microstrip line can be found in Baiocchi, *et al.* [5].

$$\alpha_c = \frac{\theta}{(1.0 - \theta)^{3/2}} \frac{G \; \sigma_n \; \mu^2 \; \lambda^3}{4.0 \; Z_0}$$

$$\cdot\left(\coth\frac{A\ G\ \sqrt{1.0-\theta}}{\lambda}+\frac{A\ G\ \sqrt{1.0-\theta}}{\lambda}\ \mathrm{csech}^2\frac{A\ G\ \sqrt{1.0-\theta}}{\lambda}\right)\omega^2$$

(3.5.1.8.1)

where

$\sigma_n$ = conductivity near critical temperature

$\lambda$ = zero temperature penetration depth

$T_c$ = critical temperature of the superconductor

$$\theta=(T\ /\ T_c\ )^4 \tag{3.5.1.8.2}$$

and the value of $G$ can be found with the relations in [29]:

$$G=\frac{1.0}{\mu}\sum_j\frac{\partial L}{\partial n}\quad(\mathrm{m}^{-1}) \tag{3.5.1.8.3}$$

where

$\frac{\partial L}{\partial n}$ = derivative of external inductance for a small recession of the $j$th conductor wall (see Chapter 1).

## REFERENCES

[1]  Abouzahra, Mohammad Deb, and Leonard Lewin, "Radiation from Microstrip Discontinuities," *IEEE Transactions on Microwave Theory and Techniques*, Vol. MTT-27, No. 8, August 1979, pp. 722–723.

[2]  Atwater, H.A., "Tests of Microstrip Dispersion Formulas," *IEEE Transactions on Microwave Theory and Techniques*, Vol. MTT-36, No. 3, pp. 619–621, March 1988.

[3]  Atwater, H.A., "Simplified Design Equations for Microstrip Line Parameters," *Microwave Journal*, November 1989, pp. 109–115, and corrections February 1990, pp. 196.

[4]  Bahl, I.J., and D.K. Trivedi, "A Designer's Guide To Microstrip Line," *Microwaves*, May 1977, pp. 174–182.

[5]     Baiocchi, O.R., *et al.*, "Effects of Superconducting Losses in Pulse Propagation on Microstrip Lines," *IEEE Microwave and Guided Wave Letters*, Vol. 1, No. 1, January 1991, pp. 2–4.

[6]     Barsotti, Edward L., *et al.*, "A Simple Method to Account for Edge Shape in the Conductor Loss in Microstrip," *IEEE Transactions on Microwave Theory and Techniques*, Vol. MTT-39, No. 1, January 1991, pp. 98–106.

[7]     Bhat, Bharathi, and Shiban K. Koul, "Unified Approach to Solve a Class of Strip and Microstrip-Like Transmission Lines," *IEEE Transactions on Microwave Theory and Techniques*, Vol. MTT-30, No. 5, May 1982, pp. 679–686.

[8]     Bedair, S.S., and M.J. Sobhy, "Characteristics of the Thick Shielded Microstrip," *IEE Proceedings*, Vol. 129, Pt. H, No. 3, June 1982, pp. 135–137.

[9]     Bogatin, Eric, "Design Rules for Microstrip Capacitance," *IEEE Transactions on Components, Hybrids, and Manufacturing Technology*, Vol. CHMT-11, No. 3, September 1988, pp. 253–259.

[10]    Bryant, T.G., and J.A. Weiss, "Parameters of microstrip transmission lines and of coupled pairs of microstrip lines," *IEEE Transactions on Microwave Theory and Techniques*, Vol. MTT-16, No. 12, December 1968, pp. 1021–1027.

[11]    Chudobiak, W.J., "Dispersion in Microstrip," *IEEE Transactions on Microwave Theory and Techniques*, Vol. MTT-19, No. 9, September 1971, pp. 783–784.

[12]    Denlinger, Edgar J., "Radiation from Microstrip Resonators," *IEEE Transactions on Microwave Theory and Techniques*, Vol. MTT-17, No. 4, April 1969, pp. 235–236.

[13]    Denlinger, Edgar J., "Losses of Microstrip Lines," *IEEE Transactions on Microwave Theory and Techniques*, Vol. MTT-28, No. 6, June 1980, p. 513–522.

[14]    Faraji-Dana, Reza, and Y. Leonard Chow, "The Current Distribution and AC Resistance of a Microstrip Structure," *IEEE Transactions on Microwave Theory and Techniques*, Vol. MTT-38, No. 9, September 1990, pp. 1268–1277. (Distribution of ground plane current as a function of frequency.)

[15]    Faraji-Dana, Reza, and Y. Leonard Chow, "The AC Resistance of a Microstripline and Its Ground Plane," *1989 IEEE MTT-S Symposium Digest*, pp. 325–327.

[16]    Gardiol, F. E., "Design and Layout of Microstrip Structures," IEE Proceedings, Vol. 135, No. 3, June 1988, pp. 145–157. (This is an overview article with a comprehensive bibliography to the current state-of-the-art in microstrip line design.)

[17]    Geshiro, Masahiro, *et al.*, "Analysis of Slotlines and Microstrip Lines on Anisotropic Substrates," *IEEE Transactions on Microwave Theory and Techniques*, Vol. MTT-39, No. 1, January 1991, pp. 64–69.

[18]    Getsinger, William J., "Microstrip Dispersion Model," *IEEE Transactions on Microwave Theory and Techniques*, Vol. MTT-21, No. 1, pp. 34–39, January 1973.

[19]    Goldfarb, Marc E., "Losses in GaAs Microstrip," *1990 IEEE MTT-S International Microwave Symposium Digest*, pp. 563–565. (Compares Touchstone, SuperCompact, and HP85150 loss models to measured data.)

[20]    Gupta, K.C., "Striplines and Microstriplines," Notes for "Fundamentals of RF Design," at RF Technology Expo '85, January 23-25, 1985.

[21]    Hammerstad, E.O., and F. Bekkadal, *Microstrip Handbook*, The University of Trondheim, ELAB report STF44 A74169, February 1975.

[22]    Hammerstad, E.O. and Ø. Jensen, "Accurate Models for Microstrip Computer-Aided Design," *1980 IEEE MTT-S International Microwave Symposium Digest*, pp. 407–409.

[23]    Hertling, D.R., and R.K. Feeney, "Microstrip Analysis and Design and Various Substrate Materials," *RF Design*, June 1988, pp. 38–42.

[24]    Hogan, Sean, "Closed Form Transmission Line Equation," presented at *New England Circuits Association*, Andover, MA, May 11, 1988.

[25]    Holloway, Albert L., "Generalized Microstrip on a Dielectric Sheet," *IEEE Transactions on Microwave Theory and Techniques*, Vol. MTT-36, No. 6, pp. 939–951, June 1988.

[26]    Homentcovschi, Dorel, "An Analytical Approach to the Analysis of Dispersion Characteristics of Microstrip Lines," *IEEE Transactions on Microwave Theory and Techniques*, Vol. MTT-39, No. 4, April 1991, pp. 740–743.

[27]    Kimura, Atsuo, and Jeffrey Frey, "Calculation and Measurements of the Frequency Dependence of Effective Dielectric Constant of Microstrip Lines," *Cornell Microwave Report* 71-5, September 1971, pp. 205–217.

[28]    Kirschning, M., and R.H. Jansen, "Accurate Model for Effective Dielectric Constant of Microstrip with Validity up to Millimetre-Wave Frequencies," *Electronic Letters*, Vol. 18, March 1982, pp. 272–273.

[29]    Lee, Hai-Young, and Tatsuo Itoh, "Phenomenological Loss Equivalence Method for Planar Quasi-TEM Transmission Lines with a Thin Normal Conductor or Superconductor," *IEEE Transactions on Microwave Theory and Techniques*, Vol. MTT-37, No. 12, December 1989, pp. 1904–1909.

[30]    Lending, Ralph D., "New Criteria For Microwave Component Surfaces," *Proceedings of the National Electronics Conference*, Vol. 11, 1955, pp. 391–401.

[31] Lev, James J., "Synthesize and Analyze Microstrip Lines," *Microwaves & RF*, January 1985, pp. 111–116. (This reference compares several microstrip references for $Z_0$ and $\varepsilon_{eff}$.)

[32] Livernois, Thomas G., "Characteristic Impedance and Electromagnetic Field Distribution in Metal-Insulator-Semiconductor Microstrip," *IEEE Transactions on Microwave Theory and Techniques*, Vol. MTT-38, No. 11, November 1990, pp. 1740–1743.

[33] March, Steven, "Microstrip Packaging: Watch the Last Step," *Microwaves*, December 1981, pp. 83–94.

[34] Michalski, Krzysztof A., and Dalian Zheng, "Rigorous Analysis of Open Microstrip Lines of Arbitrary Cross-Section in Bound and Leaky Regions," *1989 IEEE MTT-S Symposium Digest*, pp. 787–790.

[35] Mirshekar-Syahkal, D., "An Accurate Determination of Dielectric Loss Effect in Monolithic Microwave Integrated Circuits Including Microstrip and Coupled Microstrip Lines," *IEEE Transactions on Microwave Theory and Techniques*, Vol. MTT-31, No. 11, November 1983, pp. 950–954.

[36] Morgan, Samuel P., Jr., "Effect of Surface Roughness on Eddy Current Losses at Microwave Frequencies," *Journal of Applied Physics*, Vol. 20, No. 4, 1949, pp. 352–362.

[37] Nishiki, Sadayuki, and Shuomi Yuki, "Transmission Loss of Thick-Film Microstriplines," *IEEE Transactions on Microwave Theory and Techniques*, Vol. MTT-30, No. 7, July 1982, pp. 1104–1107.

[38] Owens, R.P., and M.H.N. Potok, "Analytical Methods for Calculating the Characteristic Impedance of Finite-Thickness Microstrip Lines," *International Journal of Electronics*, Vol. 41, No. 4, 1976, pp. 399–403.

[39] Pramanick, Protap, and Prakash Bhartia, "Phase Velocity Dependence of the Frequency-Dependent Characteristic Impedance of Microstrip," *Microwave Journal*, June 1989, pp. 180–184.

[40] Pramanick, Protap, and Prakash, Bhartia, "An Accurate Description of Dispersion in Microstrip," *Microwave Journal*, December 1983, pp. 89–92.

[41] Pucel, Robert A., *et al.*, "Losses in Microstrip," *IEEE Transactions on Microwave Theory and Techniques*, Vol. MTT-16, No. 6, June 1968, pp. 342–350 and corrections December 1968, p. 1064.

[42] Pucel, Robert A., and Daniel J. Massé, "Microstrip Propagation on Magnetic Substrates–Part I: Design Theory," *IEEE Transactions on Microwave Theory and Techniques*, Vol. MTT-20, No. 5, May 1972, pp. 304–308.

[43]     Pucel, Robert A., and Daniel J. Massé, "Microstrip Propagation on Magnetic Substrates–Part II: Experiment," *IEEE Transactions on Microwave Theory and Techniques*, Vol. MTT-20, No. 5, May 1972, pp. 309–313.

[44]     Pues, H.F., and A.R. Van de Capelle, "Approximate Formulas for Frequency Dependence of Microstrip Parameters," *Electronics Letters*, Vol. 16, No. 23, November 6, 1980, pp. 870–872.

[45]     Railton, C.J., and Joseph P. McGeehan, "An Analysis of Microstrip with Rectangular and Trapezoidal Conductor Cross Section," *IEEE Transactions on Microwave Theory and Techniques*, Vol. MTT-38, No. 8, August 1990, pp. 1017–1022.

[46]     Railton, C.J., and J.P. McGeehan, "Characterisation of Microstrip Open-End with Rectangular and Trapezoidal Cross-Section," *Electronics Letters*, Vol. 26, No. 11, May 24, 1990, pp. 685–686.

[47]     Rizzoli, Vittorio, "Highly Efficient Calculation of Shielded Microstrip Structures in the Presence of Undercutting," *IEEE Transactions on Microwave Theory and Techniques*, Vol. MTT-27, No. 2, February 1979, pp. 150–157.

[48]     Ross, R.F.G, and M.J. Howes, "Simple Formulas for Microstrip Lines," *Electronics Letters*, Vol. 12, No. 16, August 1976, p. 410.

[49]     Saad, T., ed., *The Microwave Engineer's Handbook*, "Surface Roughness vs. Surface Resistance at 3000 MHz," by R. Lending, Artech House, Norwood, MA, p. 186.

[50]     Sanderson, A.E., "Effect of Surface Roughness on Propagation of the TEM Mode," *Advances in Microwaves*, Volume 7, Leo Young, Ed., Academic Press, New York, 1971, pp. 2–57.

[51]     Schneider, M.V., "Microstrip Lines for Microwave Integrated Circuits," *Bell Systems Technical Journal*, Vol. 48, No. 5, May-June 1969, p. 1421-1444.

[52]     Schneider, M.V., "Dielectric Loss in Integrated Microwave Circuits," *The Bell System Technical Journal*, September 1969, pp. 2325-2332.

[53]     Schneider, M.V., *et al.*, "Microwave and Millimeter Wave Hybrid Integrated Circuits for Radio Systems," *The Bell System Technical Journal*, July-August 1969, pp. 1703-1726.

[54]     Schroeder, W., and I. Wolff, "A New Hybrid Mode Boundary Integral Method for Analysis of MMIC Waveguides with Complicated Crossection," *1989 IEEE MTT-S Symposium Digest*, pp. 711–714. (Has plot of change in $Z_0$ and $\varepsilon_{eff}$ as a function of etch angle for coupled CPW with trapezoidal traces.)

[55]     Schumacher, Walter, "Stromverteilung auf der Grundfläche der Mikrostrip-Leitung und deren Auswirkung auf die ohmsche Leitungsdämpfung," *AEÜ*, Band 33, Heft 5, 1979, pp. 207–212. (Closed-form equations for the ground plane losses)

[56]    Simpson, Ted L., and Bangjuh Tseng, "Dielectric Loss in Microstrip Lines," *IEEE Transactions on Microwave Theory and Techniques*, Vol. MTT-24, No. 2, February 1976, pp. 106–108.

[57]    Smith, S.P., and S.H. Al-Charachafchi, "On the Dependence of Radiation from Microstrip Circuits upon Substrate Parameters," *19th European Microwave Conference Proceedings*, September 4–7, 1990, London, England, pp. 1220–1225.

[58]    Sobol, Harold, "Applications of Integrated Circuit Technology to Microwave Frequencies," *Proceedings of the IEEE*, Vol. 59, No. 8, August 1971, pp. 1200–1211. (Losses due to surface roughness from unpublished Sperry data, discussion of lumped elements.)

[59]    Tam, Alan, "Principles of Microstrip Design," *RF Design*, June 1988, p. 2934.

[60]    Tanaka, Hiroyuki, and Fumiaki Okada, "Precise Measurements of Dissipation Factor in Microwave Printed Circuit Boards," *IEEE Transactions on Microwave Theory and Techniques*, Vol. MTT-38, No. 2, April 1989, pp. 509–514. (This article demonstrates separation of conductor losses from dielectric losses. It has an interesting section comparing oxygen-free rolled copper foil conductors to electrodeposited copper conductors. Scanning electron microscope photos show anisotropy of conductivity.)

[61]    Tripathi, Vijai K., "A Dispersion Model for Coupled Microstrips," *IEEE Transactions on Microwave Theory and Techniques*, Vol. MTT-34, No. 1, January 1986, pp. 66–71.

[62]    Vakanas, Loizos P., *et al.*, "A Parametric Study of the Attenuation Constant of Lossy Microstrip Lines," *IEEE Transactions on Microwave Theory and Techniques*, Vol. MTT-38, No. 8, August 1990, pp. 1137–1139.

[63]    Van der Pauw, Leo J., "The Radiation of Electromagnetic Power by Microstrip Configurations," *IEEE Transactions on Microwave Theory and Techniques*, Vol. MTT-25, No. 9, September 1977, pp. 719–725.

[64]    van Deventer, T. Emilie, *et al.*, "An Integral Equation Method for the Evaluation of Conductor and Dielectric Losses in High-Frequency Interconnects," *IEEE Transactions on Microwave Theory and Techniques*, Vol. MTT-37, No. 12, December 1989, pp. 1964–1972. (Analyzes effect of ground plane resistivity, surface roughness, skin effect in microstrip line.)

[65]    van Heuven, J.H.C., "Conduction and Radiation Losses in Microstrips," *3rd European Microwave Conference Proceedings*, Vol. 2, September 4–7, 1973, Brussels, Belgium, B.7.4.

[66]    Weiss, J. A., *et al.*, "MSTRIP2: Parameters of Microstrip Transmission Lines and of Coupled Pairs of Lines—1978 Version and its Applications," MIT Lincoln Laboratory, Technical Report 600, June 4, 1982.

[67]    Welch, J.D., and H.J. Pratt, "Losses in Microstrip Transmission Systems for Integrated Microwave Circuits," *NEREM Record*, 1966, pp. 100–101.

[68]    Wheeler, Harold A., "Transmission-Line Properties of a Strip on a Dielectric Sheet on a Plane," *IEEE Transactions on Microwave Theory and Techniques*, Vol. MTT-25, No. 8, August 1977, pp. 631–647.

[69]    Williams, Dylan F., and David W. Paananen, "Suppression of Resonant Modes in Microwave Packages," *1989 IEEE MTT-S Symposium Digest*, pp. 1263–1265.

[70]    Williams, Dylan F., "Damping of Resonant Modes of a Rectangular Metal Package," *IEEE Transactions on Microwave Theory and Techniques*, Vol. MTT-37, No. 1, January 1989, pp. 253–256.

[71]    Young, Leo, and H. Sobol, eds., *Advances in Microwaves*, Volume 8, Academic Press, New York, 1974.

### 3.5.2   Microstrip Line with Truncated Ground Plane and Dielectric

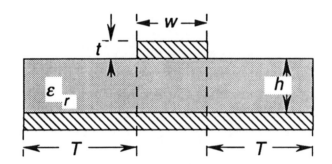

**Figure  3.5.2.1:**        **Microstrip  Line  with  Truncated  Ground  Plane  and Dielectric**

It is desirable to know the applicable range of the standard microstrip line formulas for finite width grounds and dielectrics. Figure 2 of [1] indicates that $T/w \geq 2$ has < 3% effect on $Z_0$. Figure 5 of [2] plots $\varepsilon_{eff}$ as a function of $T/w$.   As expected $\varepsilon_{eff}$ decreases with $T/w$ reaching a final value approximately 18% less than $T = \infty$ (for $\varepsilon_r = 10.2$, $w/h = 0.73$, 2 GHz). A value of $T/w > 2$ has little effect on $\varepsilon_{eff}$.

This effect is a function of frequency. At dc the ground current will be spread equally through the ground cross-section; however, as frequency increases it will concentrate into a strip directly below the signal conductor.

REFERENCES

[1]    Smith, Charles E., and Ray-Sun Chang, "Microstrip Transmission Line with Finite-Width Dielectric and Ground Plane," *IEEE Transactions on Microwave Theory and Techniques*, Vol. MTT-33, No. 9, September 1985, pp. 835–839.

[2]    Thorburn, Michael, *et al.*, "Computation of Frequency-Dependent Propagation Characteristics of Microstriplike Propagation Structures with Discontinuous Layers," *IEEE Transactions on Microwave Theory and Techniques*, Vol. MTT-38, No. 2, February 1990, pp. 148–153.

### 3.5.3    Microstrip Line with Truncated Dielectric

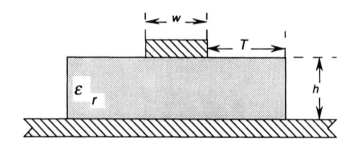

**Figure  3.5.3.1:        Microstrip  Line  with  Truncated  Dielectric**

It is desirable to know the range of standard microstrip line equations in the presence of finite width dielectrics.  Surprisingly, Figure 2 of [2] indicates for $T / w > 0.5$, $Z_0$ is affected $< 0.5\%$.  This is because most of the field lines are concentrated in the dielectric directly under the microstrip line.

It should pointed out that this effect is a function of frequency.  At dc the ground current will be spread equally through the ground cross-section; however, as frequency increases it will concentrate into a strip directly below the signal conductor.

REFERENCES:

[1]    Rong, A., and S. Li, "Frequency Dependent Transmission Characteristics of Microstrip Lines on the Finite Width Substrate or Near a Substrate Edge," *Electronics Letters*, Vo. 26, No. 12, June 7, 1990, pp. 782–783.

[2]    Smith, Charles E., and R.S. Chang, "Microstrip Transmission Line with Finite-Width Dielectric," *IEEE Transactions on Microwave Theory and Techniques*, Vol. MTT-28, February 1980, pp. 90–94.

[3]    Thorburn, Michael, *et al.*, "Computation of Frequency-Dependent Propagation Characteristics of Microstriplike Propagation Structures with Discontinuous

Layers," *IEEE Transactions on Microwave Theory and Techniques*, Vol. MTT-38, No. 2, February 1990, pp. 148–153.

### 3.5.4   Embedded or Buried Microstrip Line

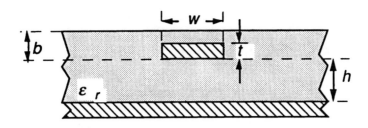

**Figure   3.5.4.1:      Embedded or Buried Microstrip Line**

This structure is encountered in multilayer PCBs, compensated couplers, and even in single-layer PCBs.  Some vendors apply a thin coat of epoxy to the PCB to make the dielectric flush with the upper microstrip line surface.  This buries surface imperfections (loose glass fibers) remaining after processing.

Many simple approximations are available in the literature.  These are included here due to their frequent occurrence in the PCB industry literature.  This is not meant to imply their usefulness.  The following approximation equation is given by [5] and others:

$$Z_0 = Z_0' - b\%$$   (3.5.4.1)

$b$   (mils)

The impedance $Z_0'$ is calculated as for normal microstrip line.  The formula then implies that the impedance is reduced 1% per mil of dielectric over the line.  In Coombs [4], it is stated that the presence of upper dielectric material reduces $Z_0$ by 22%.  This rule and equation (3.5.4.1) are often quoted by PCB vendors and are in fact not very good estimators.

An equation which appears to have the proper form is found in [6].  The equations are:

$$Z_0 = \frac{Z_{0,microstrip}\sqrt{\varepsilon_{eff,microstrip}}}{\sqrt{\varepsilon_{eff,buried}}}$$   (3.5.4.2)

where,

$Z_0$ = characteristic impedance of a (not buried, $b = 0$) microstrip line

$\varepsilon_{eff,microstrip}$ = effective relative dielectric constant of a (not buried, $b = 0$) microstrip line

$$\varepsilon_{eff,buried} = \varepsilon_{eff} \ e^{(-2.0 \ b \ / \ h)} + \varepsilon_r \left[1.0 - e^{(-2.0 \ b \ /h)}\right] \qquad (3.5.4.3)$$

The figures below plot an example of the resulting $\varepsilon_{eff}$ and $\varepsilon_{eff,buried}$ for the microstrip line and embedded microstrip line configurations respectively. Note that the extra layer of dielectric material has more effect on narrow strips. In this example, for $b \ / \ h < .1$ there is < 1% effect on the line's $\varepsilon_{eff}$ and, for $b \ / \ h > 1.0$, $\varepsilon_{eff}$ is within 5% of $\varepsilon_r$. Static results of CALIF are also shown for $w \ / \ h = 1.0$. Narrow traces have more fringing capacitance and are effected more by the presence of an overlay than wider traces.

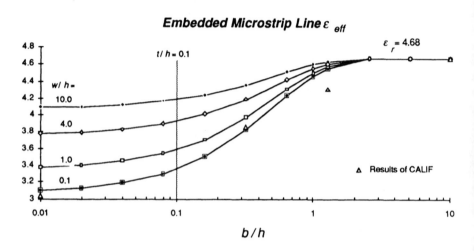

**Figure 3.5.4.2.1:   Comparison of Microstrip Line and Embedded Microstrip Line $\varepsilon_{eff}$'s**

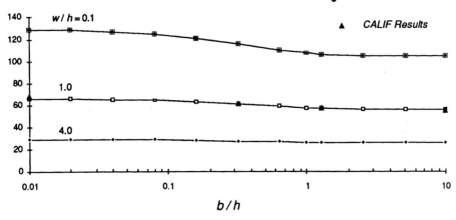

**Figure 3.5.4.2.2:** Embedded Microstrip Line $Z_0$

In Chang and Klein [3], data is presented for a line covered with a high permittivity line. The dependency on $h_2$ appears to be of the form $(1.0 - e^{-x})$, similar to the results of Buntschuh [2]. With equal dielectric constants the value of $\varepsilon_{eff}$ was nearly $\varepsilon_r$ for $h_2 / h_1 = 2.0$.

## REFERENCES

[1] Bogatin, Eric, "Design Rules for Microstrip Capacitance," *IEEE Transactions on Components, Hybrids, and Manufacturing Technology*, Vol. CHMT-11, No. 3, September 1988, pp. 253–259.

[2] Buntschuh, Charles, "High Directivity Microstrip Couplers Using Dielectric Overlays," *1975 IEEE MTT-S International Microwave Symposium Digest*, pp. 125–127.

[3] Chang, K., and J. Klein, "Dielectrically Shielded Microstrip (DSM) Lines," *Electronics Letters*, Vol. 23, No. 10, May 7, 1987, pp. 535–537.

[4] Coombs, C.F., *Printed Circuits Handbook*, McGraw-Hill, 1979.

[5] Heard, Chris, Internal Document, Teradyne Connection Systems, 1988.

[6] Hogan, Sean, "Closed Form Transmission Line Equation," presented at *New England Circuits Association*, Andover, MA, May 11, 1988.

### 3.5.5   Nonhomogeneous Dielectric Embedded Microstrip Line

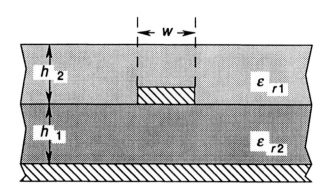

**Figure 3.5.5.1:**        Nonhomogeneous Dielectric Embedded Microstrip
Line or Microstrip Line with Overlay

This structure results when we cover a microstrip line with solder mask. Inner layer microstrip lines are sometimes used in multilayer PCBs where the dielectric constants are not equal. If $\varepsilon_{r,1} = \varepsilon_{r,2}$ use the equations of the previous section.

In lieu of design equations we can bound the problem. We know that

$$\varepsilon_{eff,microstrip} < \varepsilon_{eff,embedded} < \varepsilon_{eff,2}$$

where $\varepsilon_{eff,2}$ is calculated with the technique described in the Introduction.

REFERENCES

[1]    Buntschuh, Charles, "High Directivity Microstrip Couplers Using Dielectric Overlays," *1975 IEEE MTT-S International Microwave Symposium Digest*, pp. 125–127.

[2]    Chang, K., and J. Klein, "Dielectrically Shielded Microstrip (DSM) Lines," *Electronics Letters*, Vol. 23, No. 10, May 7, 1987, pp. 535–537.

[3]    Callarotti, Roberto C., and Augusto Gallo, "On the Solution of a Microstripline with Two Dielectrics," *IEEE Transactions on Microwave Theory and Techniques*, Vol. MTT-32, No. 4, April 1984, pp. 333–339.

### 3.5.6    Shielded Microstrip Line

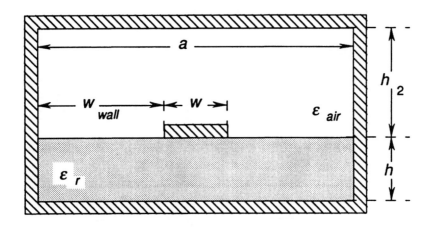

**Figure 3.5.6.1:**    Shielded Microstrip Line

Shielding lowers $\varepsilon_{effective}$ and $Z_0$ and creates resonances.  Neglect the effects of shield for $W_{wall} > 5.0 \times w$ and $h_2 > 5.0 \times h$.

REFERENCES

[1]    Bedair, S.S., "Predict Enclosure Effects on Shielded Microstrip," *Microwaves & RF*, July 1985, pp. 97–100.

[2]    Bedair, S.S., and M.I. Sobhy, "Accurate Formulas for Computer-Aided Design of Shielded Microstrip Circuits," *IEE Proceedings*, Vol. 127, Pt. H, No. 6, December 1980, pp. 305–308.

[3]    Bedair, S.S., and M.I. Sobhy, "Tolerance Analysis of Shielded Microstrip Lines," *IEEE Transactions on Microwave Theory and Techniques*, Vol. MTT-32, No. 5, May 1984, pp. 544–547.

[4]    Fikioris, John G., *et al.*, "Exact Solutions for Shielded Printed Microstrip Lines by the Carleman-Vekua Method," *IEEE Transactions on Microwave Theory and Techniques*, Vol. MTT-37, No. 1, January 1989, pp. 21–33.

[5]    Hassan, Essam E., "Simple Solution of Dispersion Characteristics of Shielded Stripline," *IEE Proceedings*, Vol. 134, Pt. H, No. 6, December 1987, pp. 566–568.

[6]    Itoh, Tatso, and Raj Mittra, "A Technique for Computing Dispersion Characteristics of Shielded Microstrip Lines," *IEEE Transactions on Microwave Theory and Techniques*, Vol. MTT-22, No. 10, October 1974, pp. 896–898.

[7]    Leong, M.S., *et al.*, "Capacitance of Shielded Microstripline:  Closed Form Quasi-TEM Results Suitable for Computer-Aided Design," *IEE Proceedings*, Vol. 134, Pt. H, No. 4, August 1987, pp. 393–396.

[8]    March, Steven, "Microstrip Packaging:  Watch the Last Step," *Microwaves*, December 1981, pp. 83–94.

[9]    March, Steven L., "Empirical Formulas for the Impedance and Effective Dielectric Constant of Covered Microstrip for Use in the Computer-Aided Design of Microwave Integrated Circuits," *11th European Microwave Conference Proceedings*, 1981, pp. 671–676.

[10]   Wolff, Edward A., and Roger Kaul, *Microwave Engineering and Systems Applications*, John Wiley and Sons, New York, 1988.

[11]   Yamashita, Eikichi, and Kuzuhiko Atsuki, "Strip Line with Rectangular Outer Conductor and Three Dielectric Layers," *IEEE Transactions on Microwave Theory and Techniques*, Vol. MTT-18, No. 5, May 1970, pp. 238–244.

### 3.5.7   Edge-Compensated Microstrip Line (ECM Line)

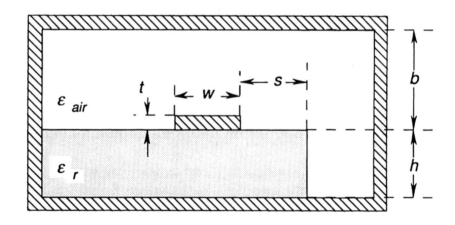

**Figure   3.5.7.1:**     **Edge-Compensated Microstrip Line**

Yamashita *et al.* [3] show a technique for solving and correcting the effects of a shielded microstrip line near a dielectric truncation.  The line is corrected by increasing the width calculated for the usual microstrip line, $w_0$, by a factor $w / w_0$ and using a modified $\varepsilon_{eff}$.  They also give simplified approximation formulas for GaAs and alumina microstrip lines.  For GaAs ($\varepsilon_r = 12.9$)

$$\frac{w}{w_0} = 1.0367 - 0.18228\, u + 0.17147\, u^2 + 0.13289\, u^3 - 0.080565\, u^4$$
$$- 0.081078\, u^5 - 0.015946\, u^6 \tag{3.5.7.1}$$

$$\varepsilon_{eff} = \left(0.35346 - 0.025010\, u + 0.022373\, u^2 + 0.018016\, u^3 \right.$$
$$\left. - 0.010253\, u^4 - 0.010717\, u^5 - 0.00211593\, u^6 \right)^{-2} \qquad (3.5.7.2)$$

For alumina ($\varepsilon_r = 9.7$)

$$\frac{w}{w_0} = 1.0162 - 0.098654\, u + 0.13236\, u^2 + 0.033574\, u^3 - 0.070952\, u^4$$
$$- 0.024336\, u^5 + 0.00022125\, u^6 \qquad (3.5.7.3)$$

$$\varepsilon_{eff} = \left(0.39554 - 0.017792\, u + 0.022627\, u^2 + 0.0068338\, u^3 \right.$$
$$\left. - 0.012018\, u^4 - 0.0046957\, u^5 + 0.0019738\, u^6 \right)^{-2} \qquad (3.5.7.4)$$

where

$$u = \log \frac{s}{w_0} \qquad (3.5.7.5)$$

Equations are curve-fit polynomials that are within 1% of the theoretical results for

$$0.1 \leq s / w_0 \leq 5.0$$

## REFERENCES

[1] Rong, A., and S. Li, "Frequency Dependent Transmission Characteristics of Microstrip Lines on the Finite Width Substrate or Near a Substrate Edge," *Electronics Letters*, Vol. 26, No. 12, June 7, 1990, pp. 782–783.

[2] Thorburn, Michael, *et al.*, "Computation of Frequency-Dependent Propagation Characteristics of Microstriplike Propagation Structures with Discontinuous Layers," *IEEE Transactions on Microwave Theory and Techniques*, Vol. MTT-38, No. 2, February 1990, pp. 148–153.

[3] Yamashita, Eikichi, *et al.*, "Characterization of Microstrip Lines Near a Substrate Edge and Design Formulas of Edge-Compensated Microstrip Lines," *IEEE Transactions on Microwave Theory and Techniques*, Vol. MTT-37, No. 5, May 1989, pp. 890–896.

### 3.5.8   Covered Microstrip Line

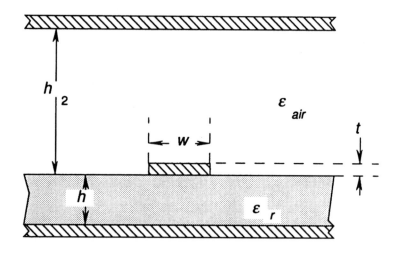

**Figure  3.5.8.1:**      Covered  Microstrip  Line

The impedance of a microstrip line of a given set of dimensions is lowered by the presence of a cover at a finite distance $h_2$. A large number of sources [1–8, 10] all describe this configuration. March [7] corrects a number of errors and compares his results with the "exact" MSTRIP analysis.

The technique is to calculate $Z_0$ for a completely air dielectric and a shield located at $h_2 = \infty$ and then correct this value for the presence of the cover with $\Delta Z_0$ and $\varepsilon_{eff}$.

$$Z_0 = \frac{Z_0\,(\varepsilon_r = 1,\ h_2 = \infty) - \Delta Z_0\,(\varepsilon_r = 1)}{\sqrt{\varepsilon_{eff}}} \qquad (3.5.8.1)$$

The equation for $Z_0$ is after Hammerstad and Jensen [6]:

$$Z_0\,(\varepsilon_r = 1,\ h_2 = \infty) = \frac{\eta_0}{2.0\,\pi}\,\ln\left\{\frac{f\,(w'/h)}{w'/h} + \sqrt{\left[1.0 + \left(\frac{2.0\,h}{w'}\right)^2\right]}\,\right\} \qquad (3.5.8.2)$$

where

$$f(w'/h) = 6.0 + (2.0\,\pi - 6.0)\,e^{-\left(\frac{30.666}{w'/h}\right)^{0.7528}} \qquad (3.5.8.3)$$

These equations agree with MSTRIP to within ±0.1% for $0.1 \le w/h \le 6.0$ and to within ±0.2% for $6.0 \le w/h \le 10.0$.

$$w' = w + \frac{t \left\{ 1.0 + \ln 4.0 - 0.5 \ln \left[ (t/h)^2 + \left( \frac{t}{\pi \, w} \right)^2 \right] \right\}}{\pi} \qquad (3.5.8.4)$$

which is Wheeler's width correction [10] for finite strip thickness presented earlier. The value of $\Delta Z(\varepsilon_r = 1)$ is calculated with corrected versions of Bahl's equations [1]:

$$\Delta Z(\varepsilon_r = 1) = P\,Q \qquad (3.5.8.5)$$

where

$$P = 270.0 \left[ 1.0 - \tanh \left( 1.192 + 0.706 \sqrt{1.0 - h_2/h} - \frac{1.389}{1.0 + h_2/h} \right) \right]$$

$$(3.5.8.6)$$

$$Q = 1.0109 - \tanh^{-1} \left[ \frac{0.012 \, w/h + 0.177 \, (w/h)^2 - 0.027 \, (w/h)^3}{(1.0 + h_2/h)^2} \right]$$

$$(3.5.8.7)$$

The effective dielectric constant corrects for the presence of the dielectric.

$$\varepsilon_{eff} = \frac{\varepsilon_r + 1}{2} + q \frac{\varepsilon_r - 1}{2} \qquad (3.5.8.8)$$

An adjusted value of $q$ is used to calculate $\varepsilon_{eff}$. March [7] modifies equations of Bahl [1] by defining a $q$ made up of three pieces:

$$q = q_c \, (q_\infty - q_T) \qquad (3.5.8.9)$$

where

$q_\infty$ = filling factor with no cover and zero-thickness strip

$q_T$ = correction for finite strip thickness

$q_c$ = correction for effect of the shield

March recommends using $q_\infty$ of [7] for $w/h \geq 5$

$$q_\infty = \left( 1.0 + \frac{10.0 \, h}{w} \right)^j \qquad (3.5.8.10)$$

$$j = a\,(w\,/\,h)\,b(\varepsilon_r) \tag{3.5.8.11}$$

$$a(w\,/\,h) = 1.0 + \frac{1.0}{49.0}\ln\left\{\frac{(w\,/\,h)^2\,\left[(w\,/\,h)^2 + (1.0\,/\,52.0)^2\right]}{(w\,/\,h)^4 + 0.432}\right.$$

$$\left. + \frac{1.0}{18.7}\ln\left[1.0 + \left(\frac{w}{18.1\,h}\right)^3\right]\right\} \tag{3.5.8.12}$$

$$b(\varepsilon_r) = -0.564\left(\frac{\varepsilon_r - 0.9}{\varepsilon_r + 3.0}\right)^{0.053} \tag{3.5.8.13}$$

These equations have a stated accuracy of ±0.2% for

$$0.01 \le w\,/\,h \le 100$$

$$1 \le \varepsilon_r \le 128$$

Stated accuracy is better than 0.2% agreement with Bryant and Weiss [3] for

$$1 \le \varepsilon_r \le 30$$

$$0.05 \le w\,/h \le 20$$

The strip thickness correction is

$$q_T = \frac{2.0\,t\,\ln 2}{\pi\,h\,\sqrt{w\,/\,h}} \tag{3.5.8.14}$$

and the correction for the shield is ($h_2\,/\,h \ge 1$)

$$q_c = \tanh\left(1.043 + 0.121\,h_2\,/\,h - \frac{1.164\,h}{h_2}\right) \tag{3.5.8.15}$$

The equations were verified with MSTRIP calculations and the error was found to be less than 1% for the conditions above and $1 < h_2\,/\,h < \infty$. March also recommends the dispersion model of [5] for this structure based on exact calculations. To correct for frequency dependencies, we modify the calculated $\varepsilon_{eff}\,(f = 0)$

$$\varepsilon_{eff}\,(f) = \varepsilon_r - \frac{\varepsilon_r - \varepsilon_{eff}\,(0)}{1.0 + G\,(f\,/\,f_p)^2} \tag{3.5.8.16}$$

where

$$f_p = \frac{Z_0}{2.0 \ \mu_0 \ h} \tag{3.5.8.17}$$

$$G = \frac{\pi^2 \ (\varepsilon_r - 1.0)}{12.0 \ \varepsilon_{eff}(0)} \sqrt{\frac{2.0 \ \pi \ Z_0}{\eta_0}} \tag{3.5.8.18}$$

To calculate the frequency-dependent characteristic impedance, $Z_0(f)$, we convert the calculated zero-frequency impedance to an equivalent parallel-plate transmission line.

$$Z_0 \ (f) = \frac{120.0 \ \pi \ h}{w_{eff} \ (f) \ \sqrt{\varepsilon_{eff} \ (f)}} \tag{3.5.8.19}$$

$$w_{eff}(f) = w + \frac{w_{eff}(0) - w}{1.0 + (f / f_g)^2} \tag{3.5.8.20}$$

$$w_{eff}(0) = \frac{120.0 \ \pi \ h}{Z_0(f = 0) \ \sqrt{\varepsilon_{eff}(0)}} \tag{3.5.8.21}$$

where $Z_0(f = 0)$ is the previously calculated zero-frequency value of the characteristic impedance, and the frequency where dispersion due to higher mode propagation begins is

$$f_g = \frac{c}{2.0 \ w_{eff}(f) \ \sqrt{\varepsilon_{eff}(f) - 1.0}} \tag{3.5.8.22}$$

### REFERENCES

[1]     Bahl, I.J., "Use Exact Methods for Microstrip Design," *Microwaves*, Vol. 19, December 1978, pp. 61–62.

[2]     Bedair, S.S., and M.I. Sobhy, "Tolerance Analysis of Shielded Microstrip Lines," *IEEE Transactions on Microwave Theory and Techniques*, Vol. MTT-32, May 1984, No. 5, pp. 544–547.

[3]     Bryant, T.G., and J.A. Weiss, "Parameters of microstrip transmission lines and of coupled pairs of microstrip lines," *IEEE Transactions on Microwave Theory and Techniques*, Vol. MTT-16, December 1968, pp. 1021–1027.

[4]     Ermert, Helmut, "Guiding and Radiation Characteristics of Planar Waveguides," *Microwaves, Optics and Acoustics*, Vol. 3, No. 2, March 1979, pp. 59–62.

[5]     Getsinger, William J., "Microstrip Dispersion Model," *IEEE Transactions on Microwave Theory and Techniques*, Vol. MTT-21, No. 1, January 1973, pp. 34–39.

[6]　Hammerstad, E. and Ø. Jensen, "Accurate Models for Microstrip Computer-Aided Design," *1980 IEEE MTT-S International Microwave Symposium Digest*, pp. 407–409.

[7]　March, Steven L., "Empirical Formulas for the Impedance and Effective Dielectric Constant of Covered Microstrip for use in the Computer-Aided Design of Microwave Integrated Circuits," *11th European Conference on Microwaves*, 1981, pp. 671–676. (March points out a number of errors in earlier articles dealing with shielded configurations.)

[8]　March, Steven, "Microstrip Packaging: Watch the Last Step," *Microwaves*, December 1981, pp. 83–94.

[9]　Pues, H.F., and A.R. Van de Capelle, "Approximate Formulas for Frequency Dependence of Microstrip Parameters," *Electronics Letters*, Vol. 16, No. 23, November 6, 1980, pp. 870–872.

[10]　Wheeler, Harold A., "Transmission-Line Properties of a Strip on a Dielectric Sheet on a Plane," *IEEE Transactions on Microwave Theory and Techniques*, Vol. MTT-25, No. 8, August 1977, pp. 631–647.

## 3.6   STRIPLINE STRUCTURES

The term stripline was originally used to describe any planar transmission line based on strip conductors, including microstrip line. Its meaning in recent years has come to be more specific and now generally refers to a structure having a strip sandwiched between ground planes. Harold Wheeler developed a planar transmission line (two coplanar strips) which could be rolled up in 1936 and a stripline-like structure in 1942. The first use of "flat strip coaxial transmission line" (later to be known as stripline) was by V.H. Rumsey and H.W. Jamieson for an antenna power divider in WWII [2]. In 1949 Robert M. Barrett applied this new structure to printed-circuit boards launching what is now one of the most popular transmission line structures. Stripline was the name coined for a ground-signal-ground transmission line with an air dielectric at AIL around 1952. Tri-Plate was simultaneously developed at Sanders Associates, Inc. (stripline with solid dielectric) and a handbook of design techniques was published in 1956 [1].

REFERENCES:

[1]   Ayer, D.R., and C.A. Wheeler, "The Evolution of Strip Transmission Line," *Microwave Journal*, Vol. 12, No. 5, May 1969, pp. 31–40.

[2]   Barrett, Robert M., "Microwave Printed Circuits—A Historical Survey," *IRE Transactions on Microwave Theory and Techniques*, MTT-3, No. 2, March 1955, pp. 1–9.

[3]   Howe, Harlan, Jr., "Microwave Integrated Circuits—An Historical Perspective," *IEEE Transactions on Microwave Theory and Techniques*, Vol. MTT-32, No. 9, September 1984, pp. 991–996.

### 3.6.1   Zero-Thickness Centered Stripline

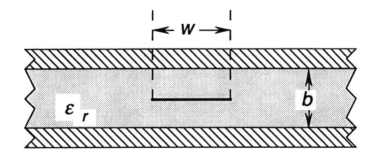

Figure 3.6.1.1:     Zero-Thickness Stripline

Equations are exact for the zero-thickness case; however see 3.6.2 for a more realistic finite-thickness trace. The zero-thickness calculation is often used to check the limiting case ($t = 0$) of finite-thickness equations and as a term in finite-thickness calculations.

$$Z_0 = \frac{\eta_0}{4.0 \sqrt{\varepsilon_r}} \frac{K(k)}{K(k')} \tag{3.6.1.1}$$

$$k = \text{sech}\left(\frac{\pi w}{2.0 \, b}\right) \tag{3.6.1.2}$$

$$k' = \tanh\left(\frac{\pi w}{2.0 \, b}\right) \tag{3.6.1.3}$$

REFERENCE

[1]    Cohn, Seymour B., "Characteristic Impedance of the Shielded-Strip Transmission Line," *Transactions of the IRE*, Vol. MTT-2, July 1954, pp. 52-57.

### 3.6.2    Centered Stripline, Tri-Plate®[2] or Sandwich Line

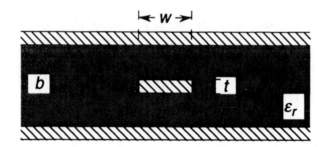

**Figure 3.6.2.1:  Centered Stripline**

Note in Figure 3.6.2.1 and this equation, that the dielectric thickness is $b$; some references refer to this as $h$ or $2.0 \, h$. The equations are [9]

$$Z_0 = \frac{\eta_0}{2.0 \, \pi \sqrt{\varepsilon_r}} \ln \left\{ 1.0 + \right.$$

$$\left. \frac{4.0(b-t)}{\pi w'} \left[ \frac{8.0 \, (b-t)}{\pi w'} + \sqrt{\left(\frac{8.0 \, (b-t)}{\pi w'}\right)^2 + 6.27} \right] \right\} \quad (\Omega) \tag{3.6.2.1}$$

---

[2]Triplate® is a registered trademark of Sanders Associates.

where

$$b = 2.0\,h + t \tag{3.6.2.2}$$

$$w' = w + \frac{\Delta w}{t}\,t \tag{3.6.2.3}$$

$$\frac{\Delta w}{t} = \frac{1.0}{\pi} \ln\left[ \frac{e}{\sqrt{\left(\dfrac{1}{2.0\,(b-t)\,/\,t+1}\right)^2 + \left(\dfrac{1/4\pi}{w/t+1.1}\right)^m}} \right] \tag{3.6.2.4}$$

The above is sometimes written:

$$\frac{\Delta w}{t} = \frac{1.0}{\pi}\left\{ 1.0 - 0.5 \ln\left[\left(\frac{1.0}{2.0\,(b-t)\,/\,t+1.0}\right)^2 + \left(\frac{1.0\,/\,4\pi}{w\,/\,t+1.1}\right)^m\right]\right\}$$

$$\tag{3.6.2.5}$$

and

$$m = \frac{6.0}{3.0 + \dfrac{2.0\,t}{(b-t)}} \tag{3.6.2.6}$$

The stated error is less than 1.5%. Nauwelaers and Van de Capelle [6] propose a thickness correction that is not dependent on the line width.

$$\frac{\Delta w}{t} = \frac{\ln\left(\dfrac{5.0\,(b-t)}{t}\right)}{3.2} \tag{3.6.2.7}$$

The approximation is valid for

$$t\,/\,b < 0.1$$

$$w\,/\,t > 2.5$$

$$w\,/\,b > 0.1$$

Accuracy of the resulting $Z_0$ is claimed to be better than 0.5%.

The effects of sidewalls can be ignored when they are $>1.5\,b$ away. If this condition is not met, use the equations of Section 3.6.4. The errors of these equations are $<0.5\%$ for

$$w'\,/\,b < 10.0.$$

Losses are calculated with the incremental inductance rule. First the skin depth, $\delta$ is calculated:

$$\delta = \sqrt{\frac{2.0}{\omega \, \mu \, \sigma}} \quad \text{(m)} \tag{3.6.2.8}$$

where $\omega$ is the frequency in radians / second, $\mu$ is the permeability in H / m, and $\sigma$ is the conductivity in mho / m. The power factor $(1.0 / Q)$ is then found from the characteristic impedance of the structure when conductor dimensions are increased by $\delta$:

$$p = 1.0 - \frac{Z_0(w + \delta, h + \delta, t + \delta)}{Z_0(w, b, t)} \tag{3.6.2.9}$$

The accuracy of $p$ is better than $\pm 0.05$ for $t / b < 0.5$, and $\pm 0.015$ for $t / b < 0.25$. Recently, an exact analysis of conductor losses was presented in [8]. This solution is not closed-form and shows that the error in the above technique is less than 6%.

Dielectric losses are calculated with

$$\alpha_d = k_0 \frac{\sqrt{\varepsilon_r} \, \tan \delta}{2.0} \tag{3.6.2.10}$$

where $k_0$ is the wavenumber in free-space.

The next higher mode will propagate in stripline when

$$\lambda_c = \sqrt{\varepsilon_r} \left( \frac{2.0 \, w}{b} + \frac{4.0 \, d}{b} \right) b \quad \text{(units of } b) \tag{3.6.2.11}$$

### REFERENCES

[1]   Bahl, I.J., and Ramesh Garg, "A Designer's Guide to Stripline Circuits," *Microwaves,* January 1978, pp. 90–96.

[2]   Cohn, Seymour B., "Characteristic Impedance of the Shielded-Strip Transmission Line," *Transactions of the IRE,* Vol. MTT-2, July 1954, pp. 52-57.

[3]   Cohn, Seymour B., "Problems in Strip Transmission Lines," *IRE Transactions on Microwave Theory and Techniques,* Vol. MTT-3, No. 3, March 1955, pp. 119–126.

[4]   Fouladian, Jamshid, and Protap Pramanick, "Generalized Analytical Equations for Strip Transmission Lines," *Microwave Journal,* November 1989, pp. 165–168.

[5]   Gupta, K.C., "Striplines and Microstriplines," Notes for "Fundamentals of RF Design," at RF Technology Expo '85, January 23-25, 1985.

[6]   Nauwelaers, B., and A. Van de Capelle, "Characteristic Impedance of Stripline," *Electronics Letters,* Vol. 23, No. 18, August 27, 1987, pp. 930–931.

[7]    Packard, Karle S., "Optimum Impedance and Dimensions for Strip Transmission Line," *IEEE Transactions on Microwave Theory and Techniques*, Vol. 5, October 1957, pp. 244–247.

[8]    Rawal, Sujata, and David R. Jackson, "An Exact TEM Calculation of Loss in a Stripline of Arbitrary Dimensions," *IEEE Transactions on Microwave Theory and Techniques*, Vol. MTT-39, No. 4, April 1991, pp. 694–699.

[9]    Wheeler, Harold A., "Transmission-Line Properties of a Strip Line Between Parallel Planes," *IEEE Transactions on Microwave Theory and Techniques*, Vol. MTT-26, No. 11, pp. 866–876, November 1978.

### 3.6.3   Off-Center Stripline

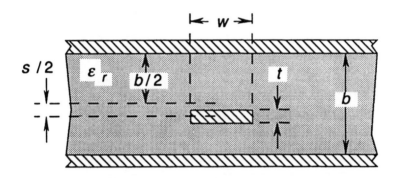

**Figure  3.6.3.1:**       Off–Center  Stripline

The off-center stripline structure is found mainly in multilayer boards. Two situations can make its use desirable. In order to make the board symmetric relative to the center-line (to avoid warpage) it is sometimes necessary to place the stripline off-center relative to its surrounding ground planes. It is also common to run two layers of striplines (*dual orthogonal stripline*) between a single pair of ground planes to increase density. To minimize coupling, the two trace layers are run perpendicular to each other. Each of these layers is off-center relative to the ground planes. See Kiang *et al.* [8] for more information on the analysis of dual orthogonal striplines.

Rao and Das [9] and Chang [1] analyze the off-center stripline using conformal mapping techniques. Chang's solution is limited to wide lines ($w \gtrsim b$) and the solution of Rao and Das' equations is somewhat involved. Equations of Robrish [10] extend Cohn [2] to a wider range of $w/b$.

For $\dfrac{w}{b-t} < 0.35$ (narrow strips) these equations are

$$Z_0 = \frac{\eta_0}{2.0 \ \pi \ \sqrt{\varepsilon_r}} \ \cosh^{-1} A \tag{3.6.3.1}$$

where

$$A = \sin\left(\frac{\pi \ c_l}{b}\right) \coth\left(\frac{\pi \ d_0}{2.0 \ b}\right) \tag{3.6.3.2}$$

$$c_l = \frac{b - s}{2.0} \tag{3.6.3.3}$$

The line width $w$ is replaced with an equivalent round wire diameter, $d_0$. A curve for this purpose is available in [3] and others[3]. A simple curve has been fit to the data for CAE use; the fit error is less than 0.35%:

$$\frac{d_0}{w} = 0.5008 + 1.0235 \ x - 1.0230 \ x^2 + 1.1564 \ x^3 - 0.4749 \ x^4 \tag{3.6.3.4}$$

$$x = \frac{t}{w} \tag{3.6.3.5}$$

where $w \geq t$ (reverse their positions if this condition is not met).

For $\frac{w}{b - t} > 0.35$ (wide strips)

$$Z_0 = \frac{\eta_0}{\sqrt{\varepsilon_r}} \ \frac{1.0}{\dfrac{w \ / \ b}{\gamma} + \dfrac{w \ / \ b}{\beta - \gamma} + \dfrac{2.0 \ C_f'}{\varepsilon_0 \ \varepsilon_r}} \quad (\Omega) \tag{3.6.3.6}$$

where

$$C_f' = \frac{\varepsilon_r \ \varepsilon_0}{\pi} \left\{ 2.0 \ \ln\left[\frac{1.0}{\gamma \ (\beta - \gamma)}\right] + \frac{1.0}{\gamma \ (\beta - \gamma)} \left[ F\left(\frac{t}{2.0 \ b}\right) - F(c_l \ / \ b) \right] \right\} \tag{3.6.3.7}$$

$$F(x) = (1.0 - 2.0 \ x) \ [(1.0 - x) \ \ln \ (1.0 - x) - x \ \ln \ x] \tag{3.6.3.8}$$

$$\gamma = \frac{c_l}{b} - \frac{t}{2.0 \ b} \tag{3.6.3.9}$$

---

[3]The source of this equivalence relation is C. Flammer, "Equivalent Radii of Thin Cylindrical Antennas with Arbitrary Cross Sections," *Stanford Research Institute Technical Report*, March 15, 1950.

$$\beta = 1.0 - \frac{t}{b} \tag{3.6.3.10}$$

for $\dfrac{w}{b-t} < \dfrac{t}{b}$ replace $w/b$ with

$$\left(\frac{w}{b}\right)_{eff} = \frac{w}{b} + \left(1.0 - \frac{t}{b}\right)^{8.0} \left[\frac{K(k')}{K(k)} - \frac{2.0}{\pi} \ln 2.0 - \frac{w}{b}\right] \tag{3.6.3.11}$$

$$k = \operatorname{sech}\left(\frac{\pi\ w}{2.0\ b}\right) \tag{3.6.3.12}$$

$$k' = \tanh\left(\frac{\pi\ w}{2.0\ b}\right) \tag{3.6.3.13}$$

Agreement with exact analytic solutions is better than 0.5%. The accuracy of these relations was compared to numerical solutions and found to be better than ±2% for

$$0.2 < \frac{cl}{b} < 0.8$$

$$w > t$$

$$t/b < 0.2$$

Two other approaches to this problem are also found in the literature. Howe uses a parallel plate and fringing capacitance calculation while Hogan calculates a correction based on equivalent widths.

These equations from Howe [6] have been rearranged to show the underlying parallel plate capacitance calculation.

$$Z_0 = \frac{\eta_0}{\sqrt{\varepsilon_r}\,\dfrac{C'}{\varepsilon}} \tag{3.6.3.14}$$

where

$$\frac{C'}{\varepsilon} = \frac{C_{p1}}{\varepsilon} + \frac{C_{p2}}{\varepsilon} + \frac{2.0\ C_{f_1}}{\varepsilon} + \frac{2.0\ C_{f_2}}{\varepsilon} \tag{3.6.3.15}$$

$$\frac{C_{p1}}{\varepsilon} = \frac{w}{\dfrac{b}{2} - \dfrac{s}{2} - \dfrac{t}{2}} = \frac{2.0\left(\dfrac{w}{b}\right)}{1.0 - \dfrac{s}{b} - \dfrac{t}{b}} \tag{3.6.3.16}$$

$$\frac{C_{p2}}{\varepsilon} = \frac{w}{\frac{b}{2} + \frac{s}{2} - \frac{t}{2}} = \frac{2\left(\frac{w}{b}\right)}{1.0 + \frac{s}{b} - \frac{t}{b}} \tag{3.6.3.17}$$

$$\frac{C_{f_1}}{\varepsilon} = \frac{1.0}{\pi}\left(\frac{2.0}{1.0 - t/(b-s)}\ln\left[\frac{1.0}{1.0 - t/(b-s)} + 1.0\right]\right.$$

$$\left. - \left[\frac{1.0}{1.0 - t/(b-s)} - 1.0\right]\ln\left\{\frac{1.0}{[1.0 - t/(b-s)]^2} - 1.0\right\}\right) \tag{3.6.3.18}$$

$$\frac{C_{f_2}}{\varepsilon} = \frac{1.0}{\pi}\left(\frac{2.0}{1.0 - t/(b+s)}\ln\left[\frac{1.0}{1.0 - t/(b+s)} + 1.0\right]\right.$$

$$\left. - \left[\frac{1.0}{1.0 - t/(b+s)} - 1.0\right]\ln\left\{\frac{1.0}{[1.0 - t/(b+s)]^2} - 1.0\right\}\right) \tag{3.6.3.19}$$

For $\dfrac{w}{b-t} < 0.35$, substitute

$$\frac{w_n}{b} = \frac{0.07\,(1 - t/b) + w/b}{1.2} \tag{3.6.3.20}$$

for $\dfrac{w}{b}$ in the equations for $C_{p_1}/\varepsilon$ and $C_{p_2}/\varepsilon$.

The term $C$ is the incremental capacitance of the line and $C_{p_1}$ and $C_{p_2}$ represent the parallel-plate capacitances of each side and $C_{f_1}$ and $C_{f_2}$ represent the fringing field capacitance of the edges. A limitation of these equations is that the results of this technique differ substantially from centered stripline equations when the strip is centered. Hogan's [7] equations offer a solution which remedies this problem.

Hogan's solution modifies Wheeler's centered stripline equations and suggests additional corrections for the trapezoidal cross-section caused by the etching process of PCBs. A different figure is required because of substantially different dimensioning.

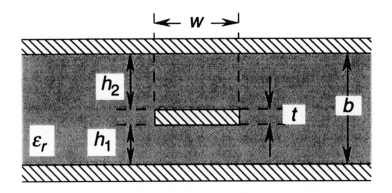

**Figure 3.6.3.2:**      **Alternate Off-Center Stripline Labeling**

The equations are:

$$Z_0 = \frac{1.0}{\sqrt{\varepsilon_r}} \left[ Z_{0,ctr}(1.0, w, b, t) - \Delta Z_{0,air} \right] \tag{3.6.3.21}$$

where

$Z_{0,ctr}(\varepsilon_r, w, b, t) =$ the characteristic impedance of a centered stripline (3.6.2.1)

$$Z_{0,air} = 2.0 \left[ \frac{Z_{0,ctr}(1.0, w, h_1, t)\, Z_{0,ctr}(1.0, w, h_2, t)}{Z_{0,ctr}(1.0, w, h_1, t) + Z_{0,ctr}(1.0, w, h_2, t)} \right]$$

$$\tag{3.6.3.22}$$

$$\Delta Z_{0,air} = \frac{0.26\,\pi}{8.0} Z_{0,air}^2 \left( 0.5 - \frac{h_1 + t/2}{h_1 + h_2 + t} \right)^{2.2} \left( \frac{t + w}{h_1 + h_2 + t} \right)^{2.9}$$

$$\tag{3.6.3.23}$$

$$h_2 > h_1 \tag{3.6.3.23a}$$

For trapezoidal cross-section traces with the widest dimension closest to a ground plane (Figure 3.6.3.3), replace equations (3.6.3.20) and (3.6.3.21) with:

$$Z_{0,air} = 2.0 \left[ \frac{Z_{0,ctr}(1.0, w_{top}, h_1, t)\, Z_{0,ctr}(1.0, w_{bottom}, h_2, t)}{Z_{0,ctr}(1.0, w_{top}, h_1, t) + Z_{0,ctr}(1.0, w_{bottom}, h_2, t)} \right]$$

$$\tag{3.6.3.24}$$

$$\Delta Z_{0,air} = \frac{0.26\ \pi}{8.0}\ Z_{0,air}^2 \left(0.5 - \frac{h_1 + t/2}{h_1 + h_2 + t}\right)^{2.2} \left(\frac{t + w_{eff}}{h_1 + h_2 + t}\right)^{2.9}$$

$$(3.6.3.25)$$

where

$$w_{eff} = \frac{w_{top} + w_{bottom}}{2.0}$$

$$(3.6.3.26)$$

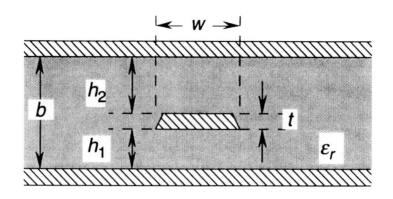

**Figure  3.6.3.3:**        **Trapezoidal  Off-Center  Stripline**

For trapezoidal cross-section traces with the narrowest dimension closest to a ground plane (Figure 3.6.3.4) replace $w$ in the equations above with $w_{eff}$ giving:

$$Z_{0,air} = 2.0 \left[\frac{Z_{0,ctr}(1.0,\ w_{eff},\ h_1,\ t)\ Z_{0,ctr}(1.0,\ w_{eff},\ h_2,\ t)}{Z_{0,ctr}(1.0,\ w_{eff},\ h_1,\ t) + Z_{0,ctr}(1.0,\ w_{eff},\ h_2,\ t)}\right]$$

$$(3.6.3.27)$$

$$\Delta Z_{0,air} = \frac{0.26\ \pi}{8.0}\ Z_{0,air}^2 \left(0.5 - \frac{h_1 + t/2}{h_1 + h_2 + t}\right)^{2.2} \left(\frac{t + w_{eff}}{h_1 + h_2 + t}\right)^{2.9}$$

$$(3.6.3.28)$$

where

$$w_{eff} = \frac{\dfrac{w_{top} + w_{bottom}}{2.0} + \dfrac{2.0\ w_{top}\ w_{bottom}}{w_{top} + w_{bottom}}}{2.0}$$

$$(3.6.3.29)$$

or

$$w_{eff} = \frac{(w_{top} + w_{bottom})^2 + 4.0\ w_{top}\ w_{bottom}}{4.0\ (w_{top} + w_{bottom})}$$

$$(3.6.3.30)$$

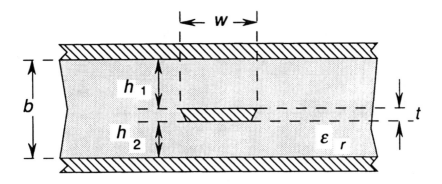

**Figure 3.6.3.4:** Trapezoidal Off-Center Stripline

Note that these equations are symmetric and the choice of $w_{top}$ or $w_{bottom}$ is relative to $h_1$ and $h_2$ and therefore arbitrary.

REFERENCES

[1] Chang, W.H., "Analytical IC Metal-Line Capacitance Formulas," *IEEE Transactions on Microwave Theory and Techniques*, Vol. MTT-24, No. 9, September 1976, pp. 608–611.

[2] Cohn, Seymour B., "Problems in Strip Transmission Lines," *IRE Transactions on Microwave Theory and Techniques*, Vol. MTT-3, No. 3, March 1955, pp. 119–126.

[3] Cohn, Seymour B., "Characteristic Impedance of the Shielded-Strip Transmission Line," *IRE Transactions on Microwave Theory and Techniques,* Vol. MTT-2, No. 3, July 1954, pp. 52–57.

[4] Fouladian, Jamshid, and Protap Pramanick, "Generalized Analytical Equations for Strip Transmission Lines," *Microwave Journal*, November 1989, pp. 165–168.

[5] Guckel, H., "Characteristic Impedances of Generalized Rectangular Transmission Lines," *IEEE Transactions on Microwave Theory and Techniques*, Vol. MTT-13, No. 3, May 1965, pp. 270–274.

[6] Howe, Jr., Harlan, *Stripline Circuit Design*, Artech House, Norwood, MA, 1974. (Howe's graphs appear <u>not</u> to include the $w_n / b$ correction he recommends in Equation 2-13 for narrow traces. Also, be careful because the $s$ term in Howe's equations and in Fig. 2–6 is not the same $s$ used in the corresponding graphs (Figures 2-7 through 2-9))

[7] Hogan, Sean, "Closed Form Transmission Line Equation," May 11, 1988.

[8]   Kiang, Jean-Fu, *et al.*, "Propagation Properties of Stripines Periodically Loaded with Crossing Strips," *IEEE Transactions on Microwave Theory and Techniques*, Vol. MTT-37, No. 4, April 1989, pp. 776–786.

[9]   Rao, J.S., and B.N. Das, "Analysis of Asymmetric Stripline by Conformal Mapping," *IEEE Transactions on Microwave Theory and Techniques*, Vol. MTT-24, No. 4, April 1979, pp. 299–303.

[10]   Robrish, Peter, "An Analytic Algorithm for Unbalanced Stripline Impedance," *IEEE Transactions on Microwave Theory and Techniques*, Vol. MTT-38, No. 8, August 1990, pp. 1011–1016.

### 3.6.4   Shielded Stripline

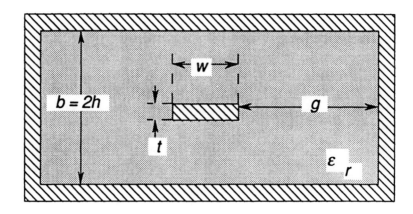

**Figure 3.6.4.1:**      **Shielded Stripline**

Generally, the stripline configuration is considered to have ground planes extending to infinity. This configuration corrects for the effect of side walls at a finite distance. The equations of [1] are:

$$Z_0 = \frac{30.0\,\pi}{\sqrt{\varepsilon_r}\left[\dfrac{w/b}{1.0 - t/b} + \dfrac{2.0\,C_f(t/b)}{\pi\,C_f(0)}\ln\left(1.0 + \coth\dfrac{\pi\,g}{b}\right)\right]} \quad (\Omega) \quad (3.6.4.1)$$

$$C_f(t/b) = \frac{\varepsilon}{\pi}\left[\frac{b}{(b-t)}\ln\frac{(2.0\,b - t)}{t} + \ln\frac{t\,(2.0\,b - t)}{(b-t)^2}\right] \quad (3.6.4.2)$$

$$C_f(0) = \frac{2.0\,\varepsilon \ln 2.0}{\pi} \quad (3.6.4.3)$$

Souza [2] describes a conformal mapping procedure that yields the equations for a shielded stripline that is off center.

## REFERENCES

[1]    Niehenke, Edward C., "Additional Notes Microwave Transmission Media Microstrip Discontinuities," Course Notes.

[2]    Souza, J.R., "Conformal Mapping Analysis of Shielded Striplines," *Electronics Letters*, Vol. 23, No. 12, June 4, 1987, pp. 637–638.

[3]    Yamashita, Eikichi, and Kuzuhiko Atsuki, "Strip Line with Rectangular Outer Conductor and Three Dielectric Layers," *IEEE Transactions on Microwave Theory and Techniques*, Vol. MTT-18, No. 5, May 1970, p. 238–244.

## 3.7 INVERTED MICROSTRIP LINE STRUCTURES

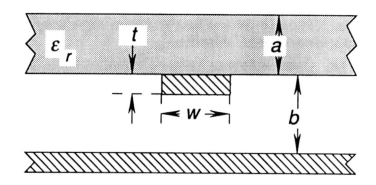

**Figure 3.7.1.1:**     Inverted Microstrip Line

Inverted microstrip is used for precision low-loss lines. It allows a wider range of achievable impedances. In addition, the air dielectric is essentially lossless and dominates the dispersion characteristics.

The equations of [1] are valid for $\varepsilon_r \leq 3.8$. For dielectrics with $\varepsilon_r \leq 20$, use the more involved equations of [2]. These closed-form equations are the results of curve-fitting data generated by an exact variational analysis.

$$Z_0 = \frac{\eta_0}{2.0 \; \pi \; \sqrt{\varepsilon_{eff}}} \ln \left[ \frac{f(u)}{u} + \sqrt{1.0 + \frac{4.0}{u^2}} \right] \; (\Omega) \tag{3.7.1.1}$$

$$\varepsilon_{eff} = (1.0 + f_1 \, f_2)^2 \tag{3.7.1.2}$$

where

$$u = \frac{w}{b} \tag{3.7.1.3}$$

$$f(u) = 6.0 + (2 \, \pi - 6.0) \, e^{-(30.666/u)^{0.7528}} \tag{3.7.1.4}$$

$$f = \ln \left( \varepsilon_r \right) \tag{3.7.1.5}$$

$$f_1 = \sqrt{\varepsilon_r} - 1 \tag{3.7.1.6}$$

$$f_2 = \cfrac{1}{\sum\limits_{j=0}^{3} c_i \left(\cfrac{w}{b}\right)^i} \qquad (3.7.1.7)$$

$$c_i = \cfrac{1}{\sum\limits_{j=0}^{3} d_{ij} \left(\cfrac{b}{a}\right)^j} \qquad (3.7.1.8)$$

$$d_{00} = (2359.401 - 97.1644\,f - 5.706 \times f^2 + 11.4112 \times f^3) \times 10^{-3} \qquad (3.7.1.9)$$

$$d_{01} = (4855.9472 - 3408.5207\,f + 15296.73\,f^2 - 2418.1785\,f^3) \times 10^{-5} \quad (3.7.1.10)$$

$$d_{02} = (1763.34 + 961.0481\,f - 2089.28\,f^2 + 375.8805\,f^3) \times 10^{-5} \qquad (3.7.1.11)$$

$$d_{03} = (-556.0909 - 268.6165\,f + 623.7094 \times f^2 + \; - 119.1402 \times f^3) \times 10^{-6}$$
$$(3.7.1.12)$$

$$d_{10} = (219.066 - 253.0864\,f + 208.7469\,f^2 - 27.3285\,f^3) \times 10^{-3} \qquad (3.7.1.13)$$

$$d_{11} = (915.5589 + 338.4033\,f - 253.2933\,f^2 + 40.4745\,f^3) \times 10^{-3} \qquad (3.7.1.14)$$

$$d_{12} = (-1957.379 - 1170.936\,f + 1480.857\,f^2 - 347.6403\,f^3) \times 10^{-5} \quad (3.7.1.15)$$

$$d_{13} = (486.7425 + 279.8323\,f - 431.3625\,f^2 + 108.824\,f^3) \times 10^{-6} \qquad (3.7.1.16)$$

$$d_{20} = (5602.767 + 4403.356\,f - 4517.034\,f^2 + 743.2717\,f^3) \times 10^{-5} \qquad (3.7.1.17)$$

$$d_{21} = (-2823.481 - 1562.782\,f + 3646.15\,f^2 - 823.4223\,f^3) \times 10^{-5} \qquad (3.7.1.18)$$

$$d_{22} = (255.893 + 158.5529\,f + \; - 3235.485\,f^2 + 919.3661\,f^3) \times 10^{-6} \quad (3.7.1.19)$$

$$d_{23} = (-147.0235 + 62.4342\,f + 887.5211\,f^2 - 270.7555\,f^3) \times 10^{-7} \qquad (3.7.1.20)$$

$$d_{30} = (-3170.21 - 1931.852\,f + 2715.327\,f^2 - 519.342\,f^3) \times 10^{-6} \qquad (3.7.1.21)$$

$$d_{31} = (596.3251 + 188.1409\,f - 1741.477\,f^2 + 465.6756\,f^3) \times 10^{-6} \qquad (3.7.1.22)$$

$$d_{32} = (124.9655 + 577.5381\,f + 1366.453\,f^2 - 481.13\,f^3) \times 10^{-7} \qquad (3.7.1.23)$$

$$d_{33} = (-530.2099 - 2666.352\,f - 3220.096\,f^2 + 1324.499\,f^3) \times 10^{-9} \quad (3.7.1.24)$$

The stated error of the fit to the exact theoretical calculation is $< 0.6\%$ ($Z_0$ and $\sqrt{\varepsilon_{eff}}$) for:

$$1.0 \le \varepsilon_r \le 20.0$$

$$0.5 \le w\,/\,b \le 10.0$$

$$0.06 \le a\,/\,b \le 1.5$$

## REFERENCES

[1]   Tomar, R.S., and Prakash Bhartia, "Suspended and Inverted Microstrip Design," *Microwave Journal*, March 1986, pp. 173–177.

[2]   Tomar, R.S., and Prakash Bhartia, "New Quasi-Static Models for the Computer-Aided Design of Suspended and Inverted Microstrip Lines," *IEEE Transactions on Microwave Theory and Techniques*, Vol. MTT-35, No. 4, April 1987, pp. 453–457, and corrections No. 11, November 1987, p. 1076.

## 3.8    SUSPENDED MICROSTRIP LINE STRUCTURES

### 3.8.1    Suspended Microstrip Line

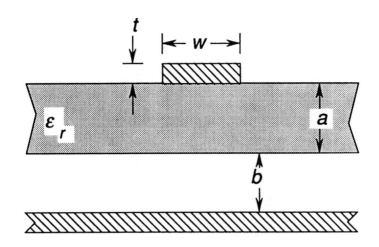

**Figure  3.8.1.1:        Suspended  Microstrip  Line**

Suspended microstrip is used for precision low-loss lines.  This configuration also allows a wider range of realizable Z's and reduced dispersion.  Again the simple equations of [4] are valid for $\varepsilon_r \leq 3.8$, otherwise use [6].

$$Z_0 = \frac{\eta_0}{2.0 \; \pi \; \sqrt{\varepsilon_{eff}}} \ln \left[ \frac{f(u)}{u} + \sqrt{1.0 + \frac{4.0}{u^2}} \; \right] \quad (\Omega) \qquad (3.8.1.1)$$

$$\varepsilon_{eff} = \frac{1.0}{(1.0 - f_1 \, f_2)^2} \quad \text{(unitless)} \qquad (3.8.1.2)$$

where

$$u = \frac{w \, / \, b}{1.0 + a \, / \, b} \qquad (3.8.1.3)$$

$$f(u) = 6.0 + (2 \, \pi - 6.0) \, e^{-(30.666 \, / \, u)^{0.7528}} \qquad (3.8.1.4)$$

$$f = \ln \left( \varepsilon_r \right) \qquad (3.8.1.5)$$

$$f_1 = 1.0 - \frac{1.0}{\sqrt{\varepsilon_r}} \tag{3.8.1.6}$$

$$f_2 = \frac{1}{\displaystyle\sum_{i=0}^{3} c_i \left(\frac{w}{b}\right)^i} \tag{3.8.1.7}$$

$$c_i = \frac{1}{\displaystyle\sum_{j=0}^{3} d_{ij} \left(\frac{b}{a}\right)^j} \tag{3.8.1.8}$$

$$d_{00} = (176.2576 - 43.1240\,f + 13.4094\,f^2 - 1.7010\,f^3) \times 10^{-2} \tag{3.8.1.9}$$

$$d_{01} = (4665.2320 - 1790.4000\,f + 291.5858\,f^2 - 8.0888\,f^3) \times 10^{-4} \tag{3.8.1.10}$$

$$d_{02} = (-3025.5070 - 141.9368\,f - 3099.4700\,f^2 + 777.6151\,f^3) \times 10^{-6} \tag{3.8.1.11}$$

$$d_{03} = (2481.5690 + 1430.3860\,f + 10095.5500\,f^2 - 2599.1320\,f^3) \times 10^{-8} \tag{3.8.1.12}$$

$$d_{10} = (-1410.2050 + 149.9293\,f + 198.2892\,f^2 - 32.1679\,f^3) \times 10^{-4} \tag{3.8.1.13}$$

$$d_{11} = (2548.7910 + 1531.9310\,f - 1027.500\,f^2 + 138.4192\,f^3) \times 10^{-4} \tag{3.8.1.14}$$

$$d_{12} = (999.3135 - 4036.7910\,f + 1762.4120\,f^2 - 298.0241\,f^3) \times 10^{-6} \tag{3.8.1.15}$$

$$d_{13} = (-1983.7890 + 8523.9290\,f - 5235.4600\,f^2 + 1145.7880\,f^3) \times 10^{-8} \tag{3.8.1.16}$$

$$d_{20} = (1954.0720 + 333.3873\,f - 700.7473\,f^2 + 121.3212\,f^3) \times 10^{-5} \tag{3.8.1.17}$$

$$d_{21} = (-3931.0900 - 1890.7190\,f + 1912.2660\,f^2 - 319.6794\,f^3) \times 10^{-5} \tag{3.8.1.18}$$

$$d_{22} = (-532.1326 + 7274.7210\,f - 4955.7380\,f^2 + 941.4134\,f^3) \times 10^{-7} \tag{3.8.1.19}$$

$$d_{23} = (138.2037 - 1412.4270\,f + 1184.2700\,f^2 - 270.0047\,f^3) \times 10^{-8} \quad (3.8.1.20)$$

$$d_{30} = (-983.4028 - 255.1229\,f + 455.8729\,f^2 - 83.9468\,f^3) \times 10^{-6} \quad (3.8.1.21)$$

$$d_{31} = (1956.3170 + 779.9975\,f - 995.9494\,f^2 + 183.1957\,f^3) \times 10^{-6} \quad (3.8.1.22)$$

$$d_{32} = (62.8550 - 3462.5000\,f + 2909.9230\,f^2 - 614.7068\,f^3) \times 10^{-8} \quad (3.8.1.23)$$

$$d_{33} = (-35.2531 + 601.0291\,f - 643.0814\,f^2 + 161.2689\,f^3) \times 10^{-9} \quad (3.8.1.24)$$

The stated error of the fit to the exact theoretical calculation is less than 0.6% ($Z_0$ and $\sqrt{\varepsilon_{eff}}$) for:

$1.0 \leq \varepsilon_r \leq 20.0$

$0.5 \leq w/b \leq 10.0$

$0.06 \leq a/b \leq 1.5$

An analysis of conductor and dielectric loss is given in Yamashita [7].

Tomar and Bhartia [5] give polynomials for the dispersion of this structure. To use these, replace $\varepsilon_{eff}$ in the equations above with $\varepsilon_{eff}(f)$ calculated below. The variable $\varepsilon_{eff}(0)$ is the static value $\varepsilon_{eff}$ calculated above.
For $\varepsilon_r = 12.9$ the equations are

$$\varepsilon_r(f) = \varepsilon_r - \frac{\varepsilon_r - \varepsilon_{eff}(0)}{1.0 + G\left(\frac{f}{f_p}\right)^2} \quad (3.8.1.25)$$

where

$$f_p = \frac{Z}{2.0\,\pi\,\mu_0\,(a+b)} \frac{a}{0.064} \quad (3.8.1.26)$$

$a, b = (\text{cm})$

$$G = C_0 + C_1 Z + C_2 Z^2 \quad (3.8.1.27)$$

and

$$C_0 = 0.0194 - 0.2398\left(\frac{a}{b}\right) + 0.8977\left(\frac{a}{b}\right)^2 - 0.9924\left(\frac{a}{b}\right)^3 + 0.3468\left(\frac{a}{b}\right)^4 \quad (3.8.1.28)$$

$$C_1 = -0.0008 + 0.0096 \left(\frac{a}{b}\right) - 0.0346 \left(\frac{a}{b}\right)^2 + 0.0384 \left(\frac{a}{b}\right)^3 - 0.0135 \left(\frac{a}{b}\right)^4 \quad (3.8.1.29)$$

$$C_2 = 0.0004 \left(\frac{a}{b}\right)^2 - 0.0004 \left(\frac{a}{b}\right)^3 + 0.0001 \left(\frac{a}{b}\right)^4 \qquad (3.8.1.30)$$

The dispersion relations are valid for

$$1.0 \leq w / b \leq 10.0$$

$$20 \text{ GHz} < f < 100 \text{ GHz}$$

The accuracy is within ±2% when compared with exact theoretical calculations.

REFERENCES

[1]  Bedair, S.S., "Closed Form Expressions for the Computer-Aided Design of Suspended Microstrip Circuits for Millimeter Wave Applications," *IEE Proceedings*, Vol. 134, Pt. H, No. 4, August 1987, pp. 386–388.

[2]  Schneider, M.V., "Microstrip Lines for Microwave Integrated Circuits," *Bell Systems Technical Journal*, Vol. 48, No. 5, May-June 1969, p. 1421-1444.

[3]  Shu, Yong-hui, Xiao-xia Qi, and Yun-ji Wang, "Analysis Equations for Shielded Suspended Substrate Microstrip Line and Broadside-Coupled Stripline," *IEEE MTT-S International Microwave Symposium Digest*, 1987, pp. 693–696.

[4]  Tomar, R.S., and Prakash Bhartia, "Suspended and Inverted Microstrip Design," *Microwave Journal*, March 1986, pp. 173–177.

[5]  Tomar, R.S., and Prakash Bhartia, "Modelling the Dispersion in a Suspended Microstripline," *1987 IEEE MTT-S Symposium Digest,* pp. 713–715.

[6]  Tomar, R.S., and Prakash Bhartia, "New Quasi-Static Models for the Computer-Aided Design of Suspended and Inverted Microstrip Lines," *IEEE Transactions on Microwave Theory and Techniques*, Vol. MTT-35, No. 4, April 1987, pp. 453–457, and corrections No. 11, November 1987, p. 1076.

[7]  Yamashita, Eikichi, and Kuzuhiko Atsuki, "Strip Line with Rectangular Outer Conductor and Three Dielectric Layers," *IEEE Transactions on Microwave Theory and Techniques*, Vol. MTT-18, No. 5, May 1970, p. 238–244.

## 3.8.2 Shielded Suspended Substrate Stripline

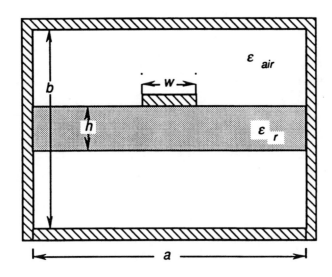

**Figure 3.8.2.1:** Shielded Suspended Substrate Stripline

The equations for this structure are broken into two ranges.
For narrow strips ($0 < w < a / 2.0$):

$$Z_0 = \frac{\eta_0}{2.0 \; \pi \sqrt{\varepsilon_{eff}}} \left\{ V + R \; \ln \left[ \frac{6.0}{w / b} + \sqrt{1.0 + \frac{4.0}{(w / b)^2}} \; \right] \right\} \; (\Omega) \quad (3.8.2.1)$$

$$\varepsilon_{eff} = \frac{1.0}{\left[ 1.0 + \left( E - F \ln \frac{w}{b} \right) \ln \left( \frac{1.0}{\sqrt{\varepsilon_r}} \right) \right]^2} \quad (3.8.2.2)$$

where

$$V = 1.7866 - 0.2035 \frac{h}{b} + 0.4750 \frac{a}{b} \quad (3.8.2.3)$$

$$R = 1.0835 + 0.1007 \frac{h}{b} - 0.09457 \frac{a}{b} \quad (3.8.2.4)$$

$$E = 0.2077 + 1.2177 \frac{h}{b} - 0.08364 \frac{a}{b} \quad (3.8.2.5)$$

$$F = 0.03451 - 0.1031 \frac{h}{b} + 0.01742 \frac{a}{b} \tag{3.8.2.6}$$

and for wide strips $(a/2 < w < a)$

$$Z_0 = \frac{\eta_0}{\sqrt{\varepsilon_{eff}}} \left[ V + \frac{R}{\frac{w}{b} + 1.3930 + 0.6670 \ln \left( \frac{w}{b} + 1.444 \right)} \right] \quad (\Omega) \tag{3.8.2.7}$$

$$\varepsilon_{eff} = \frac{1.0}{\left[ 1.0 + \left( E - F \ln \frac{w}{b} \right) \ln \left( \frac{1.0}{\sqrt{\varepsilon_r}} \right) \right]^2} \tag{3.8.2.8}$$

where

$$V = -0.6301 - 0.07082 \frac{h}{b} + 0.2470 \frac{a}{b} \tag{3.8.2.9}$$

$$R = 1.9492 + 0.1553 \frac{h}{b} - 0.5123 \frac{a}{b} \tag{3.8.2.10}$$

$$E = 0.4640 + 0.9647 \frac{h}{b} - 0.2063 \frac{a}{b} \tag{3.8.2.11}$$

$$F = -0.1424 + 0.3017 \frac{h}{b} - 0.02411 \frac{a}{b} \tag{3.8.2.12}$$

The above relations are valid for:

$$1.0 \le a/b \le 2.5$$

$$1.0 \le \varepsilon_r \le 4.0$$

$$0.1 < h/b < 0.5$$

The wide strip equations agree with finite-differential techniques to $\pm 3\%$. Narrow strip agreement is $\pm 2\%$.

## REFERENCES

[1]    Gish, Dennis L., and Odell Graham, "Characteristic Impedance and Phase Velocity of a Dielectric Supported Air Strip Transmission Line with Side Walls," *IEEE Transactions on Microwave Theory and Techniques*, MTT-18, No. 3, March 1970, pp. 131–148.

[2]     Shu, Yong-hui, Xiao-xia Qi, and Yun-ji Wang, "Analysis Equations for Shielded Suspended Substrate Microstrip Line and Broadside-Coupled Stripline," *IEEE MTT-S International Microwave Symposium Digest*, 1987, pp. 693–696.

[3]     Yamashita, Eikichi, and Kuzuhiko Atsuki, "Strip Line with Rectangular Outer Conductor and Three Dielectric Layers," *IEEE Transactions on Microwave Theory and Techniques*, Vol. MTT-18, No. 5, May 1970, p. 238–244.

## 3.9  OTHER STRUCTURES

### 3.9.1  Cylindrical Conductor in a Corner

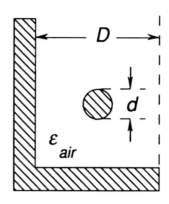

**Figure 3.9.1.1:**     **Cylindrical Conductor in a Corner**

In [1], Wheeler gives solutions for a number of "wire-in-a-shield" problems.  For this configuration:

$$Z_0 = \frac{\eta_0}{4.0\,\pi}\,\ln\left\{\left(\frac{D}{d}\right)^2 + \sqrt{\left[\left(\frac{D}{d}\right)^2 - 1.0\right]^2 + \left(\frac{D}{d}\right)^2 - 1.0}\right\} \qquad (3.9.1)$$

The stated accuracy is about 1%.

REFERENCES

[1]  Wheeler, Harold A., "Transmission-Line Properties of a Round Wire in a Polygon Shield," *IEEE Transactions on Microwave Theory and Techniques*, Vol. MTT-27, No. 8, August 1979, pp. 717–721.

## 3.9.2  Slabline

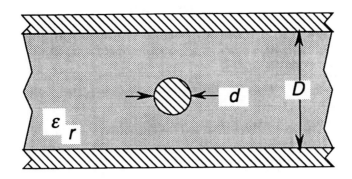

**Figure  3.9.2.1:**      Slabline

The slabline is a transmission line where a thin wire replaces a printed trace, *e.g.*, for an inductor.  A simple, frequently encountered relation is

$$Z_0 = \frac{\eta_0}{2.0 \, \pi \, \sqrt{\varepsilon_r}} \ln\left(\frac{4.0 \, D}{\pi \, d}\right) \quad (\Omega) \tag{3.9.2.1}$$

The stated error is less than 1.0% for $\frac{D}{d} \geq 2.0$ with better accuracy as $Z_0$ increases ($d \,/ \, D \to 0$).

White [5, p. 520] has a more accurate equation to be used for synthesis.  For $d$ approaching $D$, Wheeler's [4] equations are more accurate than the above:

$$Z_0 = \frac{15.0}{\sqrt{\varepsilon_r}} \ln\left[1.0 + \frac{k_0 \, g}{2.0} + \sqrt{\left(\frac{k_0 g}{2}\right)^2 + 2.0 \, g}\,\right] \quad (\Omega) \tag{3.9.2.2}$$

where

$$w\,/\,h = \frac{2.0}{\left[1.0 + \dfrac{(e^{8.0 \, \pi \, r} - 1.0)^2}{k_0(e^{8.0 \, \pi \, r} - 1.0) + 2.0}\right]^{0.25} - 1.0} \tag{3.9.2.3}$$

$$k_0 = \left(\frac{4}{\pi}\right)^4 \tag{3.9.2.4}$$

$$g = (1.0 + 2.0 \, h \,/\, 2.0)^4 - 1.0 \tag{3.9.2.5}$$

Stated accuracy is better than 0.5% for $Z_0$.

REFERENCES

[1]    Cohn, Seymour B., "Characteristic Impedance of the Shielded-Strip Transmission Line," *IRE Transactions on Microwave Theory and Techniques*, Vol. MTT-2, July 1954, pp. 52-57.

[2]    Liao, Samuel Y., *Microwave Circuit Analysis and Amplifier Design*, Prentice-Hall, Englewood Cliffs, NJ, 1987.

[3]    Wheeler, Harold A., "Transmission-Line Properties of a Round Wire in a Polygon Shield," *IEEE Transactions on Microwave Theory and Techniques*, Vol. MTT-27, No. 8, August 1979, pp. 717–721.

[4]    Wheeler, Harold A., "Transmission-Line Properties of a Strip Line Between Parallel Planes," *IEEE Transactions on Microwave Theory and Techniques*, Vol. MTT-26, November 1978, No. 11, pp. 866–876.

[5]    White, Joseph F., *Microwave Semiconductor Engineering*, Van Nostrand Reinhold, New York, 1982.

[6]    Wolff, Edward A., and Roger Kaul, *Microwave Engineering and Systems Applications*, John Wiley and Sons, New York, 1988.

### 3.9.3   Wire over Dielectric over Ground Plane

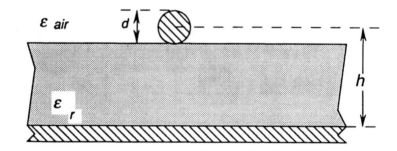

**Figure  3.9.3.1:**     **Wire over Dielectric over Ground Plane**

This configuration is useful when a line is too fine for standard PCB etching.  See Chapter 11 for available wire gauges.

$$Z_0 = \frac{\eta_0}{2.0\,\pi\,\sqrt{\varepsilon_{eff}}}\,\cosh^{-1}\left(\frac{2.0\,h}{d}\right) \approx \frac{\eta_0}{2.0\,\pi\,\sqrt{\varepsilon_{eff}}}\,\ln\left(\frac{4.0\,h}{d}\right) \qquad (3.9.3.1)$$

The approximate equation above is also commonly encountered and is valid for $h/d \gg d$. The accuracy of $Z_0$ is stated to be better than 0.024% in [2] and 1% in [3].

To calculate $\varepsilon_{\text{eff}}$, use the equation for microstrip line (3.5.1.2) replacing the microstrip line width, $w$, with an equivalent width calculated from the wire diameter, $d$. A curve for this purpose is available in [1] and others. We have fit a simple curve to the date for CAE use; the fit error is less than 0.35%:

$$\frac{d_0}{w} = 0.5008 + 1.0235\,x - 1.0230\,x^2 + 1.1564\,x^3 - 0.4749\,x^4 \tag{3.9.3.2}$$

$$x = \frac{t}{w} \tag{3.9.3.3}$$

where $w \geq t$ (reverse their positions if this condition is not met).

REFERENCES

[1]   Cohn, Seymour B., "Characteristic Impedance of the Shielded-Strip Transmission Line," *IRE Transactions on Microwave Theory and Techniques*, Vol. MTT-2, No. 3, July 1954, pp. 52-57.

[2]   Hilberg, Wolfgang, *Electrical Characteristics of Transmission Lines*, Artech House, Norwood, MA, 1979.

[3]   Wheeler, Harold A., "Transmission-Line Properties of a Round Wire in a Polygon Shield," *IEEE Transactions on Microwave Theory and Techniques*, Vol. MTT-27, No. 8, August 1979, pp. 717–721.

### 3.9.4   Bond Wire

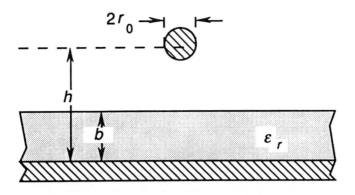

**Figure 3.9.4.1:**      **Wire above Grounded Dielectric**

The wire over grounded dielectric is encountered where we have a bond wire connecting a circuit substrate to the package leads, the case, or another substrate. Caverly [1], uses conformal mapping to derive

$$Z_0 = \frac{\eta_0}{2.0 \ \pi \ \sqrt{\varepsilon_{eff}}} \ \cosh^{-1}\left( \frac{1.0 - u_1^2}{2.0 \ R_1} + \frac{R_1}{2.0} \right) \quad (\Omega) \tag{3.9.4.1}$$

where

$$R_1 = \frac{2.0}{(4.0 \ h \ / \ r_0 - r_0 \ / \ h)} \tag{3.9.4.2}$$

$$u_1 = \frac{1.0}{(2.0 \ h \ / \ r_0)^2 - 1.0} \tag{3.9.4.3}$$

Caverly calculates $\varepsilon_{eff}$ using an equivalent resistive network and SPICE. For calculations we will use the closed-form $\varepsilon_{eff}$ of Kuester and Chang [2]:

$$\varepsilon_{eff} = \frac{\ln\left(\dfrac{2.0 \ h}{r_0}\right)}{\ln\left[ \dfrac{2.0 \ (h - b)}{r_0} + \dfrac{2.0 \ b}{r_0 \ \varepsilon_r} \right]} \quad \text{(unitless)} \tag{3.9.4.4}$$

The equation for $\varepsilon_{eff}$ is valid for quasi-TEM propagation or

$$\frac{\varepsilon_r \ (h - b)}{\lambda_0} \ln\left[ \frac{2.0 \ \lambda_0 \ (h - b)}{r_0} \right] < 1.0 \tag{3.9.4.5}$$

where

$\lambda_0$ = free-space wavelength in units of $h, b, r_0$

Mondal [3] derives equations for a bond wire arc as shown in Figure 3.9.4.2.

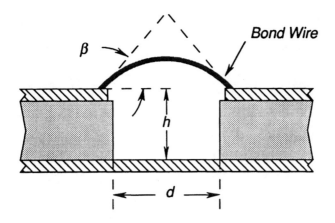

**Figure 3.9.4.2:**     **Bond Wire**

The equations for this structure are

$$L = 10.0 \int_{0.0}^{d/2.0} \ln[p(x)]\, dx \quad (\text{nH})$$

(3.9.4.6)

where

*d, h, r* are all in inches

$$p(x) = \frac{2.0}{r}\left[\sqrt{\left(\frac{d\operatorname{cosec}\beta}{2.0}\right)^{2.0} - x^2} + \left(h - \frac{d}{2.0}\cot\beta\right)\right]$$

(3.9.4.7)

And the capacitance is calculated with

$$C = \frac{1.4337}{\ln p(x)} \quad (\text{pF}/\text{in})$$

(3.9.4.8)

where

*l* = the length of the bond wire

REFERENCES

[1]    Caverly, Robert H., "Characteristic Impedance of Integrated Circuit Bond Wires," *IEEE Transactions on Microwave Theory and Techniques*, Vol. MTT-34, No. 9, September 1986, pp. 982–984.

[2]     Kuester, Edward F., and David C. Chang, "Propagating Modes Along a Thin Wire Located Above a Grounded Dielectric Slab," *IEEE Transactions on Microwave Theory and Techniques*, Vol. MTT-25, No. 12, December 1977, pp. 1065–1069.

[3]     Mondal, Jyoti P., "Octagonal Spiral Inductor Measurement and Modelling for MMIC Applications," *International Journal of Electronics*, Vol. 68, No. 1, 1990, pp. 113–125.

### 3.9.5   Multiwire®[4]

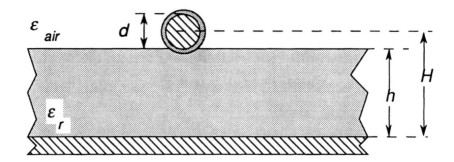

**Figure   3.9.5.1:       Multiwire®**

Multiwire® is a process that routes wires over an adhesive layer on a ground plane using a numerically controlled machine. The grid used for routing is usually 0.050 in. It can provide very quick prototypes. Multiwire allows line lengths to be monitored and controlled and easily rerouted to avoid coupling. Lines may be criss-crossed to form a psuedo-twisted-pair structure. After the wires are routed, the board processing continues similar to a standard PCB, including plated through holes.

The dielectric is a mixture of G-10 and adhesives that have a combined dielectric constant of about 5.0. Available wire sizes are 34, 38, and 41 awg.

Shibata and Terakado [4] examine the effect of different adhesive layer dielectric constants ($\varepsilon_r$ = 1.0, 2.65, and 5.0) and the positioning of the wire in the adhesive on $Z_0$ (Figure 6a), $v_p$ (Figure 6b), and $\varepsilon_{eff}$ (Figure 7). The reference analyzes three cases: (1) wire riding on top of dielectric, (2) wire halfway embedded in dielectric, and (3) wire fully embedded in dielectric. An approximation is given [4]:

$$Z_0 = 17.08 + 34.83 \ln \left( \frac{2.0\ h}{d} \right) \ (\Omega) \tag{3.9.5.1}$$

Accuracy is stated to be better than 0.35% for

---

[4]Multiwire® is a registered trademark of the Multiwire Division of the Kollmorgen Corporation. Similar products are produced by other manufacturers .

$\varepsilon_r = 5.0$

$$0.01 \leq \frac{d}{2.0 \, h} \leq 0.3$$

A controlled impedance environment may also be simulated by creating a pseudo-ground plane from a grid of grounded wires beneath the conductors. Burr [2] gives the following empirical relations for the characteristic impedance:

$$Z_0 = \frac{41.2}{\varepsilon_{eff}} \ln \left(1.88 \, H\right) \; (\Omega) \qquad (3.9.5.2)$$

$$\varepsilon_{eff} = 0.475 \, \varepsilon_r + 0.67 \qquad (3.9.5.3)$$

Sensitivity analysis shows that for 50 $\Omega$ lines the adhesive thickness needs to be monitored in order to achieve ±10%. Higher impedances do not require tight tolerancing of $h$, and at 90 $\Omega$ the line dimensions are not critical.

The dielectric constant of the wire's insulation is approximately 3.5. Bends are of constant diameter and do not require compensation. Where wires cross, they are flattened which introduces a capacitance between the lines of approximately 0.026 pF (parallel plate equation).

## REFERENCES

[1] Buck, Thomas J., "Multiwire—Its Benefits in Sub-Nanosecond Chip Carrier Applications," presented at Wescon/82.

[2] Burr, R.P., "High Frequency Propagation Characteristics of Multiwire Circuit Boards," Multiwire® Development Report #35, May 15, 1972.

[3] Plonski, J. Philip, "Implementing Controlled Impedance Techniques in Discrete Wiring," presented at Wescon/81.

[4] Shibata, Hisashi, and Ryuiti Terakado, "Characteristics of Transmission Lines with a Single Wire for a Multiwire Circuit Board," *IEEE Transactions on Microwave Theory and Techniques*, Volume 32, No. 4, April 1984, pp. 360–364.

### 3.9.6 Slotline

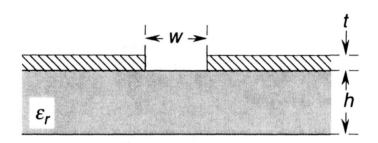

**Figure 3.9.6.1:** Slotline

Slotline is not a TEM structure; however, its single-sided, planar nature make it of interest for modern fabrication techniques. This structure requires a high dielectric constant material ($\varepsilon_r \geq 10$) to confine the fields and minimize radiation. The equations for design are from [3] for $0.02 \leq w/h \leq 0.2$

$$Z_0 = 72.62 - 15.283 \ln \varepsilon_r + \frac{50.0 \,(w/h - 0.02)(w/h - 0.1)}{w/h}$$

$$+ 0.434 \ln (w/h \times 10^2) \left(44.28 - 8.503 \ln \varepsilon_r\right)$$

$$- \left[0.139 \ln \varepsilon_r - 0.11 + w/h \,(0.465 \ln \varepsilon_r + 1.44)\right]$$

$$\times \left(11.4 - 2.636 \ln \varepsilon_r - h/\lambda \times 10^2\right)^2 \tag{3.9.6.1}$$

$$\varepsilon_{\mathit{eff}} = \left(\frac{\lambda_g}{\lambda_0}\right)^{-2}$$

$$= \left[0.923 - 0.195 \ln \varepsilon_r + 0.2 \, w/h\right.$$

$$\left. - (0.126 \, w/h + 0.020) \ln (h/\lambda_0 \times 10^2)\right]^{-2} \tag{3.9.6.2}$$

and for $0.2 \leq w/h \leq 1.0$

$$Z_0 = 113.19 - 23.257 \ln \varepsilon_r + 1.25 \, w/h \,(114.59 - 22.531 \ln \varepsilon_r)$$

$$+ 20.0 \,(w/h - 0.2)(1.0 - w/h)$$

$$- \left[ 0.15 + 0.0999 \ln \varepsilon_r + w / h \ (-0.79 + 0.899 \ln \varepsilon_r) \right]$$

$$\times \left\{ \left[ 10.25 - 2.171 \ln\varepsilon_r + w/h \ (2.1 - 0.617 \ln \varepsilon_r) - h/\lambda \times 10^2 \right]^2 \right\}$$

$$(3.9.6.3)$$

$$\varepsilon_{eff} = \left( \frac{\lambda_g}{\lambda_0} \right)^{-2}$$

$$= \left[ 0.987 - 0.210 \ln \varepsilon_r + w / h \ (0.111 - 0.0022 \ \varepsilon_r) \right.$$

$$\left. - (0.0525 + 0.0408 \ w / h - 0.00139 \ \varepsilon_r) \ \ln(h / \lambda \times 10^2) \right]^{-2}$$

$$(3.9.6.4)$$

Equations are curve fits to data calculated by the technique of [2] and are accurate to 2.0% for

$$9.7 \le \varepsilon_r \le 20.0$$

$$0.02 \le w / h \le 1.0$$

$$0.01 \le h / \lambda \le (h / \lambda)_0$$

The $TE_{10}$ mode cutoff frequency is given by

$$(h / \lambda)_0 = \frac{0.25}{\sqrt{\varepsilon_r - 1.0}} \qquad\qquad (3.9.6.5)$$

For lower dielectric constants ($2.2 \le \varepsilon_r \le 3.8$) Janaswamy and Schaubert [7] give (for $0.0015 \le w / \lambda_0 \le 0.075$):

$$Z_0 = 60.0 + 3.69 \sin \left[ \frac{(\varepsilon_r - 2.22) \ \pi}{2.36} \right] + 133.5 \ln (10.0 \ \varepsilon_r) \sqrt{w / \lambda_0}$$

$$+ 2.81 \left[ 1.0 - 0.011 \ \varepsilon_r \ (4.48 + \ln \varepsilon_r) \right] (w / h) \ln (100.0 \ h / \lambda_0)$$

$$+ 131.1 \ (1.028 - \ln \varepsilon_r) \sqrt{h / \lambda_0}$$

$$+ 12.48 \, (1.0 + 0.18 \ln \varepsilon_r) \frac{w/h}{\sqrt{\varepsilon_r - 2.06 + 0.85 \, (w/h)^2}} \qquad (3.9.6.6)$$

$$\varepsilon_{eff} = \left(\frac{\lambda_g}{\lambda_0}\right)^{-2}$$

$$= \left\{ 1.045 - 0.365 \ln \varepsilon_r + \frac{6.3 \, (w/h) \, \varepsilon_r^{0.945}}{238.64 + 100.0 \, w/h} \right.$$

$$\left. - \left[ 0.148 - \frac{8.81 \, (\varepsilon_r + 0.95)}{100.0 \, \varepsilon_r} \right] \ln \frac{h}{\lambda_0} \right\}^{-2} \qquad (3.9.6.7)$$

For $0.075 \leq w/\lambda_0 \leq 1.0$

$$Z_0 = 133.0 + 10.34 \, (\varepsilon_r - 1.8)^2 + 2.87 \left[ 2.96 + (\varepsilon_r - 1.582)^2 \right]$$

$$\left\{ (w/h + 2.32\varepsilon_r - 0.56) \left[ (32.5 - 6.67 \, \varepsilon_r)(100.0 \, h/\lambda_0)^2 - 1.0 \right] \right\}^{1/2}$$

$$- \left( 684.45 \, h/\lambda_0 \right)(\varepsilon_r + 1.35)^2 + 13.23 \left[ (\varepsilon_r - 1.722) \, w/\lambda_0 \right]^2$$

$$(3.9.6.8)$$

$$\varepsilon_{eff} = \left(\frac{\lambda_g}{\lambda_0}\right)^{-2}$$

$$= \left\{ 1.194 - 0.24 \ln \varepsilon_r - \frac{0.621 \, \varepsilon_r^{0.835} \, (w/\lambda_0)^{0.48}}{1.344 + w/h} \right.$$

$$\left. - 0.0617 \left[ 1.91 - \frac{\varepsilon_r + 2.0}{\varepsilon_r} \right] \ln \frac{h}{\lambda_0} \right\}^{-2} \qquad (3.9.6.9)$$

The above are accurate (first set: $Z_0 \leq 2.7\%$, $\lambda_g/\lambda_0 \leq 2.2\%$; second set: $Z_0 \leq 5.4\%$, $\lambda_g/\lambda_0 \leq 2.6\%$) for

$$2.2 \leq \varepsilon_r \leq 3.8$$

$$0.006 \leq h/\lambda_0 \leq 0.060$$

For intermediate dielectric constants $(3.8 \leq \varepsilon_r \leq 9.8)$ and $0.0015 \leq w / \lambda_0 \leq 0.075$

$$Z_0 = 73.6 - 2.15\,\varepsilon_r + (638.9 - 31.37\,\varepsilon_r)\,(w / \lambda_0)^{0.6}$$

$$+ \left(36.23\,\sqrt{\varepsilon_r^2 + 41.0} - 225.0\right) \frac{w / h}{w / h + 0.876\,\varepsilon_r - 2.0}$$

$$+ 0.51\,(\varepsilon_r + 2.12)\,(w / h)\,\ln \frac{100.0\,h}{\lambda_0}$$

$$- 0.753\,\varepsilon_r\,\frac{h / \lambda_0}{\sqrt{w / \lambda_0}} \qquad (3.9.6.10)$$

$$\varepsilon_{eff} = \left(\frac{\lambda_g}{\lambda_0}\right)^{-2}$$

$$= \left\{ 0.9217 - 0.277\,\ln \varepsilon_r + 0.0322\,(w / h)\,\sqrt{\frac{\varepsilon_r}{w / h + 0.435}} \right.$$

$$\left. -0.01\,\ln (h / \lambda_0) \left[ 4.6 - \frac{3.65}{\varepsilon_r^2\,\sqrt{w / \lambda_0}\,(9.06 - 100\,w / \lambda_0)} \right] \right\}^{-2}$$

$$(3.9.6.11)$$

For $0.075 \leq w / \lambda_0 \leq 1.0$

$$Z_0 = 120.75 - 3.74\,\varepsilon_r + 50.0 \left[ \tan^{-1} (2.0\,\varepsilon_r) - 0.8 \right]$$

$$(w / h)^{\{1.11 + [0.132\,(\varepsilon_r - 27.7) / (100.0\,h / \lambda_0 + 5.0)]\}}$$

$$\ln \left[ 100\,h / \lambda_0 + \sqrt{(100\,h / \lambda_0)^2 + 1.0} \right]$$

$$+ 14.21\,(1.0 - 0.458\,\varepsilon_r)\,(100.0\,h / \lambda_0 + 5.1\,\ln \varepsilon_r - 13.1)\,(w / \lambda_0 + 0.33)^2$$

$$(3.9.6.12)$$

$$\varepsilon_{eff} = \left(\frac{\lambda_g}{\lambda_0}\right)^{-2}$$

$$= \left\{ 1.05 - 0.04 \ \varepsilon_r + 1.411 \times 10^{-2} \ (\varepsilon_r - 1.421) \right.$$

$$\ln \left[ w \ / \ h - 2.012 \ (1.0 - 0.146 \ \varepsilon_r) \right]$$

$$+ 0.111 \ (1.0 - 0.366 \ \varepsilon_r) \ \sqrt{w \ / \ \lambda_0}$$

$$\left. + 0.139 \left[ 1.0 + 0.52 \ \varepsilon_r \ \ln \ (14.7 - \varepsilon_r) \right] (h \ / \ \lambda_0) \ \ln \ (h \ / \ \lambda_0) \right\}^{-2}$$

$$(3.9.6.13)$$

The above are accurate (first set: $Z_0$ accurate to 5.4%, $\varepsilon_{eff}$ accurate to 3%; second set: $Z_0$ accurate to 5.8%, $\varepsilon_{eff}$ accurate to 3.2%) for

$$3.8 \le \varepsilon_r \le 9.8$$

Cohn [2] indicates that losses may be approximated with the losses of microstrip line. Kirschning *et al.* [8] analyzes slotline between two dielectrics.

### REFERENCES

[1]    Biswas, Animesh, and Vijai K. Tripathi, "Modeling of Asymmetric and Offset Gaps in Shielded Microstrips and Slotlines, " *IEEE Transactions on Microwave Theory and Techniques*, Vol. 38, No. 6, June 1990, pp. 818–822.

[2]    Cohn, Seymour B., "Slot Line on a Dielectric Substrate," *IEEE Transactions on Microwave Theory and Techniques*, Vol. MTT-17, No. 10, October 1969, pp. 768–778.

[3]    Garg, Ramesh, and K.C. Gupta, "Expressions for Wavelength and Impedance of a Slotline," *IEEE Transactions on Microwave Theory and Techniques*, Vol. MTT-24, No. 8, August 1976, p. 532. (Closed-form versions of Cohn [2])

[4]    Geshiro, Masahiro, *et al.*, "Analysis of Slotlines and Microstrip Lines on Anisotropic Substrates," *IEEE Transactions on Microwave Theory and Techniques*, Vol. MTT-39, No. 1, January 1991, pp. 64–69.

[5]    Heinrich, Wolfgang, "The Slot Line in Uniplanar MMIC's: Propagation Characteristics and Loss Analysis," *1990 IEEE MTT-S International Microwave Symposium Digest*, pp. 167–170.

[6]    Itoh, T., and R. Mittra, "Dispersion Characteristics of Slot Lines," *Electronics Letters*, July 1, 1971, Vol. 7, No. 13, pp. 364–365.

[7]    Janaswamy, R., and D.H. Schaubert, "Characteristic Impedance of a Wide Slotline on Low-Permittivity Substrates," *IEEE Transactions on Microwave Theory and Techniques*, Vol. MTT-34, No. 8, August 1986, pp. 900–902.

[8] Kirschning, M., *et al.*, "Sandwich Slot Line," *IEEE Transactions on Microwave Theory and Techniques*, Vol. MTT-19, No. 9, September 1971, pp. 773–774.

[9] Knorr, Jeffrey B., and Klaus-Dieter Kuchler, "Analysis of Coupled Slots and Coplanar Strips on Dielectric Substrate," *IEEE Transactions on Microwave Theory and Techniques*, Vol. MTT-23, No. 7, July 1975, pp. 541–548.

[10] Knorr, Jeffrey B., "Slot-Line Transitions," *IEEE Transactions on Microwave Theory and Techniques*, Vol. MTT-6, No. 5, May 1974, pp. 548–554.

[11] Lee, J.J., "Slotline Impedance," *IEEE Transactions on Microwave Theory and Techniques*, Vol. MTT-39, No. 4, April 1991, pp. 666–672.

[12] Mariani, Elio A., *et al.*, "Slot Line Characteristics," *IEEE Transactions on Microwave Theory and Techniques*, Vol. MTT-17, No. 12, December 1969, pp. 1091–1096.

### 3.9.7    Finline

Again, this is not a TEM structure but because of its planar nature it is very useful for fabricating circuits. It is also easy to interface to waveguide because of its fully shielded structure. Propagation is a hybrid mode that is useful mainly in the 30 to 100 GHz range. There are several different forms of finline. Metallization can be on a single side of the dielectric, both sides, or diagonal. The finline may be captivated in the shield and grounded or isolated.

### *3.9.7.1    Unilateral Finline*

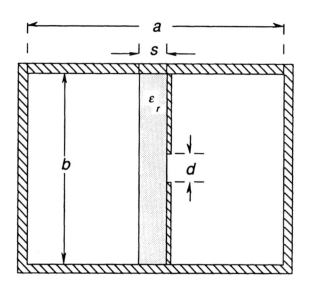

Figure   3.9.7.1:        Unilateral   Finline

Unilateral finline closed-form equations are in Praminick and Bhartia [10]

$$Z_0 = \frac{240.0 \ \pi^2 \ (P \ X + Q) \frac{b}{a}}{\sqrt{\varepsilon_{eff}(f)} \ (0.385 \ X + 1.762)^2} \qquad (3.9.7.1.1)$$

$$\varepsilon_{eff}(f) = k_e - \left(\frac{\lambda_0}{\lambda_{ca}}\right)^2 \qquad (3.9.7.1.2)$$

where for $d / b > 0.3$

$$P = -0.763 \left(\frac{b}{\lambda_0}\right)^2 + 0.58 \left(\frac{b}{\lambda_0}\right) + 0.0775 \left[\ln\left(\frac{a}{s}\right)\right]^2 - 0.668 \left(\ln \frac{a}{s}\right) + 1.262$$

$$(3.9.7.1.3)$$

$$Q = 0.372 \left(\frac{b}{\lambda}\right) + 0.914 \tag{3.9.7.1.4}$$

and for $d / b \leq 0.3$

$$P = 0.17 \left(\frac{b}{\lambda}\right) + 0.0098 \tag{3.9.7.1.5}$$

$$Q = 0.138 \left(\frac{b}{\lambda}\right) + 0.873 \tag{3.9.7.1.6}$$

where

$$X = \ln \csc \left(\frac{\pi\, d}{2.0\, b}\right) \tag{3.9.7.1.7}$$

For thin dielectric and low $\varepsilon_r$ [11] gives

$$k_e \approx k_c = \left(\frac{\lambda_{cd}}{\lambda_{ca}}\right)^2 \tag{3.9.7.1.8}$$

$$\lambda_{cd} = 2.0 \,(a - s) \sqrt{1.0 + N \left[1.0 - \frac{s}{a}(-0.0769\, \varepsilon_r + 1.231)\, F\left(\frac{s}{a}\right) X_a\right]}$$

$$(3.9.7.1.9)$$

$$\lambda_{ca} = 2.0\, a \sqrt{1.0 + \frac{4.0}{\pi} \frac{b}{a}\left(1.0 + 0.2 \sqrt{\frac{b}{a}}\,\right) \ln \csc \left(\frac{\pi\, d}{2.0\, b}\right)} \tag{3.9.7.1.10}$$

$$N = \frac{4.0}{\pi} \frac{b}{a - s}\left(1.0 + 0.2 \sqrt{\frac{b}{a - s}}\,\right) \tag{3.9.7.1.11}$$

$$X_a = \ln \csc \left(\frac{\pi\, d}{2.0\, b}\right) + \varepsilon_r\, (G_d + G_a) \tag{3.9.7.1.12}$$

$$G_a = \eta_a \ \tan^{-1}\left(\frac{1.0}{\eta_a}\right) + \ln \sqrt{1.0 + \eta_a{}^2} \qquad\qquad (3.9.7.1.13)$$

$$G_d = \eta_d \ \tan^{-1}\left(\frac{1.0}{\eta_d}\right) + \ln \sqrt{1.0 + \eta_d{}^2} \qquad\qquad (3.9.7.1.14)$$

$$\eta_a = \frac{\eta_d \, d}{b} \qquad\qquad (3.9.7.1.15)$$

$$\eta_d = \frac{s / a}{\dfrac{b \, d}{a \, b}} \qquad\qquad (3.9.7.1.16)$$

where for $d / b \le 0.5$

$$F\!\left(\frac{s}{a}\right) = -25.1223 + 31.524 \ \ln \frac{a}{s} - 12.504 \left(\ln \frac{a}{s}\right)^2 + 1.9454 \left(\ln \frac{a}{s}\right)^3 \qquad (3.9.7.1.17)$$

and for $0.5 \le d / b \le 0.75$

$$F\!\left(\frac{s}{a}\right) = -33.934 + 42.451 \ \ln \frac{a}{s} - 17.057 \left(\ln \frac{a}{s}\right)^2 + 2.5885 \left(\ln \frac{a}{s}\right)^3 \qquad (3.9.7.1.18)$$

and for $0.75 \le d / b \le 1.0$

$$F\!\left(\frac{s}{a}\right) = -48.0487 + 59.2846 \ \ln \frac{a}{s} - 23.77 \left(\ln \frac{a}{s}\right)^2 + 3.47 \left(\ln \frac{a}{s}\right)^3 \qquad (3.9.7.1.19)$$

The accuracy of the equation for $Z_0$ is $\pm 2.0\%$ for $s / a \le 0.05$ and $\pm 3\%$ for $s / a > 0.05$.

### 3.9.7.2    *Finite Thickness Off-center Unilateral Finline*

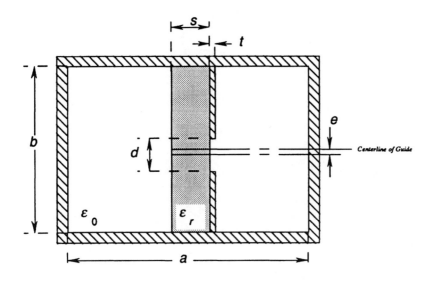

**Figure 3.9.7.2:    Finite Thickness, Off-Center Finline**

Pramanick *et al.* [14] include the effects of finite thickness metallization and an off-center fin as shown in Figure 3.9.7.2. The equations are

$$Z_0 = \frac{120.0 \, \pi^2 \, d}{\lambda_{ca} \sqrt{\varepsilon_{eff} \, (f)} \, (f_1 + q \, f_2)} \left[ \frac{k_e(f) - 1.0}{k_c - 1.0} \right] \qquad (3.9.7.2.1)$$

$$\varepsilon_{eff} \, (f) = k_e(f) - \left( \frac{\lambda_0}{\lambda_{ca}} \right)^2 \qquad (3.9.7.2.2)$$

where

$$f_1 = \frac{2.0 \, d}{\lambda_{ca}} \cos^2 \left( \frac{\pi \, t}{\lambda_{ca}} \right) + \frac{\pi \, t}{2.0 \, \lambda_{ca}} + \frac{\sin^2 \left( \frac{\pi \, t}{\lambda_{ca}} \right)}{4.0} \qquad (3.9.7.2.3)$$

$$f_2 = \frac{\pi \, (a - t)}{2.0 \, \lambda_{ca}} - \frac{\sin^4 \left[ \frac{\pi \, (a - t)}{\lambda_{ca}} \right]}{4.0} \qquad (3.9.7.2.4)$$

$$q = \left[ \frac{\frac{d}{b} \cos^2 \left( \frac{\pi t}{\lambda_{ca}} \right)}{\sin^2 \left( \frac{a - t}{\lambda_{ca}} \right)} \right] \tag{3.9.7.2.5}$$

$$k_e(f) = \varepsilon_r - \frac{\varepsilon_r - k_c}{1.0 + \left[ \frac{s}{a} \sqrt{\frac{f - f_c}{f}} \right]^{(1.0 + 0.06 \, b / d)}} \tag{3.9.7.2.6}$$

$$k_c = 1.0 + \frac{s}{a} (a_1 X + b_1) (\varepsilon_r - 1.0) \tag{3.9.7.2.7}$$

$$\lambda_{ca} = 2.0 \, (a - t) \sqrt{1.0 + N \, X + \left( 2.75 + 0.2 \frac{t}{a} \right) \left( \frac{t}{a - t} \right) b / d} \tag{3.9.7.2.8}$$

$$f_c = \frac{c}{\lambda_{ca}}$$

$$N = \frac{4.0}{\pi} \left( 1.0 + 0.2 \sqrt{\frac{b}{a - t}} \right) \left( \frac{b}{a - t} \right) \tag{3.9.7.2.9}$$

$$a_1 = p_0 + p_1 \ln \frac{a}{s} + p_2 \left( \ln \frac{a}{s} \right)^2 \tag{3.9.7.2.10}$$

$$b_1 = q_0 + q_1 \ln \frac{a}{s} + q_2 \left( \ln \frac{a}{s} \right)^2 \tag{3.9.7.2.11}$$

and

$$p_0 = 2.455 - 69.366 \left( \frac{t}{a} \right) + 1826.189 \left( \frac{t}{a} \right)^2 \tag{3.9.7.2.12}$$

$$p_1 = 2.5646 + 54.2675 \left( \frac{t}{a} \right) - 1381.6 \left( \frac{t}{a} \right)^2 \tag{3.9.7.2.13}$$

$$p_2 = 0.7837 - 18.954 \left( \frac{t}{a} \right) + 466.3185 \left( \frac{t}{a} \right)^2 \tag{3.9.7.2.14}$$

$$q_0 = 2.1697 + 3.638 \left(\frac{t}{a}\right) + 240.356 \left(\frac{t}{a}\right)^2 \qquad (3.9.7.2.15)$$

$$q_1 = 3.4661 - 7.3661 \left(\frac{t}{a}\right) - 180.6 \left(\frac{t}{a}\right)^2 \qquad (3.9.7.2.16)$$

$$q_2 = -0.6149 + 3.11624 \left(\frac{t}{a}\right) + 19.605 \left(\frac{t}{a}\right)^2 \qquad (3.9.7.2.17)$$

For a centered fin gap use

$$X = -\ln\left[\sin\left(\frac{0.5 \ \pi \ d}{b}\right)\right] \qquad (3.9.7.2.18)$$

For an off-center fin gap replace this with

$$X = \ln\left\{\csc\left(\frac{0.5 \ \pi \ d}{b}\right)\csc\left[0.5 \ \pi \ (1.0 - 2.0 \ e \ / \ b)\right]\right\} \qquad (3.9.7.2.19)$$

The equations are valid for

$\varepsilon_r \le 4.0$

$t \, / \, a \le 0.02094$

$1.0 \, / \, 64.0 \le s \, / \, a \le 1.0 \, / \, 8.0$

$1.0 \, / \, 32.0 \le d \, / \, b \le 1.0 \, / \, 2.0$

Results are valid to within $\pm 0.8\%$ of the complex power calculations.

### 3.9.7.3    *Bilateral Finline*

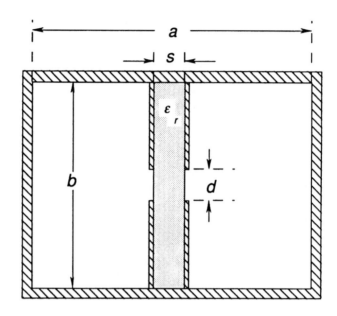

**Figure  3.9.7.3:**    **Bilateral  Finline**

Bilateral finline closed-form equations are in Praminick and Bhartia [10]

$$Z_0 = \frac{240.0 \ \pi^2 \ (P \ X + Q) \ \dfrac{b}{a - s}}{\sqrt{\varepsilon_{eff}(f)} \ (0.385 \ X + 1.762)^2}$$
(3.9.7.3.1)

$$\varepsilon_{eff}(f) = k_e - \left(\frac{\lambda_0}{\lambda_{ca}}\right)^2$$
(3.9.7.3.2)

where

$$P = 0.01 \left(\frac{b}{\lambda_0}\right)^2 + 0.097 \left(\frac{b}{\lambda_0}\right) + 0.04095$$
(3.9.7.3.3)

$$Q = 0.0031 \left(\frac{b}{\lambda_0}\right) + 0.89$$
(3.9.7.3.4)

$$X = X_b - G_d \ (\varepsilon_r - 1.0)$$
(3.9.7.3.5)

$$X_b = \ln \csc\left(\frac{\pi\, d}{2.0\, b}\right) + \varepsilon_r\, G_d \qquad (3.9.7.3.6)$$

$$G_d = \eta_d \arctan\left(\frac{1.0}{\eta_d}\right) + \ln\left(1.0 + \eta_d{}^2\right)^{1/2} \qquad (3.9.7.3.7)$$

$$\eta_d = \frac{(s\,/\,a)}{(b\,/\,a)\,(d\,/\,b)} \qquad (3.9.7.3.8)$$

For low $\varepsilon_r$ and thin dielectric

$$k_e \approx k_c = \left(\frac{\lambda_{cd}}{\lambda_{ca}}\right)^2 \qquad (3.9.7.3.9)$$

where

$$\lambda_{cd} = 2.0\,(a - s)\,\sqrt{1.0 + N\,X_b} \qquad (3.9.7.3.10)$$

$X_b$ is defined above and

$$N = \frac{4.0}{\pi}\frac{b}{a - s}\left(1.0 + 0.2\,\sqrt{\frac{b}{a - s}}\,\right) \qquad (3.9.7.3.11)$$

To calculate $\lambda_{ca}$, set $\varepsilon_r = 1.0$ in the equations for $\lambda_{cd}$. The accuracy of $\varepsilon_{eff}$ is stated to be within $\pm 0.8\%$ for $s\,/\,a \le 0.125$ and $Z_0$ is accurate to $\pm 2\%$ for $d\,/\,b \le 0.2$.

### 3.9.7.4    *Antipodal Finline*

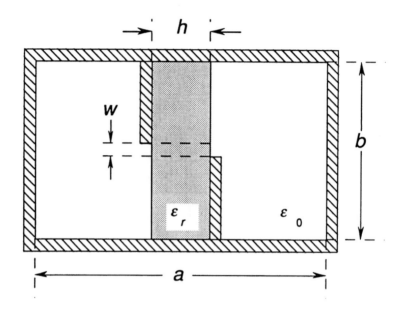

**Figure 3.9.7.4:**    Antipodal Finline

Antipodal finline provides a wider range of impedance than the previous finline structures. For sufficiently large gap $w$, and thin dielectric the equations of unilateral finline can be used for design. Also see [1] for design curves.

REFERENCES

[1]    Bhat, Bharathi, and Shiban K. Koul, *Analysis, Design and Applications of Fin Lines,* Artech House, Norwood, MA, 1987.

[2]    Button, Kenneth J., *ed., Infrared and Millimeter Waves*, Vol. 13, Millimeter Components and Techniques, Part IV, "Integrated Fin-Line Components for Communication, Radar and Radiometer Applications," by Wolfgang Menzel, pp. 77–121 and Vol. 9, Millimeter Components and Techniques, Part I, "A Comparative Study of Millimeter-Wave Transmission Lines," by Tatsuoh Itoh and Juan Rivera, Academic Press, New York, 1983.

[3]    El Hennaway, H., *et al.,* "Impedance Transformation in Fin-Line," *IEE Proceedings*, Part H, Vol. 129, No. 6, April 1983, pp. 342–350.

[4]    Hoefer, Wolfgang J.R., and Miles N. Burton, "Analytical Expressions for the Parameters of Finned and Ridged Waveguides," *1982 IEEE MTT-S International Microwave Symposium Digest*, pp. 311–313.

[5]     Hofmann, Holger, "Dispersion of Planar Waveguides for Millimeter-Wave Application," *Archiv Für Elektronik und Übertragungstechnik*, Band 31, Heft 1, 1977, pp. 40–44.

[6]     Kitazawa, Toshihide, and Raj Mittra, "Analysis of Finline with Finite Metallization Thickness," *IEEE Transactions on Microwave Theory and Techniques*, Vol. MTT-32, No. 11, November 1984, pp. 1484–1487.

[7]     Meier, Paul J., "Integrated Fin-Line Millimeter Components," *1981 IEEE MTT-S Symposium Digest*, pp. 195–197.

[8]     Meier, Paul J., "Integrated Fin-Line Millimeter Components," *IEEE Transactions on Microwave Theory and Techniques*, Vol. MTT-22, No. 12, December 1974, pp. 1209–1216.

[9]     Piotrowski, J.K., "Accurate and Simple Formulas for Dispersion in Fin-Lines," *1984 IEEE MTT-S Symposium Digest*, pp. 333–335.

[10]    Pramanick, Protap, and Prakash Bhartia, "Accurate Analysis Equations and Synthesis Technique for Unilateral Finlines," *IEEE Transactions on Microwave Theory and Techniques*, Vol. MTT-33, No. 1, January 1985, pp. 24–30.

[11]    Pramanick, Protap, and Prakash Bhartia, "CAD Models for Millimeter-Wave Suspended Substrate Microstrip Lines and Fin-lines," *1985 IEEE MTT-S International Microwave Symposium Digest*, St. Louis, Missouri, pp. 453-456.

[12]    Pramanick, Protap, and Prakash Bhartia, "Modeling of the Apparent Characteristic Impedance of Finned-Waveguide and Finlines," *1986 IEEE MTT-S Symposium Digest*, pp. 225–228.

[13]    Pramanick, Protap, and Prakash Bhartia, "A New Model for the Apparent Characteristic Impedance of Finned Wavegide and Finlines," *IEEE Transactions on Microwave Theory and Techniques*, Vol. MTT-34, No. 12, December 1986, p. 1437–1441.

[14]    Pramanick, P., *et al.*, "Computer Aided Design Models for Unilateral Finlines with Finite Metallization Thickness and Arbitrarily Located Slot Widths," *1987 IEEE MTT-S Symposium Digest*, pp. 703–706.

[15]    Saad, A.M.K., and G. Begemann, "Electrical Performance of Finlines of Various Configurations," *Microwaves, Optics and Acoustics*, Vol. 1, No. 2, January 1977, pp. 81–88.

[16]    Saad, A.M.K., "Analysis of Fin-Line Tapers and Transitions," *IEE Proceedings*, Part H, Vol. 130, No. 3, April 1983, pp. 230–235.

[17]    Sharma, Arvind K. and Wolfgang J.R. Hoefer, "Empirical Expressions for Fin-Line Design," *IEEE Transactions on Microwave Theory and Techniques*, Vol. MTT-31, No. 4, April 1983, pp. 350–356.

[18]    Vahldieck, Rüdiger, "Accurate Hybrid-Mode Analysis of Various Finline Configurations Including Multilayered Dielectrics, Finite Metallization Thickness, and Substrate Holding Grooves," *IEEE Transactions on Microwave Theory and Techniques*, Vol. MTT-32, No. 11, November 1984, pp. 1454–1460.

[19]    Yang, Hung-Yu, and Nicólaos G. Alexópoulos, "Characterization of the Finline Step Discontinuity on Anisotropic Substrates," *IEEE Transactions on Microwave Theory and Techniques*, Vol. MTT-35, No. 11, November 1987, pp. 956–963.

### 3.9.8 Others

Other useful types of transmission lines not covered in this book include image guide, H-guide (100 to 200 GHz), groove guide (100 to 300 GHz), and microguide.

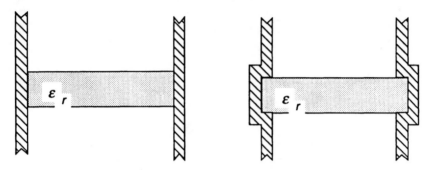

**Figure 3.9.8.1:**     **H-Guide and Groove Guide**

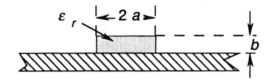

**Figure 3.9.8.2:**     **Image Guide**

REFERENCES

[1]     Button, Kenneth J., ed., *Infrared and Millimeter Waves*, Vol. 9, Millimeter Components and Techniques, Part I, "A Comparative Study of Millimeter-Wave Transmission Lines," by Tatsuoh Itoh and Juan Rivera, Academic Press, New York, 1983.

[2]     Wolff, Edward A., and Roger Kaul, *Microwave Engineering and Systems Applications*, John Wiley and Sons, New York, 1988.

## 3.10  WAVEGUIDES

The analysis of waveguides is often used when analyzing planar transmission line discontinuities. An equivalence can be established between a planar line discontinuity and a waveguide discontinuity allowing existing waveguide models to be transformed. The analysis of waveguides is a very large topic in and of itself and will not be covered in depth here. There are a number of excellent references. See Marcuvitz [2], for example.

### 3.10.1 Rectangular Waveguide

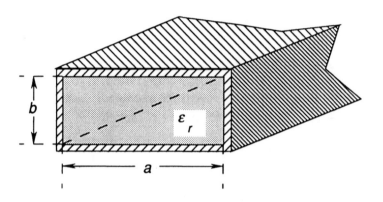

Figure 3.10.1.1:    Rectangular Waveguide

Waveguide has a major difference from most of the other transmission lines described so far—it does not pass dc. As such, it behaves somewhat like a high-pass filter. There are many *modes* or solutions of the wave equations in the guide and each has an associated *cutoff wavelength* or *cutoff frequency* . A wave can propagate only when the frequency is greater than the cutoff frequency. Because there is a single conductor, the modes of propagation cannot be *TEM*. The field patterns or modes are named $TE_{m,n}$ or $TM_{m,n}$. The subscript $m$ represents the number of half wavelengths in the $x$ direction and $n$ represents the number of half wavelengths in the $y$ direction (where $z$ is length). *TE* (*transverse electric*) waves, also known as $H$ waves have a nonzero longitudinal magnetic field component. *TM* (*transverse magnetic*) waves, also known as $E$ waves have a nonzero longitudinal electric field component. *TE* modes may have only one of $m$, $n$ equal to zero. *TM* modes may have neither equal to zero. Conventionally, $a > b$ as shown in Figure 3.10.1 above. This is assumed throughout this section.

A *degenerate mode* occurs whenever multiple field patterns have the same cutoff frequency. The lowest frequency that can actually propagate in the waveguide (the $TE_{1,0}$ mode) is known as the *dominant mode* and occurs when:

$$\lambda_c = 2.0 \, a \qquad\qquad\qquad (3.10.1.1)$$

The wavelength in the guide is:

$$\lambda_g = \frac{\lambda_0}{\sqrt{\varepsilon_r \, \mu_r \left[ 1.0 - \left( \frac{\lambda_0}{\lambda_{c_{m,n}}} \right)^2 \right]}}$$

(3.10.1.2)

The cutoff frequencies of both $TE_{m,n}$ and $TM_{m,n}$ modes are:

$$\omega_{c_{m,n}} = \frac{c_0}{\sqrt{\mu_r \, \varepsilon_r}} \sqrt{\left( \frac{m \, \pi}{a} \right)^2 + \left( \frac{n \, \pi}{b} \right)^2} \quad \text{(rad / s)}$$

(3.10.1.3)

The guide losses of the $TE$ modes ($n \neq 0$) are:

$$\alpha_{c_{TE_{m,n}}} = \frac{2.0 \, R_s}{b \, \eta \, \sqrt{1.0 - (f_c / f)^2}} \left\{ \left( 1.0 + \frac{b}{a} \right) \left( \frac{f_c}{f} \right)^2 \right.$$

$$\left. + \left[ 1.0 - \left( \frac{f_c}{f} \right)^2 \right] \left[ \frac{(b / a) \left[ (b / a) \, m^2 + n^2 \right]}{(b^2 \, m^2 / a^2) + n^2} \right] \right\}$$

(3.10.1.4)

and the losses for the $TE_{m,0}$ mode are

$$\alpha_{c_{TE_{m,0}}} = \frac{2.0 \, R_s}{b \, \eta \, \sqrt{1.0 - (f_c / f)^2}} \left[ 1.0 + \frac{2.0 \, a}{b} \left( \frac{f_c}{f} \right)^2 \right]$$

(3.10.1.5)

For the $TM$ modes the attenuation is calculated with

$$\alpha_{c_{TM_{m,n}}} = \frac{2.0 \, R_s}{b \, \eta \, \sqrt{1.0 - (f_c / f)^2}} \frac{(b / a)^3 \, m^2 + n^2}{(b^2 \, m^2 / a^2) + n^2}$$

(3.10.1.6)

The dielectric losses are

$$\alpha_d = \frac{k_0^2 \, \varepsilon_r \, \tan \delta}{2.0 \, \beta}$$

(3.10.1.7)

where

$$k_0^2 = k_c^2 + \beta^2$$

(3.10.1.8)

$$\beta = \frac{2.0 \, \pi}{\lambda_g}$$

(3.10.1.9)

$$k_c = \frac{2.0 \; \pi \sqrt{\mu_r \, \varepsilon_r}}{\lambda_c} \tag{3.10.1.10}$$

For the commonly used $TE_{1,0}$ mode the impedance is

$$Z_{TE_{1,0}} = \frac{\eta_0}{\sqrt{1.0 - \left(\frac{\lambda}{2.0 \, a}\right)^2}} \tag{3.10.1.11}$$

The phase and group velocities are given by

$$v_p \, v_g = v_c{}^2 \tag{3.10.1.12}$$

$$v_g = \frac{\lambda_0}{\lambda_g} v_c = \text{velocity of energy propagation} \tag{3.10.1.13}$$

$$v_c = \sqrt{\mu_0 \, \varepsilon_0} = \text{velocity of the } TEM \text{ mode in free space} \tag{3.10.1.14}$$

$$v_p = \frac{\omega}{\beta} = \frac{k_0}{\beta} v_c = \text{speed of a constant phase point on the wave} \tag{3.10.1.15}$$

## REFERENCES

[1]  Collin, Robert E., *Field Theory of Guided Waves*, IEEE Press, 1991.

[2]  Marcuvitz, N., *Waveguide Handbook*, McGraw-Hill, New York, 1955. Reprint with errata and preface by N. Marcuvitz. Peter Peregrinus Ltd., 1986.

[3]  Moreno, T., *ed.*, *Microwave Transmission Design Data*, Sperry Gyroscope Company, Great Neck, New York, Publication No. 23-80, 1944. (Reprint by Artech House, Norwood, MA, 1989.)

[4]  Ramo, Simon, John R. Whinnery, and Theodore Van Duzer, *Fields and Waves in Communication Electronics*, John Wiley & Sons, New York, 1984. (A classic text which has been revised several times since the first version, *Fields and Waves in Modern Radio* in 1944. Good example problems.)

[5]  Terman, F.E., *Radio Engineering*, McGraw-Hill, New York, 1947.

## 3.10.2 Circular Waveguide

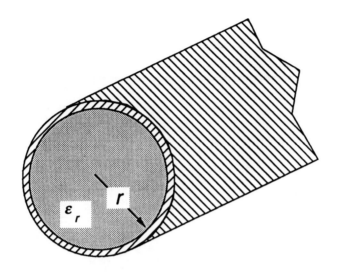

Figure 3.10.2.1:    Circular Waveguide

The $TE_{m,n}$ cutoff wavelengths for this structure are

$$\lambda_c = \frac{2 \pi a \sqrt{\varepsilon_r}}{u'_{m,n}}$$
(3.10.2.1)

The subscript $m$ indicates angular variation in the field and $n$ indicates radial field variations.  The roots of the Bessel function are given in Marcuvitz [2]:

| $u'_{m,n}$ | $m$ | | | | |
|---|---|---|---|---|---|
| $n$ | 0 | 1 | 2 | 3 | 4 |
| 1 | 3.832 | 1.841 | 3.054 | 4.201 | 5.317 |
| 2 | 7.016 | 5.331 | 6.706 | 8.015 | 9.282 |
| 3 | 10.173 | 8.536 | 9.969 | 11.346 | 12.682 |
| 4 | 13.324 | 11.706 | 13.170 | | |

Table  3.10.2.1:    Values of $u'_{m,n}$ for $TE_{m,n}$  Mode  Calculations

$TM_{m,n}$ modes are calculated in the same fashion with the following values for $u_{m,n}$.

| $u_{m,n}$ | $m$ | | | | |
|---|---|---|---|---|---|
| $n$ | 0 | 1 | 2 | 3 | 4 |
| 1 | 2.405 | 3.832 | 5.136 | 6.380 | 7.588 |
| 2 | 5.520 | 7.016 | 8.417 | 9.761 | 11.065 |
| 3 | 8.654 | 10.173 | 11.620 | 13.015 | 14.372 |
| 4 | 11.792 | 13.323 | 14.796 | | |

**Table 3.10.2.2:**    Values of $u_{m,n}$ for TM$_{m,n}$ Mode Calculations

The wavelength inside the waveguide is $\lambda_g$

$$\lambda_g = \frac{\lambda}{\sqrt{\varepsilon_r - \left(\dfrac{\lambda}{\lambda_c}\right)}} \tag{3.10.2.2}$$

The conductor losses for *TM* modes the conductor loss is

$$\alpha_c = \frac{R_s}{\eta\, a} \frac{1.0}{\sqrt{1.0 - \left(\dfrac{\lambda}{\lambda_c}\right)^2}} \tag{3.10.2.3}$$

for *TE* modes the conductor loss is

$$\alpha_c = \frac{R_s}{\eta\, a} \frac{1.0}{\sqrt{1.0 - \left(\dfrac{\lambda}{\lambda_c}\right)^2}} \left[\frac{m^2}{u'_{m,n} - m^2} + \left(\frac{\lambda}{\lambda_c}\right)^2\right] \tag{3.10.2.4}$$

Dielectric losses are

$$\alpha_d = \frac{k_0^2\, \varepsilon_r\, \lambda_g\, \tan\delta}{4.0\, \pi} \tag{3.10.2.5}$$

## REFERENCES

[1]    Collin, Robert E., *Field Theory of Guided Waves,* IEEE Press, 1991.

[2]    Marcuvitz, N.,*Waveguide Handbook*, McGraw-Hill, 1955. Reprint with errata and preface by N. Marcuvitz. Peter Peregrinus Ltd., 1986.

[3]    Moreno, T., ed., *Microwave Transmission Design Data*, Sperry Gyroscope Company, Great Neck, New York, Publication No. 23-80, 1944. (Reprint by Artech House, Norwood, MA, 1989.)

[4]     Ramo, Simon, John R. Whinnery, and Theodore Van Duzer, *Fields and Waves in Communication Electronics,* John Wiley & Sons, New York, 1984. (A classic text which has been revised several times since the first version, *Fields and Waves in Modern Radio* in 1944. Good example problems.)

# Chapter 4

# Coupled Lines

When two transmission lines are in close enough proximity for their field patterns to be disrupted (relative to their isolated patterns) it is discovered that a portion of the signal in one line is now present in the second line. This property can be desired—as when we design a directional coupler, or undesired—as in a multi-conductor cable carrying several signals. In the former case we use term the second signal *coupling*, in the latter we refer to *crosstalk*. In either case we wish to understand the source of the phenomena and the equations we will need to control it.

This chapter covers the equations for simple side-by-side transmission lines. Side-by-side lines can be used alone, in series, or in combination with other lines and components to form broadband components.

For design of symmetric, directional couplers we make use of two equations. The first determines the amount of coupling and the second designs the coupler to be matched at its ports to the impedance $Z_0$.

$$k = \frac{Z_{0,e} - Z_{0,o}}{Z_{0,e} + Z_{0,o}} \tag{4.0.1}$$

$$Z_0 = \sqrt{Z_{0,e} Z_{0,o}} \tag{4.0.2}$$

or equivalently

$$Z_{0,e} = Z_0 \sqrt{\frac{1 + k}{1 - k}} \tag{4.0.3}$$

$$Z_{0,o} = Z_0 \sqrt{\frac{1 - k}{1 + k}} \tag{4.0.4}$$

The value of $k$ is the maximum coupling, which is achieved when the electrical length of the line is $\lambda / 4$. Equation 4.0.5 is the equation as a function of electrical length $\theta$, and Figure 4.0 shows this repeating function.

181

$$\frac{V_2}{V_1} = \frac{j\,k\,\sin\,\theta}{\sqrt{1.0 - k^2}\,\cos\,\theta + j\,\sin\,\theta} \qquad (4.0.5)$$

$$k \equiv \left|\frac{V_2}{V_1}\right|_{\theta = 90° = \pi/2 \text{ radians}} \qquad (4.0.6)$$

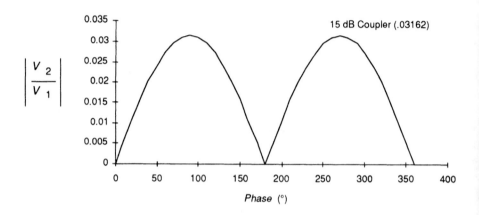

Figure 4.0:    Directional Coupler Coupled Port Response

## 4.1    COUPLED LINE ABCD MATRIX

The physical lines described in the previous chapter are two-port networks. Coupled lines are four-port networks. Coupled lines have two ABCD matrices corresponding to the even and odd modes:

$$\begin{bmatrix} \cos\theta_e & j\,Z_{0,e}\,\sin\theta_e \\ \dfrac{j\,\sin\theta_e}{Z_{0,e}} & \cos\theta_e \end{bmatrix}$$

$$\begin{bmatrix} \cos\theta_o & j\,Z_{0,o}\,\sin\theta_o \\ \dfrac{j\,\sin\theta_o}{Z_{0,o}} & \cos\theta_o \end{bmatrix}$$

where the subscript $e$ denotes the even mode parameters and $o$ denotes the odd mode.

## 4.2    COAXIAL LINES

### 4.2.1    Shielded Pair

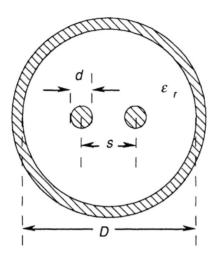

**Figure   4.2.1.1:**        Shielded  Pair

For circuits requiring differential signal transmission or low frequency shielding, the addition of an extra conductor in the standard coaxial line provides a good solution. The shielded pair may also be used for coupler, balun, or transformer designs. This structure has higher losses than coax for equivalent dimensions and dielectric.

Ramo and Whinnery [6] give the following equations:

$$Z_{0,o} = \frac{\eta_0}{\pi \sqrt{\varepsilon_r}} \left\{ \ln \left[ \frac{2s}{d} \frac{1.0 - (s/D)^2}{1.0 + (s/D)^2} \right] - \right.$$

$$\left. \frac{\left[ 1.0 + 4 (s/d)^2 \right]\left[ 1.0 - 4 (s/D)^2 \right]}{16 (s/d)^4} \right\} \tag{4.2.1.1}$$

$$Z_{0,e} = \frac{\eta_0}{2\pi \sqrt{\varepsilon_r}} \ln \left\{ \frac{s/d}{2 (s/D)^2} \left[ 1.0 - (s/D)^4 \right] \right\} \tag{4.2.1.2}$$

The above equations are valid for

$$d \ll D$$

$$d \ll s$$

Attenuation is calculated with:

$$\alpha_d = \frac{\pi \sqrt{\varepsilon_r \mu_r}}{\lambda_0} \frac{\varepsilon_1''}{\varepsilon_r} \tag{4.2.1.3}$$

$$\alpha_c = \frac{R}{2.0 \, Z_0} \tag{4.2.1.4}$$

where

$\lambda_0$ = wavelength in free space (2.3.13)

$\varepsilon_1''$ = loss factor; see (2.4.1.1.)

## REFERENCES

[1]     Fanchiotti, H., and H. Vucetich, "Angular Asymmetries in a Shielded-Pair Line," *International Journal of Electronics*, Vol. 47, No. 6, 1979, pp. 609–616.

[2]     Fanchiotti, H., *et al,* "Improved Formulae for the Capacitance of a Shielded-Pair Line," *International Journal of Electronics*, Vol. 49, No. 6, 1980, pp. 487–495.

[3]     Green, E.I., *et al.*, "The Proportioning of Shielded Circuits for Minimum High-Frequency Attenuation," *The Bell System Technical Journal*, 1936, pp. 248–283.

[4]     Liboff, Richard L., and G. Conrad Dalman, *Transmission Lines, Waveguides, and Smith Charts*, Macmillan, New York, 1985. (Has an unresolved typo and does not give even and odd mode impedances.)

[5]     Nordgard, John D., "The Capacitances and Surface-Charge Distributions of Shielded Balanced Pair," *IEEE Transactions on Microwave Theory and Techniques*, Vol. MTT-24, No. 2, February 1976, pp. 94–100.

[6]     Ramo, Simon, and John R. Whinnery, *Fields and Waves in Modern Radio*, Wiley, London, 1944.

[7]     *Reference Data For Radio Engineers*, Howard W. Sams & Co., Inc., Indianapolis, 1982. (Gives the same equations as [6] without the second term in $Z_{0,o}$.)

[8]     Zinke, O., and H. Brunswig, *Lehrbuch der Hochfrequenztechnik I*, 1 Band, 2 Auflage, Springer-Verlag, Berlin, 1973.

## 4.2.2 Shielded Twisted Pair

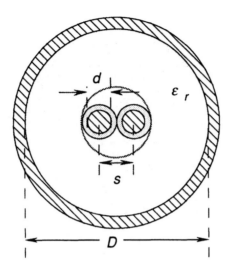

**Figure 4.2.2.1:**     Shielded Twisted Pair

Most analyses use the equations given for the shielded pair of Section 4.2.1; however, those equations are not valid for $d$ approaching $s$ or $d$ approaching $D$, which are common situations for this type of cable. Until exact equations are available, a reasonable starting point might be to use twisted-pair equations for the odd mode and coax equations (Chapter 3) for the even mode.

Marx [1] analyzes and gives the equations for proper termination of a shielded twisted pair. It is important to terminate both propagation modes to prevent reflections from being converted from one mode to the other. This helps minimize overshoot.

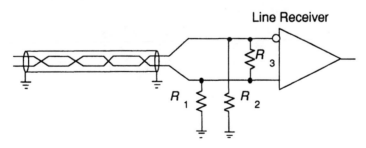

**Figure 4.2.2.2:**     **Terminating Shielded Twisted Pair**

For the circuit of Figure 4.2.2.2 the termination equations are (it is assumed that the source is similarly matched to the line):

$$R_1 = R_2 = Z_{0,e} \tag{4.2.2.2.1}$$

$$R_3 = \frac{2\,Z_{0,e}\,Z_{0,o}}{Z_{0,e} - Z_{0,o}} \tag{4.2.2.2.2}$$

and to terminate as in the circuit of Figure 4.2.2.3:

$$R_1 = R_2 = Z_{0,o} \tag{4.2.2.2.3}$$

$$R_3 = \frac{Z_{0,e} - Z_{0,o}}{2} \tag{4.2.2.2.4}$$

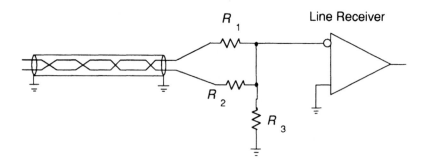

**Figure 4.2.2.3:     Terminating Shielded Twisted Pair**

REFERENCES

[1]   Marx, Kenneth P., "Propagation Modes, Equivalent Circuits, and Characteristic Terminators for Multiconductor Transmission Lines with Inhomogeneous Dielectric," *IEEE Transactions on Microwave Theory and Techniques*, MTT-21, No. 7, July 1973, pp. 450–457.

[2]   Zinke, O., and H. Brunswig, *Lehrbuch der Hochfrequenztechnik I*, 1 Band, 2 Auflage, Springer-Verlag, Berlin, 1973.

## 4.3   PAIRED LINES

### 4.3.1   Wireline™ Coupler

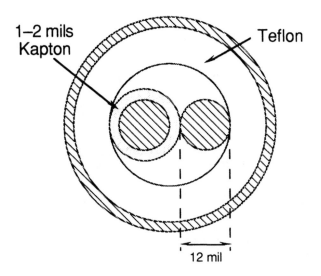

**Figure   4.3.1.1:        Wireline™ Coupler**

This structure is a shielded twisted pair used as a coupler. It can be used as a 3 dB coupler at its center frequency or narrowband (≈1 octave) as any other amount of coupling by operating off the center frequency. The shield may be foil, braid, or solid tubing. The maximum frequency is limited to about 5 GHz due to strays caused by the transition from the Wireline's leads to the PCB.

A Wireline™ coupler may be designed in two ways:
- a 3 dB coupler at a desired frequency
- an $X$ dB coupler at a desired frequency

For a 3 dB coupler, calculate the $\lambda / 4$ length at the desired center frequency using $\varepsilon_r = 2.55$.

For other couplers, because the spacing of the lines is fixed, we operate off of the center frequency of the coupling curve to get the desired coupling. If the desired operating frequency is $f_{op}$ and the desired coupling (dB) is $X$, we calculate:

$$K_2 = \sin^{-1} \left( \sqrt{\frac{10^{-X/10}}{1.0 - 10^{-X/10}}} \right) \qquad (4.3.1.1.1)$$

and

$$f_q = \text{quarter wave frequency} = f_{op} \frac{90°}{K_2} \qquad (4.3.1.1.2)$$

Rearranging we get an equation for the coupling over frequency:

$$X = -10 \log\left(\frac{\sin^2 (f/f_q)}{\sin^2 (f/f_q) - 1}\right) \quad \text{(dB)} \qquad (4.3.1.1.3)$$

### 4.3.1.1    Example:   10 dB Wireline Coupler

Design a 10 dB Wireline coupler at 372.8 MHz.
**Solution:**

$$K_2 = \sin^{-1}\left(\sqrt{-\frac{10^{-10/10}}{10^{-10/10} - 1}}\right) = \sin^{-1}\left(\sqrt{1/9}\right) = \sin^{-1}(0.3333) = 19.47°$$

$$f_q = 372.8 \text{ MHz} \times 90° / 19.47° = 1.7233 \text{ GHz (quarter-wave frequency)}$$

$$l = c/f_q/4/\sqrt{\varepsilon_r} = 3 \times 10^8 \text{ m/s} / 1.7233 \text{ GHz} / 4 / \sqrt{2.55}$$

$$\boxed{l = 0.0273 \text{ m} = 1.073 \text{ inches}}$$

### 4.3.1.2    Example:   3 dB Wireline Coupler

Design a 3 dB Wireline coupler at 402 MHz.
**Solution:**

$$f_q = 402 \text{ MHz}$$

$$l = c/f_q/4/\sqrt{\varepsilon_r} = 3 \times 10^8 \text{ m/s} / 402 \text{ MHz} / 4 / \sqrt{2.55}$$

$$\boxed{l = 0.1168 \text{ m} = 4.6 \text{ inches}}$$

#### REFERENCES

[1]    Saad, Theodore S., "Wireline™," *1987 SBMO International Microwave Symposium Proceedings*, Rio de Janeiro, pp,. 563–569.

[2]    Sage Laboratories, "Wireline® Quadrature Hybrids and Couplers."

## 4.3.2 Coupled Rectangular Coaxial Lines

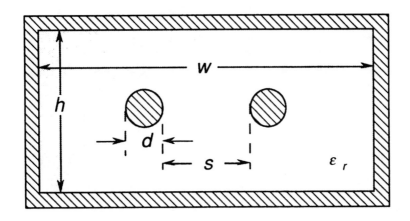

**Figure 4.3.2.1:** **Coupled Lines in Rectangular Coaxial**

This configuration is found in interdigital filters and when we have close side walls in slab line designs. Agarwal [1] modifies the (no side wall) equations of Stracca *et al.* [3] for slab line and gives

$$Z_{0,e} = \frac{\eta_0}{2.0 \, \pi \, \sqrt{\varepsilon_r}} \, \ln\left(\frac{4.0}{\pi \, \frac{d}{h} \, F \, F_e}\right)$$

$$-\frac{138}{\sqrt{\varepsilon_r}} \sum_{m=1}^{\infty} (-1)^m \log \frac{\tanh\left\{\frac{\pi}{h} \, [m \, (D + C \, / \, 2.0) - C \, / \, 2.0]\right\}}{\tanh\left\{\frac{\pi}{h} \, [m \, (D + C \, / \, 2.0) + C \, / \, 2.0]\right\}}$$

$$(4.3.2.1.1)$$

$$Z_{0,o} = \frac{\eta_0}{2.0 \, \pi \, \sqrt{\varepsilon_r}} \, \ln\left(\frac{4.0}{\pi \, \frac{d}{h} \, F \, F_o}\right)$$

$$-\frac{138}{\sqrt{\varepsilon_r}} \sum_{m=1}^{\infty} \log \frac{1.0 + \left\{ \sinh\left(\frac{\pi c}{2.0\,h}\right) \Big/ \left[(D + C/2.0)\cosh\left(\frac{\pi m}{h}\right)\right]\right\}}{1.0 - \left\{ \sinh\left(\frac{\pi c}{2.0\,h}\right) \Big/ \sinh\left[\frac{\pi m\,(D + C/2.0)}{h}\right]\right\}}$$

$$(4.3.2.1.2)$$

where

$$F_e = M_e \tanh\left(\frac{\pi c}{2.0\,h}\right) \tag{4.3.2.1.3}$$

$$F_o = M_o \coth\left(\frac{\pi c}{2.0\,h}\right) \tag{4.3.2.1.4}$$

$$M_e = 1.0 + \exp\left[(-1.8861\,x^2 + 1.4177\,x - 5.5142)y\right]$$

$$- \exp\left[(-2.7030\,x^2 + 4.6772\,x - 8.5900)y\right] \tag{4.3.2.1.5}$$

$$M_o = 1.0 + \exp\left[(3.0989\,x^2 - 8.2997\,x - 3.0843)y\right]$$

$$- \exp\left[(4.0797\,x^2 - 2.1808\,x - 8.5534)y\right] \tag{4.3.2.1.6}$$

$$F = \frac{\left[1.0 + e^{16.0\,(d/h) - 1.142}\right]}{\sqrt{1.0 - \frac{(d/h)^4}{5.905}}} \tag{4.3.2.1.7a}$$

$$C = s + d \tag{4.3.2.1.7b}$$

$$D = \frac{w - C}{2.0} \tag{4.3.2.1.7c}$$

and

$$x = d/h \tag{4.3.2.1.8}$$

The accuracy of these equations was checked by designing a 10 dB coupler. The experimental results showed good agreement with the calculation. The original equations of Stracca *et al.* are curve fits to numerical results. These equations are accurate to 1.4% for $Z_{0,e}$ and 3.2% for $Z_{0,o}$ for

$$d / h \leq 0.9$$

$$s / d \geq 0.3$$

### REFERENCES

[1] Agarwal, A., *et al.*, "Coupled Bars in Rectangular Coaxial," *Electronics Letters*, Vol. 25, No. 1, January 5, 1989, pp. 66–67. (Bars of circular cross-section.)

[2] Perlow, Stewart M., "Analysis of Edge-Coupled Shielded Strip and Slabline Structures," *IEEE Transactions on Microwave Theory and Techniques*, Vol. MTT-35, No. 5, May 1987, pp. 522–529. (Bars of rectangular and unequal cross-section.)

[3] Stracca, Giovanni B., *et al.*, "Numerical Analysis of Various Configurations of Slab Lines," *IEEE Transactions on Microwave Theory and Techniques*, Vol. MTT-34, No. 3, March 1986, pp. 359–363.

## 4.4 COUPLED COPLANAR LINES

### 4.4.1 Broadside Coupled Coplanar Waveguide

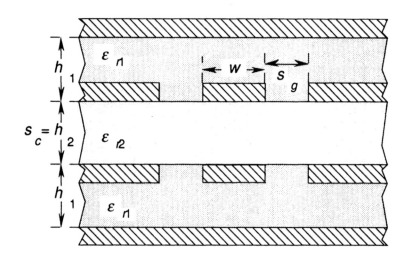

**Figure 4.4.1.1:** Broadside Coupled Coplanar Waveguide

This multiple dielectric configuration is useful in integrated circuits. Bedair and Wolff [1] gives the even and mode impedances in terms of the the even and odd mode capacitances:

$$Z_{0,o} = \frac{1.0}{c \sqrt{C_o C_o^a}} \tag{4.4.1.1}$$

$$Z_{0,e} = \frac{1.0}{c \sqrt{C_e C_e^a}} \tag{4.4.1.2}$$

$$\varepsilon_{eff,o} = \frac{C_o}{C_o^a} \tag{4.4.1.3}$$

$$\varepsilon_{eff,e} = \frac{C_e}{C_e^a} \tag{4.4.1.4}$$

$$C_o = C_{o,1} + C_{o,2} \tag{4.4.1.5}$$

$$C_{o,1} = 2.0 \; \varepsilon_0 \; \varepsilon_{r1} \frac{K(k_{o,1})}{K(k'_{o,1})} \tag{4.4.1.6}$$

$$C_{o,2} = 2.0 \; \varepsilon_0 \; \varepsilon_{r2} \frac{K(k_{o,2})}{K(k'_{o,2})} \tag{4.4.1.7}$$

$$C_e = 2.0 \; \varepsilon_0 \; \varepsilon_{r1} \frac{K(k_{o,1})}{K(k'_{o,1})} + 2.0 \; \varepsilon_0 \; \varepsilon_{r2} \frac{K(k_{e,2})}{K(k'_{e,2})} \tag{4.4.1.8}$$

$$k_{o,1} = \frac{\tanh \dfrac{\pi \; w}{4 \; h_1}}{\tanh \dfrac{\pi \; (w + 2 \; s_g)}{4 \; h_1}} \tag{4.4.1.9}$$

$$k_{o,2} = \frac{\tanh \dfrac{\pi \; w}{4 \; h_2}}{\tanh \dfrac{\pi \; (w + 2 \; s_g)}{4 \; h_2}} \tag{4.4.1.10}$$

$$k_{e,2} = \frac{\sinh \dfrac{\pi \; w}{4 \; h_2}}{\sinh \dfrac{\pi \; (w + 2 \; s_g)}{4 \; h_2}} \tag{4.4.1.11}$$

and $C_i^{\;a}$ $(i = e, o)$ are the capacitances when the dielectric is replaced with air ($\varepsilon_r = 1$). As before,

$$k_i' = \sqrt{1 - k_i^2} \tag{4.4.1.12}$$

These equations are quasi-static and are exact at zero frequency and for coplanar propagation. The authors state that these equations are useful to 40 GHz for small $s_g$. A graph of dispersion for two cases is given in Bedair and Wolff [2]; however, no general conclusions or formulas are available. For coplanar waveguide-like propagation to within 1%, $s_g / h \geq 0.56$ is recommended.

## REFERENCES

[1]   Bedair, Said S., and Ingo Wolff, "Computer Oriented Analytic Formulas for the Quasi-TEM Parameters of Broadside-Coupled Coplanar Waveguides with a General Dielectric for MMICs," *18th European Microwave Conference*

*Proceedings*, Stockholm, Sweden, 1988, pp.1139–1144. (Omits the factor of 2 in front of $C_{0,1}$, $C_{0,2}$, and in $C_e$ .)

[2]    Bedair, Said S., and Ingo Wolff, "Fast and Accurate Analytic Formulas for Calculating the Parameters of a General Broadside-Coupled Coplanar Waveguide for (M)MIC Applications," *IEEE Transactions on Microwave Theory and Techniques*, Vol. MTT-37, No. 5, May 1989, pp. 843–850.

## 4.4.2   Edge-Coupled  CPW

Edge-coupled coplanar waveguide (CPW) has all of the advantages of the planarity of the coplanar structure together with an increased possible isolation over microstrip line structures.

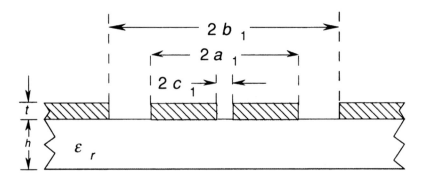

**Figure  4.4.2.1:**        **Edge-Coupled  CPW**

$$Z_{0,o} = \frac{\eta_0}{2\sqrt{\varepsilon_{eff}}} \frac{K'(k_5)}{K(k_5)} \frac{c}{754} \quad (\Omega) \tag{4.4.2.1}$$

$$Z_{0,e} = 2 Z_0 k_1 \frac{K(k_7)}{K'(k_7)} \frac{K'(k_1)}{K(k_1)} \quad (\Omega) \tag{4.4.2.2}$$

$$\varepsilon_{eff} = \sqrt{\frac{\varepsilon_r + 1}{2}} \tag{4.4.2.3}$$

where

$$k = \frac{Z_{0,e} - Z_{0,o}}{Z_{0,e} + Z_{0,o}} \tag{4.4.2.4}$$

$$k_1 = \frac{a_1}{b_1} \tag{4.4.2.5}$$

$$k_5 = \frac{c_5}{a_5} \text{ which is found by solving } \frac{F(k_1,k_5)}{K(k_1)} = 1.0 - \frac{F(k_1,c_1/a_1)}{K(k_1)} \qquad (4.4.2.6)$$

$$k_7 = \frac{c_7}{b_7} \text{ which is found by solving } F(k_1,c_7/b_7) = 1.0 - \frac{F(k_1,c_1/a_1)}{K(k_1)}$$

$$(4.4.2.7)$$

where $F(k_1, m)$ is the elliptic integral of the first kind.

In Bandrand *et al.* [1] a coupled CPW is described that is fabricated as an active device (FET) to allow DC programmability of the coupling.

Another coupling configuration, Figure 4.4.2.2, with increased isolation between strips is described in [2]. Chang models this as two isolated strips coupled by a lumped capacitance.

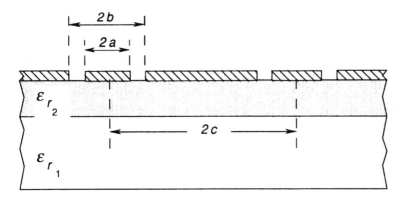

Figure 4.4.2.2:     Edge-Coupled CPW

The coupling of the configuration shown in Figure 4.4.2.2 can be calculated approximately with equations of Ghione and Naldi [3].

$$Z_{0,o} = Z_\infty \frac{K(k_8')}{K(k_8)} \frac{K(k_9)}{K(k_9')} \qquad (4.4.2.8)$$

$$Z_{0,e} = Z_\infty + \Delta Z_e \qquad (4.4.2.9)$$

where

$Z_\infty$ = impedance of an isolated CPW with dimensions $2\,a$, $2\,b$ (3.4.1.1)

$$\Delta Z_e = Z_\infty - Z_{0,o} \qquad (4.4.2.10)$$

$$k_9 = \frac{2.0 \sqrt{a\,b}}{a + b} \tag{4.4.2.11}$$

$$k_8 = \frac{k_9}{\sqrt{1.0 - \dfrac{(b - a)^2}{(2.0\,c)^2}}} \tag{4.4.2.12}$$

These equations assume loose coupling and that the substrate is infinitely thick.

REFERENCES

[1]    Bandrand, H., *et al.*, "Bias-Variable Characteristics of Coupled Coplanar Waveguide on GaAs Substrate," *Electronics Letters*, Vol. 23, No. 4, February 13, 1987, pp. 171–172.

[2]    Chang, Ching Ten, and Graham A. Garcia, "Crosstalk Between Two Coplanar Waveguides," *Archiv Für Elektronik und Übertragungstechnik*, Band 43, Heft 1, 1989, pp. 55–58.

[3]    Ghione, Giovanni, and Carlo U. Naldi, "Coplanar Waveguides for MMIC Applications: Effect of Upper Shielding Conductor Backing, Finite-Extent Ground Planes, and Line-to-Line Coupling," *IEEE Transactions on Microwave Theory and Techniques*, Vol. MTT-35, No. 3, March 1987, pp. 260–267.

[4]    Wen, Cheng P., "Coplanar-Waveguide Directional Couplers," *IEEE Transactions on Microwave Theory and Techniques*, Vol. MTT-18, No. 6, June 1970, pp. 318–322.

## 4.4.3  Edge-Coupled CPWG

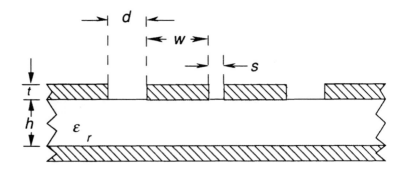

**Figure  4.4.3.1:**      **Edge-Coupled  CPWG**

$$Z_{0,o} = \frac{\eta_0}{\sqrt{\varepsilon_{eff,o}}} \left[ \frac{1.0}{2.0 \dfrac{K(k_o)}{K'(k_o)} + \dfrac{K(\beta_1)}{K'(\beta_1)}} \right] \quad (\Omega) \tag{4.4.3.1}$$

$$Z_{0,e} = \frac{\eta_0}{\sqrt{\varepsilon_{eff,e}}} \left[ \frac{1.0}{2.0 \dfrac{K(k_e)}{K'(k_e)} + \dfrac{K(\beta_1 k_1)}{K'(\beta_1 k_1)}} \right] \quad (\Omega) \tag{4.4.3.2}$$

$$\varepsilon_{eff,o} = \frac{2.0 \, \varepsilon_r \dfrac{K(k_o)}{K'(k_o)} + \dfrac{K(\beta_1)}{K'(\beta_1)}}{2.0 \dfrac{K(k_o)}{K'(k_o)} + \dfrac{K(\beta_1)}{K'(\beta_1)}} \tag{4.4.3.3}$$

$$\varepsilon_{eff,e} = \frac{2.0 \, \varepsilon_r \dfrac{K(k_e)}{K'(k_e)} + \dfrac{K(\beta_1 k_1)}{K'(\beta_1 k_1)}}{2.0 \dfrac{K(k_e)}{K'(k_e)} + \dfrac{K(\beta_1 k_1)}{K'(\beta_1 k_1)}} \tag{4.4.3.4}$$

where

$$k_o = \Lambda \, \frac{-\sqrt{\Lambda^2 - t_c^2} + \sqrt{\Lambda^2 - t_B^2}}{t_B \sqrt{\Lambda^2 - t_c^2} + t_c \sqrt{\Lambda^2 - t_B^2}} \tag{4.4.3.5}$$

$$k_e = \Lambda' \, \frac{-\sqrt{\Lambda'^2 - t'_c^2} + \sqrt{\Lambda'^2 - t'_B^2}}{t'_B \sqrt{\Lambda'^2 - t'_c^2} + t'_c \sqrt{\Lambda'^2 - t'_B^2}} \tag{4.4.3.6}$$

$$\Lambda = \frac{\sinh^2 \left[ \dfrac{\pi \, (s \, / \, 2.0 + w + d)}{2.0 \, h} \right]}{2} \tag{4.4.3.7}$$

$$t_c = \sinh^2 \left[ \frac{\pi \, (s \, / \, 2.0 + w)}{2.0 \, h} \right] - \Lambda \tag{4.4.3.8}$$

$$t_B = \sinh^2 \left( \frac{\pi \, s}{4.0 \, h} \right) - \Lambda \tag{4.4.3.9}$$

$$\Lambda' = \frac{\cosh^2\left[\dfrac{\pi\ (s\ /\ 2.0\ +\ w\ +\ d)}{2.0\ h}\right]}{2.0} \tag{4.4.3.10}$$

$$t'_c = \sinh^2\left[\frac{\pi\ (s\ /\ 2.0\ +\ w\ )}{2.0\ h}\right] - \Lambda' + 1.0 \tag{4.4.3.11}$$

$$t'_B = \sinh^2\left[\frac{\pi\ s}{4.0\ h}\right] - \Lambda' + 1.0 \tag{4.4.3.12}$$

$$\beta_1 = \sqrt{\frac{1.0 - y^2}{1.0 - k_1{}^2\ y^2}} \tag{4.4.3.13}$$

$$y = \frac{s}{s\ +\ 2.0\ w} \tag{4.4.3.14}$$

$$k_1 = \frac{s\ +\ 2.0\ w}{s\ +\ 2.0\ w\ +\ 2.0\ d} \tag{4.4.3.15}$$

To guarantee coplanar propagation,

$s + 2.0\ w + 2.0\ d \le h.$

The analysis is a conformal mapping technique, so the equations are exact for low frequency calculation.

## REFERENCES

[1]    Chang, Ching Ten, and Graham A. Garcia, "Crosstalk Between Two Coplanar Waveguides," *Archiv Für Elektronik und Übertragungstechnik*, Band 43, Heft 1, 1989, pp. 55–58. (Edge-coupled CPW's separated by ground plane.)

[2]    Hanna, Victor Fouad, "Parameters of Coplanar Directional Couplers with Lower Ground Plane," *15th European Microwave Conference Proceedings*, 1985, pp. 820–825. ([2] dropped primes in (8), corrected here.)

## 4.5    COUPLED MICROSTRIP LINES

### 4.5.1    Edge-Coupled Microstrip Lines

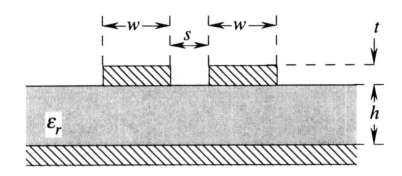

**Figure 4.5.1.1: Coupled Microstrip Line**

Kirschning and Jansen [8] analyzed this structure with a rigorous spectral-domain hybrid mode calculation. The results of this were then fit numerically. These equations are:

$$Z_{0,e}(0) = \frac{Z_0(0) \sqrt{\dfrac{\varepsilon_{eff}(0)}{\varepsilon_{eff,e}(0)}}}{1.0 - \dfrac{Z_0(0)}{\eta_0} \sqrt{\varepsilon_{eff}(0)}\, Q_4} \tag{4.5.1.1}$$

$$Z_{0,o}(0) = \frac{Z_0(0) \sqrt{\dfrac{\varepsilon_{eff}(0)}{\varepsilon_{eff,o}(0)}}}{1.0 - \dfrac{Z_0(0)}{\eta_0} \sqrt{\varepsilon_{eff}(0)}\, Q_{10}} \tag{4.5.1.2}$$

where

$$Z_0(f) = Z_0(0) \left(\frac{R_{13}}{R_{14}}\right)^{R_{17}}$$

$Z_0(0)$ = dc value (3.5.1.1) of a single, isolated strip

$\varepsilon_{eff}(f_n)$ = frequency-dependent effective relative permittivity of a single, isolated strip (3.5.1.7)

$$Q_0 = R_7 \left[ 1.0 - \frac{1.1241\, R_{12}}{R_{16}}\, e^{\left(-0.026\, f_n^{1.15656} - R_{15}\right)} \right] \tag{4.5.1.3a}$$

$$Q_1 = 0.8695\, u^{\,0.194} \tag{4.5.1.3b}$$

$$Q_2 = 1.0 + 0.7519\, g + 0.189\, g^{\,2.31} \tag{4.5.1.4}$$

$$Q_3 = 0.1975 + \left[ 16.6 + (8.4\,/\,g)^{6.0} \right]^{-0.387} + \frac{\ln\left[ \dfrac{g^{10}}{1.0 + (g\,/\,3.4)^{10}} \right]}{241} \tag{4.5.1.5}$$

$$Q_4 = \frac{\dfrac{2\, Q_1}{Q_2}}{e^{-8}\, u^{Q_3} + (2.0 - e^{-8})\, u^{-Q_3}} \tag{4.5.1.6}$$

$$Q_5 = 1.794 + 1.14 \ln \left( 1.0 + \frac{0.638}{g + 0.517\, g^{2.43}} \right) \tag{4.5.1.7}$$

$$Q_6 = 0.2305 + \frac{\ln\left[ \dfrac{g^{10}}{1.0 + (g\,/\,5.8)^{10}} \right]}{281.3} + \frac{\ln\left(1.0 + 0.598\, g^{1.154}\right)}{5.1} \tag{4.5.1.8}$$

$$Q_7 = \frac{10.0 + 190.0\, g^2}{1.0 + 82.3\, g^3} \tag{4.5.1.9}$$

$$Q_8 = e^{\left[ -6.5 - 0.95 \ln (g) - (g\,/\,0.15)^5 \right]} \tag{4.5.1.10}$$

$$Q_9 = \ln (Q_7)\left( Q_8 + 1\,/\,16.5 \right) \tag{4.5.1.11}$$

$$Q_{10} = \frac{Q_2 Q_4 - Q_5\, e^{\left[ \ln (u)\, Q_6\, u^{-Q_9} \right]}}{Q_2} \tag{4.5.1.12}$$

$$R_1 = 0.03891\, \varepsilon_r^{1.4} \tag{4.5.1.12b}$$

$$R_2 = 0.267\, u^{7.0} \tag{4.5.1.12c}$$

$$R_3 = 4.766 \exp(-3.228\, u^{0.641})$$

$$R_4 = 0.016 + (0.0514\ \varepsilon_r)^{4.524}$$

$$R_5 = (f_n\ /\ 28.843)^{12.0}$$

$$R_6 = 22.20\ u^{1.92}$$

$$R_7 = 1.206 - 0.3144\ e^{-R_1}\left[1.0 - e^{-R_2}\right] \tag{4.5.1.12d}$$

$$R_8 = 1.0 + 1.275\left\{1.0 - \exp\left[-0.004625\ R_3\ \varepsilon_r^{1.674}\ (f_n\ /\ 18.365)^{2.745}\right]\right\}$$

$$R_9 = 5.086\ R_4\ \frac{R_5}{0.3838 + 0.386\ R_4}\ \frac{e^{-R_6}}{1.0 + 1.2992\ R_5}\ \frac{(\varepsilon_r - 1.0)^6}{1.0 + 10.0\ (\varepsilon_r - 1.0)^6}$$

$$R_{10} = 0.00044\ \varepsilon_r^{2.136} + 0.0184 \tag{4.5.1.12e}$$

$$R_{11} = \frac{\left(\dfrac{f_n}{19.47}\right)^{6.0}}{\left[1.0 + 0.0962\left(\dfrac{f_n}{19.47}\right)^{6.0}\right]} \tag{4.5.1.12f}$$

$$R_{12} = \frac{1.0}{1.0 + 0.00245\ u^2} \tag{4.5.1.12g}$$

$$R_{13} = 0.9408\ \varepsilon_{eff}(f_n)^{R_8} - 0.9603$$

$$R_{14} = (0.9408 - R_9)\ \varepsilon_{eff}(0)^{R_8} - 0.9603$$

$$R_{15} = 0.707\ R_{10}\left(\frac{f_n}{12.3}\right)^{1.097} \tag{4.5.1.12h}$$

$$R_{16} = 1.0 + 0.0503\ \varepsilon_r^2\ R_{11}\left\{1.0 - \exp\left[-(u\ /\ 15)^6\right]\right\} \tag{4.5.1.12i}$$

$$R_{17} = Q_0$$

The frequency-dependent even and odd mode impedances are calculated from the above with:

$$Z_{0,e}\,(f_n) = \frac{Z_{0,e}(0)\left\{0.9408\left[\varepsilon_{eff}(f_n)\right]^{C_e} - 0.9603\right\}^{Q_0}}{\left\{(0.9408 - d_e)\left[\varepsilon_{eff}(0)\right]^{C_e} - 0.9603\right\}^{Q_0}} \qquad (4.5.1.13)$$

$$Z_{0,o}\,(f_n) = Z_0(f_n) + \frac{Z_{0,o}\,(0)\left[\dfrac{\varepsilon_{eff,o}(f_n)}{\varepsilon_{eff,o}(0)}\right]^{Q_{22}} - Z_0(f_n)\,Q_{23}}{1.0 + Q_{24} + (0.46\ g)^{2.2}\,Q_{25}} \qquad (4.5.1.14)$$

where

$$C_e = 1.0 + 1.275\left[1.0 - e^{-0.004625\ p_e\ \varepsilon_r^{1.674}\left(\frac{f_n}{18.365}\right)^{2.745}}\right]$$

$$- Q_{12} + Q_{16} - Q_{17} + Q_{18} + Q_{20} \qquad (4.5.1.15)$$

$$d_e = 5.086\ q_e\ \frac{r_e}{0.3838 + 0.386\ q_e}\ \frac{e^{-22.2\ u^{1.92}}}{1.0 + 1.2992\ r_e}\ \frac{(\varepsilon_r - 1.0)^6}{1.0 + 10.0\ (\varepsilon_r - 1.0)^6}$$

$$(4.5.1.16)$$

$$p_e = 4.766\ e^{-3.228\ u^{0.641}} \qquad (4.5.1.17)$$

$$q_e = 0.016 + \left(0.0514\ \varepsilon_r\ Q_{21}\right)^{4.524} \qquad (4.5.1.18)$$

$$r_e = \left(\frac{f_n}{28.843}\right)^{12} \qquad (4.5.1.19)$$

$$Q_{11} = 0.893\left[1.0 - \frac{0.3}{1.0 + 0.7(\varepsilon_r - 1.0)}\right] \qquad (4.5.1.20)$$

$$Q_{12} = 2.121\ \frac{(f_n/20)^{4.91}}{\left[1.0 + Q_{11}\ (f_n/20)^{4.91}\right]}\ \exp(-2.87\ g)\ g^{0.902} \qquad (4.5.1.21)$$

$$Q_{13} = 1.0 + 0.038\ (\varepsilon_r/8)^{5.1} \qquad (4.5.1.22)$$

$$Q_{14} = 1.0 + \frac{1.203\ (\varepsilon_r/15)^4}{1.0 + (\varepsilon_r/15)^4} \qquad (4.5.1.23)$$

$$Q_{15} = \frac{1.887 \, e^{-1.5 \, g^{0.84}} \, g^{Q_{14}}}{1.0 + \dfrac{0.41 \, (f_n / 15)^3 \, u^{(2 / Q_{13})}}{0.125 + u^{(1.626 / Q_{13})}}} \tag{4.5.1.24}$$

$$Q_{16} = Q_{15} \left[ 1.0 + \frac{9.0}{1.0 + 0.403 \, (\varepsilon_r - 1.0)^2} \right] \tag{4.5.1.25}$$

$$Q_{17} = 0.394 \left[ 1.0 - e^{-1.47 \, (u / 7)^{0.672}} \right] \left[ 1.0 - e^{-4.25 \, (f_n / 20)^{1.87}} \right] \tag{4.5.1.26}$$

$$Q_{18} = 0.61 \, \frac{1.0 - e^{-2.13 \, (u / 8)^{1.593}}}{1.0 + 6.544 \, g^{4.17}} \tag{4.5.1.27}$$

$$Q_{19} = \frac{0.21 \, g^4}{\left( 1.0 + 0.18 \, g^{4.9} \right) \left( 1.0 + 0.1 \, u^2 \right) \left[ 1.0 + (f_n / 24.0)^3 \right]} \tag{4.5.1.28}$$

$$Q_{20} = Q_{19} \left[ 0.09 + \frac{1.0}{1.0 + 0.1 \, (\varepsilon_r - 1.0)^{2.7}} \right] \tag{4.5.1.29}$$

$$Q_{21} = \left| 1.0 - \frac{42.54 \, g^{0.133} \, e^{-0.812 \, g} \, u^{2.5}}{1.0 + 0.033 \, u^{2.5}} \right| \tag{4.5.1.30}$$

$$Q_{22} = \frac{0.925 \, (f_n / Q_{26})^{1.536}}{1.0 + 0.3 \, (f_n / 30)^{1.536}} \tag{4.5.1.31}$$

$$Q_{23} = 1.0 + \frac{0.005 \, f_n \, Q_{27}}{\left[ 1.0 + 0.812 \, (f_n / 15.0)^{1.9} \right] \left( 1.0 + 0.025 \, u^2 \right)} \tag{4.5.1.32}$$

$$Q_{24} = 2.506 \, Q_{28} \, u^{0.894} \, \frac{\left[ \dfrac{(1.0 + 1.3 \, u) \, f_n}{99.25} \right]^{4.29}}{3.575 + u^{0.894}} \tag{4.5.1.33}$$

$$Q_{25} = \left( \frac{0.3 \, f_n^2}{10.0 + f_n^2} \right) \left[ 1.0 + \frac{2.333 \, (\varepsilon_r - 1.0)^2}{5.0 + (\varepsilon_r - 1.0)^2} \right] \tag{4.5.1.34}$$

$$Q_{26} = 30.0 - \frac{22.2 \left[ \frac{(\varepsilon_r - 1.0)}{13} \right]^{12}}{1.0 + 3.0 \left[ \frac{(\varepsilon_r - 1.0)}{13} \right]^{12}} - Q_{29} \qquad (4.5.1.35)$$

$$Q_{27} = 0.4 \, g^{0.84} \left[ 1.0 + \frac{2.5 \, (\varepsilon_r - 1.0)^{1.5}}{5.0 + (\varepsilon_r - 1.0)^{1.5}} \right] \qquad (4.5.1.36)$$

$$Q_{28} = \frac{0.149 \, (\varepsilon_r - 1.0)^3}{94.5 + 0.038 \, (\varepsilon_r - 1.0)^3} \qquad (4.5.1.37)$$

$$Q_{29} = \frac{15.16}{1.0 + 0.196 \, (\varepsilon_r - 1.0)^2} \qquad (4.5.1.38)$$

$$u = w / h \qquad (4.5.1.39)$$

$$g = s / h \qquad (4.5.1.40)$$

$$f_n = f h \qquad (4.5.1.41)$$

where $f$ is the frequency in GHz and $h$ is the substrate thickness in mm. The effective dielectric constants for the even and odd modes are calculated by beginning with the dc $\varepsilon_{eff}$ for a single strip derived in [5]:

$$\varepsilon_{eff}(0) = (3.5.1.2) \text{ and } (3.5.1.3) \qquad (4.5.1.42)$$

$$\varepsilon_{eff}(f_n) = (3.5.1.7) \qquad (4.5.1.42a)$$

where

$$v = \frac{u \, (20.0 + g^2)}{10.0 + g^2} + g \, e^{-g} \quad (4.5.1.43)$$

$$a_e(v) = 1.0 + \frac{\ln \left[ \frac{v^4 + (v / 52.0)^2}{v^4 + 0.432} \right]}{49.0} + \frac{\ln \left[ 1.0 + (v / 18.1)^3 \right]}{18.7} \qquad (4.5.1.44)$$

$$b_e(\varepsilon_r) = 0.564 \left( \frac{\varepsilon_r - 0.9}{\varepsilon_r + 3.0} \right)^{0.053} \qquad (4.5.1.45)$$

The even and odd mode effective relative dielectric constants at dc are then calculated:

$$\varepsilon_{eff,o}(0) = \left[0.5\ (\varepsilon_r + 1.0) + a_0(u,\ \varepsilon_r) - \varepsilon_{eff}(0)\right] e^{-c_0\ g^{d_0}} + \varepsilon_{eff}(0) \qquad (4.5.1.46)$$

$$\varepsilon_{eff,e}(0) = 0.5\ (\varepsilon_r + 1.0) + 0.5\ (\varepsilon_r - 1.0)\left[\left(1.0 + \frac{10.0}{v}\right)^{-a_e(v)\ b_e(\varepsilon_r)}\right] \qquad (4.5.1.47)$$

where

$$a_0(u,\ \varepsilon_r) = 0.7287\ \left[\varepsilon_{eff}(0) - 0.5\ (\varepsilon_r + 1.0)\right](1.0 - e^{-0.179\ u}) \qquad (4.5.1.48)$$

$$b_0(\varepsilon_r) = \frac{0.747\ \varepsilon_r}{0.15 + \varepsilon_r} \qquad (4.5.1.49)$$

$$c_0 = b_0(\varepsilon_r) - \left[b_0(\varepsilon_r) - 0.207\right]e^{-0.414\ u} \qquad (4.5.1.50)$$

$$d_0 = 0.593 + 0.694\ e^{-0.562\ u} \qquad (4.5.1.51)$$

These can be corrected for frequency-dependent dispersion with:

$$\varepsilon_{eff,o}(f_n) = \varepsilon_r - \frac{\varepsilon_r - \varepsilon_{eff,o}(0)}{1.0 + F_o(f_n)} \qquad (4.5.1.52)$$

$$\varepsilon_{eff,e}(f_n) = \varepsilon_r - \frac{\varepsilon_r - \varepsilon_{eff,e}(0)}{1.0 + F_e(f_n)} \qquad (4.5.1.53)$$

where

$$F_o(f_n) = P_1\ P_2\ \left[(P_3\ P_4 + 0.1844)\ f_n\ P_{15}\right]^{1.5763} \qquad (4.5.1.54)$$

$$F_e(f_n) = P_1\ P_2\ \left[(P_3\ P_4 + 0.1844\ P_7)\ f_n\right]^{1.5763} \qquad (4.5.1.55)$$

$$P_1 = 0.27488 + \left[0.6315 + \frac{0.525}{(1.0 + 0.0157\ f_n)^{20}}\right] u - 0.065683\ e^{-8.7513\ u}$$

$$(4.5.1.56)$$

$$P_2 = 0.33622\ \left(1.0 - e^{-0.03442\varepsilon_r}\right) \qquad (4.5.1.57)$$

$$P_3 = 0.0363\ e^{-4.6\ u}\left[1 - e^{-(f_n/38.7)^{4.97}}\right] \qquad (4.5.1.58)$$

$$P_4 = 1.0 + 2.751 \left[ 1.0 - e^{-(\varepsilon_r / 15.916)^8} \right] \tag{4.5.1.59}$$

$$P_5 = 0.334 \, e^{-3.3(\varepsilon_r / 15)^3} + 0.746 \tag{4.5.1.60}$$

$$P_6 = P_5 \, e^{-(f_n / 18)^{0.368}} \tag{4.5.1.61}$$

$$P_7 = 1.0 + 4.069 \, P_6 \, g^{0.479} \, e^{(-1.347 \, g^{0.595} - 0.17 \, g^{2.5})} \tag{4.5.1.62}$$

$$P_8 = 0.7168 \left[ 1.0 + \frac{1.076}{1.0 + .0576 \, (\varepsilon_r - 1.0)} \right] \tag{4.5.1.63}$$

$$P_9 = P_8 - 0.7913 \left[ 1.0 - e^{-(f_n / 20)^{1.424}} \right] \tan^{-1} \left[ 2.481 \, (\varepsilon_r / 8)^{0.946} \right] \tag{4.5.1.64}$$

$$P_{10} = 0.242 \, (\varepsilon_r - 1.0)^{0.55} \tag{4.5.1.65}$$

$$P_{11} = 0.6366 \left( e^{-3.401 \, f_n} - 1.0 \right) \tan^{-1} \left[ 1.263 \, (u / 3)^{1.629} \right] \tag{4.5.1.66}$$

$$P_{12} = P_9 + \frac{1.0 - P_9}{1.0 + 1.183 \, u^{1.376}} \tag{4.5.1.67}$$

$$P_{13} = \frac{1.695 \, P_{10}}{0.414 + 1.605 \, P_{10}} \tag{4.5.1.68}$$

$$P_{14} = 0.8928 + 0.10722 \left[ 1.0 - e^{-0.42(f_n / 20.0)^{3.215}} \right] \tag{4.5.1.69}$$

$$P_{15} = \left| 1.0 - \frac{0.8928 \, (1.0 + P_{11}) \, P_{12} \, e^{-P_{13} \, g^{1.092}}}{P_{14}} \right| \tag{4.5.1.70}$$

Equations are valid for:    $0.1 \le w / h \le 10.0$

$$0.1 \le s / h \le 10.0$$

$$1.0 \le \varepsilon_r \le 18$$

and the stated accuracy for the range $\varepsilon_r \le 12.9$ and $f_n \le 15$ is better than 1.5%.

The above equations are lengthy; they can be simplified to their dispersionless form in many applications. PCB dimensions are not usually practical for less than about 6 dB of coupling (see the section on Lange couplers for tighter couplings). Other coupling structures (*i.e.*, branch-line hybrids, Lange couplers, Wilkinson dividers, and directional couplers) can be used when tighter or wideband coupling is required.

REFERENCES

[1]     Bochtler, U. and F. Endress, "CAD Program Designs Stripline Couplers," *Microwaves & RF*, December 1986, pp. 91–95.

[2]     Bryant, T.G., and J.A. Weiss, "Parameters of Microstrip Transmission Lines and of Coupled Pairs of Microstrip Lines," *IEEE Transactions on Microwave Theory and Techniques*, Vol. MTT-16, December 1968, pp. 1021–1027.

[3]     Garg, Ramesh, and I.J. Bahl, "Characteristics of Coupled Microstriplines," *IEEE Transactions on Microwave Theory and Techniques*, Vol. MTT-27, No. 7, July 1979, pp. 700–705 and "Correction to 'Characteristics of Coupled Microstriplines,'" Vol. MTT-28, No. 3, March 1980, p. 272.

[4]     Getsinger, W.J., "Dispersion of Parallel-Coupled Microstrip," *IEEE Transactions on Microwave Theory and Techniques*, Vol. MTT-23, No. 3, 1973, pp. 144–145.

[5]     Hammerstad, E., and Ø. Jensen, "Accurate Models for Microstrip Computer-Aided Design," *1980 IEEE MTT-S International Microwave Symposium Digest*, pp. 407–409.

[6]     Hinton, J.H., "On Design of Coupled Microstrip Lines," *IEEE Transactions on Microwave Theory and Techniques*, Vol. MTT-28, No. 3, March 1980, pp. 272.

[7]     Horno, Manuel, and Ricardo Marqués, "Coupled Microstrips on Double Anisotropic Layers," *IEEE Transactions on Microwave Theory and Techniques*, Vol. 32, No. 4, April 1984, pp. 467–470.

[7a]    Jansen, Rolf, and Martin Kirschning, "Arguments and an Accurate Model for the Power-Current Formula of Microstrip Characteristic Impedance,"*Archiv Für Elektronik und Übertragungstechnik*, Band 37, Heft 3/4, 1983, pp. 108–112.

[8]     Kirschning, Manford, and Rolf H. Jansen, "Accurate Wide-Range Design Equations for the Frequency-Dependent Characteristic of Parallel Coupled Microstrip Lines," *IEEE Transactions on Microwave Theory and Techniques*, Vol. MTT-32, No. 1, January 1984, pp. 83–90.

[9]     Krage, Mark K., and George I. Haddad, "Characteristics of Coupled Microstrip Transmission Lines-I: Coupled Mode Formulation of Inhomogeneous Line," *IEEE Transactions on Microwave Theory and Techniques*, MTT-18, No. 4, April 1970, pp. 217-222.

[10]    Krage, Mark K., and George I. Haddad, "Characteristics of Coupled Microstrip Transmission Lines-II: Evaluation of Coupled-Line Parameters," *IEEE Transactions on Microwave Theory and Techniques*, MTT-18, No. 4, April 1970, pp. 222-228.

[11]    Lau, Wai Yuen, "Network Analysis Verifies Models in CAD Packages," *Microwaves & RF*, November 1989, pp. 99–110.

[12]    Lev, James J., "Calculator Program Synthesizes Microstrip Coupled Lines," *Microwaves & RF*, May 1984, pp. 215–220.

[13]    March, Steven, "Microstrip Packaging: Watch the Last Step," *Microwaves*, December 1981, pp. 83–94.

[14]    Mirshekar-Syahkal, D., "An Accurate Determination of Dielectric Loss Effect in Monolithic Microwave Integrated Circuits Including Microstrip and Coupled Microstrip Lines," *IEEE Transactions on Microwave Theory and Techniques*, Vol. MTT-31, No. 11, November 1983, pp. 950–954.

[15]    Quaglia, Antonio, "Use Coupled Lines to Achieve High Impedances in Microstrip," *Microwaves & RF*, October 1983, pp. 120–127.

[16]    Rao, B. Rama, "Effect of Loss and Frequency Dispersion on the Performance of Microstrip Directional Couplers and Coupled Line Filters," *IEEE Transactions on Microwave Theory and Techniques*, Vol. 22, No. 8, July 1974, p. 747–750.

[17]    Rosloniec, Stanislaw, "Design of Coupled Microstrip Lines by Optimization Methods," *IEEE Transactions on Microwave Theory and Techniques*, Vol. MTT-35, No. 11, November 1987, pp. 1072–1074.

[18]    Tripathi, V.K., "Loss Calculations for Coupled Transmission-Line Structures," *IEEE Transactions on Microwave Theory and Techniques*, Vol. MTT-20, No. 2, February 1972, pp. 178–180.

[19]    Zehentner, Jan, "Analysis and Synthesis of Coupled Microstrip Lines by Polynomials," *Microwave Journal*, May 1980, pp. 95–110.

## 4.5.2   Asymmetric Coupled Microstrip Line

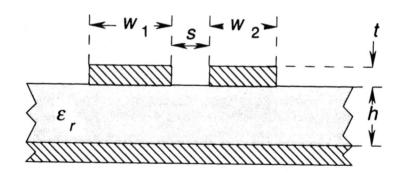

**Figure  4.5.2.1:**      **Asymmetric  Coupled  Microstrip  Lines**

This configuration has received a great deal of attenuation for its potential applications in filters, etc. Current designs are based on symmetric couplers and the asymmetry would give us an additional degree of freedom in design. Asymmetric couplers allow the coupling of power at the same time as an impedance transformation takes place.

Unfortunately, closed-form equations for microstrip line do not exist. The reader with sufficient skill may be able to adapt the curves or analyses presented in the references. EM programs may also be used.

The general relationships for coupling and impedances in asymmetric couplers can be defined [1]

$$Z_{0,e}{}^a = \frac{\sqrt{1.0 - k^2}}{G_a - k \sqrt{G_a G_b}} \tag{4.5.2.1}$$

$$Z_{0,o}{}^a = \frac{\sqrt{1.0 - k^2}}{G_a - k \sqrt{G_a G_b}} \tag{4.5.2.2}$$

$$Z_{0,e}{}^b = \frac{\sqrt{1.0 - k^2}}{G_a - k \sqrt{G_a G_b}} \tag{4.5.2.3}$$

$$Z_{0,o}{}^b = \frac{\sqrt{1.0 - k^2}}{G_a - k \sqrt{G_a G_b}} \tag{4.5.2.4}$$

$$k = \frac{Y^a_{0,o} - Y^a_{0,e}}{\sqrt{\left(Y^a_{0,o} + Y^a_{0,e}\right)\left(Y^b_{0,o} - Y^b_{0,e}\right)}} \tag{4.5.2.5}$$

where the superscript $a$ corresponds to the impedance of line 2 measured at port 1 and $b$ corresponds to measurements at port 2. For a matched condition to exist the ports are terminated with $R_a$ and $R_b$ such that

$$G_a G_b = \frac{1.0}{4.0}\left[\left(Y^a_{0,o} + Y^a_{0,e}\right)\left(Y^b_{0,o} + Y^b_{0,e}\right) - \left(Y^a_{0,o} + Y^a_{0,e}\right)^2\right] \tag{4.5.2.6}$$

$$\frac{R_b}{R_a} = \frac{Y^a_{0,o} + Y^a_{0,e}}{Y^b_{0,o} + Y^b_{0,e}} \tag{4.5.2.7}$$

The limitations on impedance transformation for a given coupling are set by the following equations

$$k \le \sqrt{\frac{R_a}{R_b}} \qquad \text{and} \qquad k \le \sqrt{\frac{R_b}{R_a}} \tag{4.5.2.8}$$

REFERENCES

[1]    Cristal, Edward G., "Coupled-Transmission-Line Directional Couplers with Coupled Lines of Unequal Characteristic Impedances," *IEEE Transactions on Microwave Theory and Techniques*, Vol. MTT-14, No. 7, July 1966, pp. 337–346.

[2]    Jansen, R.H., "Fast Accurate Hybrid Mode Computation of Non-Symmetrical Coupled Microstrip Characteristics," *7th European Microwave Conference Proceedings*, 1977, pp. 135–139.

[3]    Kal, S., *et al.*, "Normal-Mode Parameters of Microstrip Coupled Lines of Unequal Width," *IEEE Transactions on Microwave Theory and Techniques*, Vol. MTT-32, No. 2, February 1984, pp. 198–200.

[4]    Sachse, Krzysztof, "The Scattering Parameters and Directional Coupler Analysis of Characteristically Terminated Asymmetric Coupled Transmission Lines in an Inhomogeneous Medium," *IEEE Transactions on Microwave Theory and Techniques*, Vol. MTT-38, No. 4, April 1990, pp. 4117–425.

[5]    Tripathi, Vijai K., "Asymmetric Coupled Transmission Lines in an Inhomogenous Medium," *IEEE Transactions on Microwave Theory and Techniques*, Vol. MTT-23, No. 9, September 1975, pp. 734–739.

[6]    Tripathi, Vijai K., and C.L. Chang, "Quasi-TEM Parameters of Non-Symmetrical Coupled Microstrip Lines," *International Journal of Electronics*, Vol. 45, No. 2, 1978, pp. 215–223.

## 4.5.3    Covered Coupled Microstrip Line

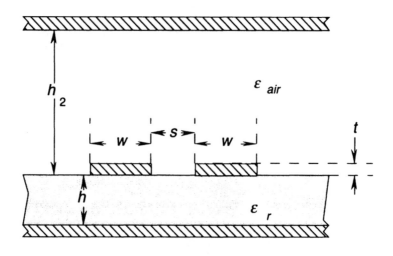

**Figure  4.5.3.1:**        **Covered  Coupled  Microstrip  Line**

March [3, 4] proposes a technique that corrects and solves a number of problems with previous solutions. The equations of Hammerstad and Jensen [2] are modified for the presence of the cover by a series of corrections. Results were verified by comparison with MSTRIP simulations. The equations of Hammerstad and Jensen are:

$$Z_{0,e}(u, g) = \frac{Z_0(u)}{1.0 - Z_0(u)\,\Phi_e(u, g)\,/\,\eta_0} \quad (\Omega) \tag{4.5.3.1}$$

$$Z_{0,o}(u, g) = \frac{Z_0(u)}{1.0 - Z_0(u)\,\Phi_o(u, g)\,/\,\eta_0} \quad (\Omega) \tag{4.5.3.2}$$

$$g = s\,/\,h \tag{4.5.3.3}$$

$$u = w\,/\,h \tag{4.5.3.4}$$

$$\varepsilon_{eff,e}(u, g, \varepsilon_r) = \frac{\varepsilon_r + 1.0}{2.0} + F_e(u, g, \varepsilon_r)\frac{\varepsilon_r - 1.0}{2.0} \tag{4.5.3.5}$$

$$\varepsilon_{eff,o}(u, g, \varepsilon_r) = \frac{\varepsilon_r + 1.0}{2.0} + F_o(u, g, \varepsilon_r)\frac{\varepsilon_r - 1.0}{2.0} \tag{4.5.3.6}$$

$$F_e(u, g, \varepsilon_r) = \left[1.0 + \frac{10.0}{\mu(u, g)}\right]^{-a(\mu)\,b(\varepsilon_r)} \tag{4.5.3.7}$$

$$F_o(u, g, \varepsilon_r) = f_0(u, g, \varepsilon_r)\left[1.0 + \frac{10.0}{\mu(u, g)}\right]^{-a(\mu)\,b(\varepsilon_r)} \tag{4.5.3.8}$$

$$\Phi_e(u, g) = \frac{\varphi(u)}{\left(\psi(g)\left\{\alpha(g)\,u^{m(g)} + [1.0 - \alpha(g)]\,u^{-m(g)}\right\}\right)} \tag{4.5.3.9}$$

$$\varphi(u) = 0.8645\,u^{0.172} \tag{4.5.3.10}$$

$$\psi(g) = 1.0 + g\,/\,1.45 + \frac{g^{2.09}}{3.95} \tag{4.5.3.11}$$

$$\alpha(g) = 0.5\,e^{-g} \tag{4.5.3.12}$$

$$m(g) = 0.2175 + \left[4.113 + \left(\frac{20.36}{g}\right)^6\right]^{-0.251} + \left(\frac{1.0}{323.0}\right)\ln\left[\frac{g^{10}}{1.0 + \left(\frac{g}{13.8}\right)^{10}}\right]$$

$$(4.5.3.13)$$

$$\phi_o(u, g) = \phi_o(u, g) - \frac{\theta(g)}{\psi(g)} \; e^{\beta(g) \, u^{-n(g)} \, \ln u}$$

$$(4.5.3.14)$$

$$\theta(g) = 1.729 + 1.175 \ln\left(1.0 + \frac{0.627}{g + 0.327 \, g^{2.17}}\right)$$

$$(4.5.3.15)$$

$$\beta(g) = 0.2306 + \frac{1.0}{301.8} \ln\left[\frac{g^{10}}{1.0 + \left(\frac{g}{3.73}\right)^{10}}\right] + \frac{1.0}{5.3} \ln\left(1.0 + 0.646 \, g^{1.175}\right)$$

$$(4.5.3.16)$$

$$n(g) = \left\{\frac{1.0}{17.7} + e^{\left[-6.424 - 0.76 \ln g - \left(\frac{g}{0.23}\right)^5\right]}\right\} \ln\left(\frac{10.0 + 68.3 \, g^2}{1.0 + 32.5 \, g^{3.093}}\right)$$

$$(4.5.3.17)$$

$$\mu(u, g) = g \, e^{-g} + u \, \frac{20.0 + g^2}{10.0 + g^2}$$

$$(4.5.3.18)$$

$$f_0(u, g, \varepsilon_r) = f_{0,1}(g, \varepsilon_r) \, e^{p(g) \ln u + q(g) \sin\left(\frac{\pi \ln u}{\ln 10.0}\right)}$$

$$(4.5.3.19)$$

$$p(g) = \frac{e^{-0.745 \, g^{0.295}}}{\cosh\left(g^{0.15}\right)}$$

$$(4.5.3.20)$$

$$q(g) = e^{-1.366 - g}$$

$$(4.5.3.21)$$

$$f_{0,1}(g, \varepsilon_r) = 1.0 - e^{\left\{-0.179 \, g^{0.15} - \frac{0.328 \, g^{r(g, \varepsilon_r)}}{\ln\left[e^{1.0} + (g/7.0)^{2.8}\right]}\right\}}$$

$$(4.5.3.22)$$

$$r(g, \varepsilon_r) = 1.0 + 0.15\left\{1.0 - \frac{e^{\left[\frac{1.0 - (\varepsilon_r - 1.0)^{2.0}}{8.2}\right]}}{1.0 + g^{-6.0}}\right\}$$

$$(4.5.3.23)$$

The values of the modal impedances and effective dielectric constants agree with MSTRIP to better than ±0.5% for

$$0.1 \le w/h \le 10.0$$

$$0.01 \le s/h$$

$$1.0 \le \varepsilon_r \le 128.0$$

March corrects for the finite trace thickness by replacing $w$ in the above equations with equivalent widths $w_e'$ and $w_o'$ for the even and odd modes, respectively.

$$w_e' = w + \Delta w_e \tag{4.5.3.24}$$

$$s_e' = s - \Delta w_e \tag{4.5.3.25}$$

$$\Delta w_e = \Delta w \left[ 1.0 - 0.5 \; e^{(-0.69 \; \Delta w / \Delta t)^2} \right] \tag{4.5.3.26}$$

$$w_o' = w + \Delta w_o \tag{4.5.3.27}$$

$$s_o' = s - \Delta w_o \tag{4.5.3.28}$$

$$\Delta w_o = \Delta w_e + \Delta t \tag{4.5.3.29}$$

$$\Delta t = \frac{2.0 \; t \; h}{s \; (\varepsilon_r + 1.0)} \tag{4.5.3.30}$$

$$\Delta w = \frac{t}{\pi} \left\{ 1.0 + \ln 4.0 - 0.5 \; \ln \left[ \left( \frac{t}{h} \right)^2 + \left( \frac{t}{\pi \; w} \right)^2 \right] \right\} \tag{4.5.3.31}$$

Finite trace thickness corrections are valid for $s \ge 2.0 \; t$. The equations have been verified for $0.1 \le u \le 10.0$ and $g \ge 0.01$.

To correct for the presence of the cover when $h_2 / h \le 39.0$

$$q_{e,c} = \tanh \left( 1.626 + 0.107 \; h_2 / h \; - \frac{1.733}{\sqrt{h_2 / h}} \right) \tag{4.5.3.32}$$

and when $h_2 / h > 39.0$

$$q_{e,c} = 1.0 \tag{4.5.3.33}$$

and the odd mode is corrected for the cover when $h_2 / h \geq 7.0$ with

$$q_{o,c} = \tanh \left[ \frac{9.575}{7.0 - h_2 / h} - 2.965 + 1.68 \, h_2 / h - 0.311 \, (h_2 / h)^2 \right]$$

(4.5.3.34)

and for $h_2 / h < 7.0$

$$q_{o,c} = 1.0$$

(4.5.3.35)

These equations are accurate to ±0.4% and typically to ±0.2%. Even mode corrections for the cover are

$$\Delta Z_{0,e}(\varepsilon_r = 1.0) = f_e(w / h, h_2 / h) \, g_e(s / h, h_2 / h)$$

(4.5.3.36)

$$f_e (w / h, h_2 / h) = 1.0 - \tanh^{-1} \left[ A + \left( B + \frac{C}{h} \frac{w}{h} \right) \frac{w}{h} \right]$$

(4.5.3.37)

$$A = \frac{-4.351}{(1.0 + h_2 / h)^{1.842}}$$

(4.5.3.38)

$$B = \frac{6.639}{(1.0 + h_2 / h)^{1.861}}$$

(4.5.3.39)

$$C = \frac{-2.291}{(1.0 + h_2 / h)^{1.90}}$$

(4.5.3.40)

$$g_e (s / h, h_2 / h) = 270.0 \left[ 1.0 - \tanh \left( D + E \sqrt{1.0 + h_2 / h} - \frac{F}{1.0 + h_2 / h} \right) \right]$$

(4.5.3.41)

$$D = 0.747 \csc \left( \frac{\pi \, x}{2.0} \right)$$

(4.5.3.42)

$$E = 0.725 \sin \left( \frac{\pi \, y}{2.0} \right)$$

(4.5.3.43)

$$F = 10^{(0.11 - 0.0947 \, s / h)}$$

(4.5.3.44)

$$x = 10^{(0.103 \, s / h - 0.159)}$$

(4.5.3.45)

$$y = 10^{(0.0492 \ s \ / \ h \ - \ 0.073)} \tag{4.5.3.46}$$

and odd mode corrections for the cover are

$$\Delta Z_{0,o}(\varepsilon_r = 1.0) = f_o(w \ / \ h, h_2 \ / \ h) \ g_o(s \ / \ h, h_2 \ / \ h) \tag{4.5.3.47}$$

$$f_o \ (w \ / \ h, h_2 \ / \ h) = \left(\frac{w}{h}\right)^J \tag{4.5.3.48}$$

$$J = \tanh\left[\frac{(1.0 + h_2 \ / \ h)^{1.585}}{6.0}\right] \tag{4.5.3.49}$$

$$g_o \ (s \ / \ h, h_2 \ / \ h) = 270.0\left[1.0 - \tanh\left(G + K \ \sqrt{1.0 + \frac{h_2}{h}} - \frac{L}{1.0 + h_2 \ / \ h}\right)\right] \tag{4.5.3.50}$$

$$G = 2.7178 - 0.796 \ \frac{s}{h} \tag{4.5.3.51}$$

for $s \ / \ h > 0.858$

$$K = \log\left[20.492 \ \left(\frac{s}{h}\right)^{0.174}\right] \tag{4.5.3.52}$$

for $s \ / \ h \leq 0.858$

$$K = 1.30 \tag{4.5.3.53}$$

for $s \ / \ h > 0.873$

$$L = 2.51 \ \left(\frac{s}{h}\right)^{-0.462} \tag{4.5.3.54}$$

and for $s \ / \ h \leq 0.873$

$$L = 2.674 \tag{4.5.3.55}$$

Dispersion can now be accounted for by the remaining relations. First we replace $\varepsilon_{eff,e}$ $u, g, \varepsilon_r)$ and $\varepsilon_{eff,o} \ (u, g, \varepsilon_r)$, which are zero-frequency relations [we will denote them $_{eff,o}(0)$ and $\varepsilon_{eff,e}(0)$] with frequency-dependent relations

$$\varepsilon_{eff,o} \ (f) = \varepsilon_r - \frac{\varepsilon_r - \varepsilon_{eff,o} \ (0)}{1.0 - G_o \ (f \ / \ f_{p,o})^2} \tag{4.5.3.56}$$

$$\varepsilon_{eff,e}\ (f) \ = \varepsilon_r - \frac{\varepsilon_r - \varepsilon_{eff,e}\ (0)}{1.0 - G_e\ (f\ /\ f_{p,e})^2} \qquad (4.5.3.57)$$

where

$$G_e = \frac{\pi^2\ (\varepsilon_r - 1.0)}{12.0\ \varepsilon_{eff,e}(0)} \sqrt{\frac{Z_{0,e}}{120.0}} \qquad (4.5.3.58)$$

$$G_o = \frac{\pi^2\ (\varepsilon_r - 1.0)}{12.0\ \varepsilon_{eff,e}(0)} \sqrt{\frac{Z_{0,e}}{120.0}} \qquad (4.5.3.59)$$

$$f_{p,e} = \frac{Z_{0,e}}{4.0\ \mu_0\ H} \qquad (4.5.3.60)$$

$$f_{p,o} = \frac{Z_{0,o}}{\mu_0\ H} \qquad (4.5.3.61)$$

The even and odd mode impedances are similarly corrected. The zero-frequency relations $Z_{0,e}\ (u, g)$ and $Z_{0,o}\ (u, g)$ are denoted $Z_{0,e}\ (0)$ and $Z_{0,o}\ (0)$. At high frequencies the fields become completely contained in the dielectric ($\varepsilon_{eff}$ approaches $\varepsilon_r$) and the impedance of the line can be related to that of a stripline. Calculate the impedances of edge-coupled striplines (see later in this chapter) with the microstrip line dimensions as follows

$$Z_{0,e\ sl}\ (w, b = 2.0\ h, t, s) \qquad (4.5.3.62)$$

$$Z_{0,o\ sl}\ (w, b = 2.0\ h, t, s) \qquad (4.5.3.63)$$

We can now calculate the frequency-corrected impedances:

$$Z_{0,e}\ (f) = Z_{0,e\ sl} - \frac{2.0\ Z_{0,e\ sl} - Z_{0,e}\ (0)}{1.0 + G_e\ (f\ /\ f_{p,e})^{1.6}} \qquad (4.5.3.64)$$

$$Z_{0,o}\ (f) = Z_{0,o\ sl} - \frac{2.0\ Z_{0,o\ sl} - Z_{0,o}\ (0)}{1.0 + G_o\ (f\ /\ f_{p,o})^{1.6}} \qquad (4.5.3.65)$$

The accuracy of these corrections is better than ±0.5% to 18 GHz.

Although these equations are quite involved, we are rewarded with a general-purpose set of relations that describe a frequently encountered structure quite accurately. In fact since most microstrip lines are packaged one might use only this relation.

REFERENCES

[1]    Fikioris, John G., *et al.*, "Exact Solutions for Shielded Printed Microstrip Lines by the Carleman-Vekua Method," *IEEE Transactions on Microwave Theory and Techniques*, Vol. MTT-37, No. 1, January 1989, pp. 21–33.

[2]    Hammerstad, E., and Ø. Jensen, "Accurate Models for Microstrip Computer-Aided Design," *1980 IEEE MTT-S International Microwave Symposium Digest*, pp. 407–409.

[3]    March, Steven L., "Empirical Formulas for the Impedance and Effective Dielectric Constant of Covered Microstrip for use in the Computer-Aided Design of Microwave Integrated Circuits," *11th European Conference on Microwaves*, 1981, pp. 671–676. (March points out a number of errors in earlier articles dealing with shielded configurations.)

[4]    March, Steven, "Microstrip Packaging: Watch the Last Step," *Microwaves,* December 1981, pp. 83–94.

[5]    Wheeler, Harold A., "Transmission-Line Properties of a Strip on a Dielectric Sheet on a Plane," *IEEE Transactions on Microwave Theory and Techniques*, Vol. MTT-25, No. 8, August 1977, pp. 631–647.

## 4.5.4   Broadside Coupled Microstrip Lines

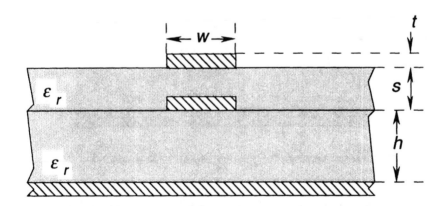

Figure  4.5.4.1:      Broadside  Coupled  Microstrip  Lines

The broadside coupled microstrip line structure is useful for tight coupling directional couplers and for differential transmission of signals.

Malone [1] reports good results with design of this structure by equating the capacitance of an equivalent broadside coupled stripline structure.  If the desired coupled line parameters are calculated for broadside coupled striplines (Figure 4.6.2.1), the dimensions of the equivalent microstrip line structure are then calculated:

$$h = \left(\frac{b_{sl} - s_{sl}}{4.0}\right) \tag{4.5.4.1}$$

$$s = s_{sl} \tag{4.5.4.2}$$

$$w = w_{sl} \tag{4.5.4.3}$$

where the dimensions of the stripline are as shown in Figure 4.6.2.1 and are indicated above with the subscript $sl$.

REFERENCES:

[1]   Malone, Hugh R., "Microstrip Overlay Coupler Suits Broadband Use," *Microwaves & RF*, July 1985, pp. 84–86.

## 4.5.5   Three Coupled Microstrip Lines

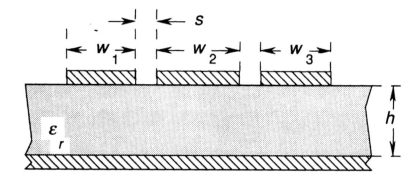

**Figure  4.5.5.1:**      **Three Coupled Microstrip Lines**

Three coupled lines have the potential to reduce space over two-line couplers where the signal is to be split into three paths. This structure has five mode impedances.

Abdallah and El-Deeb [1] present the relation of the widths for equal mode impedances in terms of a polynomial fit to numerical data. If $w_1 = w_3$, then

$$w_2 / w_1 = \sum_{n=0}^{3} B_n \left(\frac{s}{h}\right)^n \tag{4.5.5.1}$$

where

$$B_0 = 2.40599 - 3.06631 \, (w_1 \, / \, h \,) + 3.46389 \, (w_1 \, / \, h \,)^2 - 1.66833 \, (w_1 \, / \, h \,)^3$$

$$(4.5.5.2)$$

$$B_1 = -1.41242 + 0.75437 \, (w_1 \, / \, h \,) + 1.98412 \, (w_1 \, / \, h \,)^2 - 1.49878 \, (w_1 \, / \, h \,)^3$$

$$(4.5.5.3)$$

$$B_2 = 0.61033 - 0.38901 \, (w_1 \, / \, h \,) - 0.17853 \, (w_1 \, / \, h \,)^2 - 0.42474 \, (w_1 \, / \, h \,)^3$$

$$(4.5.5.4)$$

$$B_3 = -0.20601 + 1.04401 \, (w_1 \, / \, h \,) - 2.23510 \, (w_1 \, / \, h \,)^2 + 1.74802 \, (w_1 \, / \, h \,)^3$$

$$(4.5.5.5)$$

which is valid for $s \, / \, h \leq 0.8$ and $w_1 \, / \, h \leq 0.8$ and accurate to better than 1%. For lines with $w_1 \, / \, h > 0.8$, equal width lines ($w_1 = w_2$) create equal mode impedances.

Perlow and Presser [3] form a two-port from this structure by tying the outer lines' ends together. The resulting coupler can achieve tighter coupling than two-lines. Equations are derived by equating the capacitances of this structure to those of the two line coupler.

## REFERENCES

[1] Abdallah, Esmat A.F., and Nabil A. El-Deeb, "Design Parameters of Three Coupled Microstriplines," *13th European Microwave Conference Proceedings*, 1983, Nürnberg, W. Germany, pp. 333–338.

[2] Pavlidis, Dimitrios, "The Design and Performance of Three-Line Microstrip Couplers," *IEEE Transactions on Microwave Theory and Techniques*, Vol. MTT-24, No. 10, October 1976, pp. 631–640.

[3] Perlow, S.M., and A. Presser, "The Interdigitated Three-Strip Coupler," *IEEE Transactions on Microwave Theory and Techniques*, Vol. MTT-32, No. 10, October 1984, pp. 1418–1422.

[4] Tripathi, Vijai K., "On the Analysis of Symmetrical Three-Line Microstrip Circuits," *IEEE Transactions on Microwave Theory and Techniques*, Vol. MTT-25, No. 9, September 1977, pp. 726–729.

### 4.5.6    Microstrip Line-to-Stripline Aperture Coupler

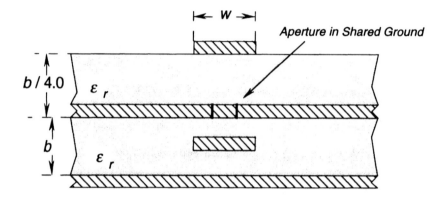

**Figure 4.5.6.1:**    **Microstripline-to-Stripline Aperture Coupler**

The aperture coupler allows us to couple between layers of a multilayer structure. This is useful when we need a small amount of signal coupled to another layer. The equations may also be useful when evaluating coupling through areas of discontinuous ground plane.

Even and odd mode impedances for this structure are not found in the literature. We do find relations for the coupling, $k$. For a circular-shaped aperture in the shared ground plane,

$$k_{dB} = 20.0 \log_{10} \left[ \frac{5.0 \, d_0^3 \left( \frac{-\varepsilon_r \, \varepsilon_{eff}}{\varepsilon_r + \varepsilon_{eff}} + 2.0 \, \sqrt{\varepsilon_r \, \varepsilon_{eff}} \right)}{9.0 \, \lambda \, b^2 \, \sqrt{F(m)}} \right] \qquad (4.5.6.1)$$

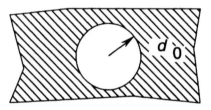

**Figure 4.5.6.2:**    **Circular Aperture Dimensions**

For a slot aperture,

$$k_{\text{dB}} = 20.0 \log_{10} \left[ \frac{20.0 \left( \dfrac{\alpha_c \, \varepsilon_r \, \varepsilon_{\text{eff}}}{\varepsilon_r + \varepsilon_{\text{eff}}} + \alpha_m \, \sqrt{\varepsilon_r \, \varepsilon_{\text{eff}}} \right)}{3.0 \, \lambda \, b^2 \, \sqrt{F(m)}} \right] \qquad (4.5.6.2)$$

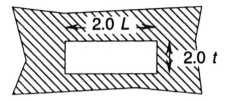

**Figure 4.5.6.3:** **Rectangular Aperture Dimensions**

where

$$\alpha_m = \frac{1.054 \, l^3}{\ln \left( 1.0 + \dfrac{0.66}{\alpha} \right)} \qquad (4.5.6.3)$$

$$\alpha_e = \frac{\pi \, \alpha^2 \, l^3 \, (1.0 - 0.5663 \, \alpha + 0.1398 \, \alpha^2)}{2.0} \qquad (4.5.6.4)$$

$$\alpha = \frac{t}{l} \qquad (4.5.6.5)$$

and for both of the equations:

$$F(m) = \frac{4.0}{b^2} \int_{-b/2}^{b/2} \int_{-\infty}^{\infty} \frac{dx \, dy}{\left| 1.0 - m^2 \cosh^2 \left[ \dfrac{\pi \, (x + jy)}{b} \right] \right|} \qquad (4.5.6.6)$$

$$m = \text{sech} \left( \frac{\pi \, w}{2.0 \, b} \right) \qquad (4.5.6.7)$$

## REFERENCES

[1]    Jogiraju, G.V., and V.M. Pandharipande, "Stripline to Microstrip Line Aperture Coupler," *IEEE Transactions on Microwave Theory and Techniques*, Vol. MTT-38, No. 4, April 1990, pp. 440–443.

## 4.6    COUPLED STRIPLINES

### 4.6.1    Nonhomogeneous Dielectric Broadside Coupled Striplines

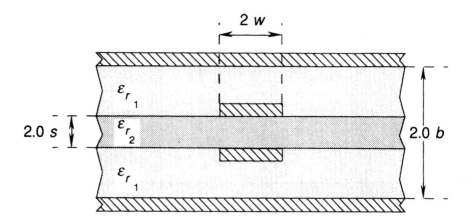

**Figure  4.6.1.1:**        Nonhomogeneous  Dielectric  Broadside  Coupled
Striplines

This configuration is described in [2] and [5].  Broadside coupling allows tight
coupling to be achieved with practical PCB dimensions.  Bahl and Bhartia [2] treat this
topology as two separate cases.  The equations are in terms of the structure's capacitance:

$$Z_{0,e} = \frac{1.0}{c_0 \sqrt{C_e\, C_e{}^a}} \tag{4.6.1.1}$$

$$Z_{0,o} = \frac{1.0}{c_0 \sqrt{C_o\, C_o{}^a}} \tag{4.6.1.2}$$

$$\varepsilon_{eff,e} = \frac{C_e}{C_e{}^a} \tag{4.6.1.3}$$

$$\varepsilon_{eff,o} = \frac{C_o}{C_o{}^a} \tag{4.6.1.4}$$

where, for $\varepsilon_{r,2} = 1.0$

$$\frac{1.0}{C_e} = \frac{1.0}{\pi \, \varepsilon_0 \, Q^2} \int_0^\infty \frac{h_e \, \tilde{f}^2(\beta) \, d(\beta b)}{\{ \tanh (\beta s) + \varepsilon_r \coth [\beta (b-s)] \} (\beta b)} \tag{4.6.1.5}$$

$$\frac{1.0}{C_o} = \frac{1.0}{\pi \, \varepsilon_0 \, Q^2} \int_0^\infty \frac{h_e \, \tilde{f}^2(\beta) \, d(\beta b)}{\{ [\tanh (\beta s)]^{-1} + \varepsilon_r \coth [\beta (b-s)] \} (\beta b)} \tag{4.6.1.6}$$

where

$$h_e = \frac{1.0}{2.0} \left\{ 1.0 + \frac{\cosh[\beta (s-t)]}{\cosh(\beta s)} \right\} \tag{4.6.1.7}$$

$$h_o = \frac{1.0}{2.0} \left\{ 1.0 + \frac{\sinh[\beta (s-t)]}{\sinh(\beta s)} \right\} \tag{4.6.1.8}$$

For $\varepsilon_{r,1} = 1.0$,

$$\frac{1.0}{C_e} = \frac{1.0}{\pi \, \varepsilon_0 \, Q^2} \int_0^\infty \frac{g_e \, \tilde{f}^2(\beta) \, d(\beta b)}{\{ \varepsilon_r [\tanh (\beta s)] + \coth [\beta (b-s)] \} (\beta b)} \tag{4.6.1.9}$$

$$\frac{1.0}{C_o} = \frac{1.0}{\pi \, \varepsilon_0 \, Q^2} \int_0^\infty \frac{g_o \, \tilde{f}^2(\beta) \, d(\beta b)}{\{ \varepsilon_r [\tanh (\beta s)]^{-1} + \coth [\beta (b-s)] \} (\beta b)} \tag{4.6.1.10}$$

where

$$\frac{\tilde{f}(\beta)}{Q} = 1.6 \left( \frac{\sin \beta \, w}{\beta \, w} \right)$$

$$+ \frac{2.4}{(\beta w)^2} \left[ \cos \beta \, w - \frac{2.0 \sin \beta \, w}{\beta \, w} + \frac{\sin^2 (\beta \, w / 2.0)}{(\beta \, w / 2.0)^2} \right] \tag{4.6.1.11}$$

$$g_e = \frac{1.0}{2.0}\left\{1.0 + \frac{\cosh\left[\beta\ (b - s - t)\right]}{\cosh\left[\beta\ (b - s)\right]}\right\} \qquad (4.6.1.12)$$

$$g_o = \frac{1.0}{2.0}\left\{1.0 + \frac{\sinh\left[\beta\ (b - s - t)\right]}{\sinh\left[\beta\ (b - s)\right]}\right\} \qquad (4.6.1.13)$$

and $C_o{}^a$ and $C_e{}^a$ are the odd and even mode capacitances with air replacing the dielectric in the equations. Integrals are evaluated numerically.

Conductor losses are calculated with the incremental inductance rule:

$$\alpha_{c,e} = \frac{0.0116\ R_s\ \sqrt{\varepsilon_{eff,e}}}{Z_{0,e}{}^a}\left[\frac{\partial Z_{0,e}{}^a}{\partial(s\ /\ b)}\ (1.0 - s\ /\ b)\right.$$

$$\left.- \frac{\partial Z_{0,e}{}^a}{\partial(w\ /\ b)}\ (1.0 + w\ /\ b) - \frac{\partial Z_{0,e}{}^a}{\partial(t\ /\ b)}\ (0.5 + t\ /\ b)\right] \quad \text{(dB / length)}$$

$$(4.6.1.14)$$

$$\alpha_{c,o} = \frac{0.0116\ R_s\ \sqrt{\varepsilon_{eff,o}}}{Z_{0,o}{}^a}\left[\frac{\partial Z_{0,o}{}^a}{\partial(s\ /\ b)}\ (1.0 - s\ /\ b)\right.$$

$$\left.- \frac{\partial Z_{0,o}{}^a}{\partial(w\ /\ b)}\ (1.0 + w\ /\ b) - \frac{\partial Z_{0,o}{}^a}{\partial(t\ /\ b)}\ (0.5 + t\ /\ b)\right] \quad \text{(dB / length)}$$

$$(4.6.1.15)$$

Dielectric losses are calculated with

$$\alpha_{d,e} = 27.3\ \frac{\varepsilon_r}{\sqrt{\varepsilon_{eff,e}}}\ \frac{\varepsilon_{eff,e} - 1.0}{\varepsilon_r - 1.0}\ \frac{\tan\ \delta}{\lambda_0} \qquad (4.6.1.16)$$

$$\alpha_{d,o} = 27.3\ \frac{\varepsilon_r}{\sqrt{\varepsilon_{eff,o}}}\ \frac{\varepsilon_{eff,o} - 1.0}{\varepsilon_r - 1.0}\ \frac{\tan\ \delta}{\lambda_0} \qquad (4.6.1.17)$$

The effect of an open end as a discontinuity in this structure is reported in [7]. See Chapter 5 for more information.

REFERENCES

[1]    Barnes, William J., and James L. Allen, "Modified Broadside-Coupled Strips in a Layered Dielectric Medium," *International Journal of Electronics*, Vol. 40, No. 4, 1976, pp. 377–391. (Analyzes nonhomogenous dielectrics that are broken up vertically as well the horizontal layering shown above.)

[2]    Bhal, I.J., and Prakash Bhartia, "Characteristics of Inhomogeneous Broadside-Coupled Striplines," *IEEE Transactions on Microwave Theory and Techniques*, Vol. MTT-28, No. 6, June 1980, pp. 529–535.

[3]    Bhartia, Prakash, and Protap Pramanick, "Computer-Aided Design Models for Broadside-Coupled Striplines and Millimeter-Wave Suspended Substrate Microstrip Lines," *IEEE Transactions on Microwave Theory and Techniques*, Vol. MTT-36, No. 11, November 1988, pp. 1476–1481 and errata Vol. MTT-37, No. 10, p. 1658.

[4]    Cohn, Seymour, "Characteristic Impedances of Broadside-Coupled Strip Transmission Lines," *IEEE Transactions on Microwave Theory and Techniques*, Vol. MTT-8, No. 11, November 1960, pp. 633–637.

[5]    Dalley, James E., "A Strip-Line Directional Coupler Utilizing a Non-Homogeneous Dielectric Medium," *IEEE Transactions on Microwave Theory and Techniques*, MTT-17, No. 9, September 1969, pp. 706–712.

[6]    Jones, E.M.T., and J.T. Bolljahn, "Coupled Strip-Transmission-Line Filters and Directional Couplers, *IRE-MTT*, Vol. MTT-4, No. 2, April 1956, pp. 75–81.

[7]    Khoul, Shiban K., and Bharathi Bhat, "Equivalent Circuit of Broadside-Coupled Microstrip Open Ends," *1985 IEEE MTT-S International Microwave Symposium Digest*, St. Louis, Missouri, pp. 497–498.

## 4.6.2    Broadside Coupled Striplines

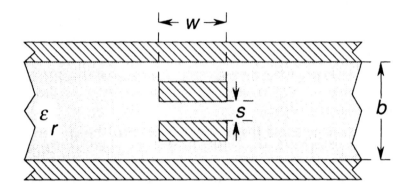

**Figure  4.6.2.1:**        Broadside  Coupled  Striplines

This structure is described in [7] and [5]; however, broader range equations are available in Bahl and Bhartia [1]:

$$Z_{0,e} = \frac{\eta_0}{\sqrt{\varepsilon_r}} \frac{K'(k)}{K(k)} \ (\Omega) \qquad\qquad (4.6.2.1)$$

$$Z_{0,o} = \frac{296.1}{\sqrt{\varepsilon_r} \frac{b}{s} \tanh^{-1}(k)} \ (\Omega) \qquad\qquad (4.6.2.2)$$

where $k$ is the solution of

$$\frac{w}{b} = \frac{1.0}{\pi} \left[ \ln \left( \frac{1.0 + R}{1.0 - R} \right) - \frac{s}{b} \ln \left( \frac{1.0 + R / k}{1.0 - R / k} \right) \right] \qquad (4.6.2.3)$$

and

$$R = \sqrt{\frac{\left( \frac{k}{s} \frac{b}{s} - 1.0 \right)}{\left( \frac{b}{k} \frac{1}{s} - 1.0 \right)}} \qquad (4.6.2.4)$$

The accuracy is "virtually exact" for $w/s \geq 0.35$. These equations are more exact for high dielectric constant materials. Bahl and Bhartia [1] also do a sensitivity analysis of this structure and show that the sensitivity to dimensional changes increases with coupling.

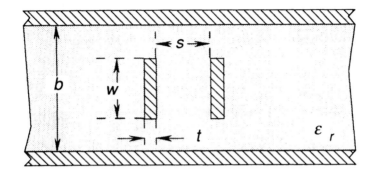

**Figure 4.6.2.2:** Vertical Plate Broadside Coupled Striplines

Another broadside coupled stripline configuration is described in Cohn, [2] and shown in the figure above. This configuration is less often encountered due to the difficulty of fabricating it with standard photolithographic techniques. The relevant design equations are

$$Z_{0,e} = \frac{\eta_0}{2.0\sqrt{\varepsilon_r}} \frac{K(k)}{K(k')} \tag{4.6.2.5}$$

$$Z_{0,o} = \frac{295.9}{\sqrt{\varepsilon_r}} \frac{1.0}{\frac{b}{s}\cos^{-1}k + \ln\frac{1.0}{k}} \tag{4.6.2.6}$$

where

$$\frac{w}{b} = \frac{2.0}{\pi}\left[ \frac{k'}{k}\tan^{-1}\left(\sqrt{\frac{1.0 - \frac{k}{k'}\frac{s}{b}}{1.0 + \frac{k'}{k}\frac{s}{b}}}\right)\right.$$

$$\left. - \frac{s}{b}\tanh^{-1}\left(\sqrt{\frac{1.0 - \frac{k}{k'}\frac{s}{b}}{1.0 + \frac{k'}{k}\frac{s}{b}}}\right)\right] \tag{4.6.2.7}$$

which is accurate for

$$w/s > 1.0$$

and when $s/b \ll 1.0$ they are valid for smaller $w/s$. For thick strips the effect of $t$ is compensated with equations of [3] by increasing the zero-thickness capacitance:

$$C'_{0,o}\left(\frac{w}{b},\frac{s}{b},\frac{t}{s}\right)=C'_{0,o}\left(\frac{w}{b},\frac{s}{b},0.0\right)+4.0\,\Delta C'_2\left(\frac{t}{s}\right)\tag{4.6.2.8}$$

for $t \ll s$

$$C'_{0,e}\left(\frac{w}{b},\frac{s}{b},\frac{t}{s}\right)=C'_{0,e}\left(\frac{w}{b},\frac{s'}{b},0.0\right)\tag{4.6.2.9}$$

and for $t \gg s$

$$C'_{0,e}\left(\frac{w}{b},\frac{s}{b},\frac{t}{s}\right)=C_0\left(\frac{w'}{b},\frac{t'}{w'}\right)\tag{4.6.2.10}$$

where

$$s' = s + 2.0\,t\tag{4.6.2.11}$$

$$w' = s + 2.0\,t\tag{4.6.2.12}$$

$$t' = w\tag{4.6.2.13}$$

$$\Delta C'_2(t/s)=\frac{\varepsilon}{2.0\,\pi}\left[\left(1.0+\frac{t}{s}\right)\ln\left(1.0+\frac{t}{s}\right)-\frac{t}{s}\ln\frac{t}{s}\right]\tag{4.6.2.14}$$

$C_0\left(\frac{w'}{b},\frac{t'}{w'}\right)$ = capacitance of a single strip of width $w'$, thickness $t'$, and having

ground plane spacing $b$. (4.6.2.15)

## REFERENCES

[1]  Bahl, I.J., and P. Bhartia, "The Design of Broadside-Coupled Stripline Circuits," *IEEE Transactions on Microwave Theory and Techniques*, Vol. MTT-29, No. 2, February 1981, pp. 165–168.

[2]  Cohn, Seymour, "Characteristic Impedances of Broadside-Coupled Strip Transmission Lines," *IEEE Transactions on Microwave Theory and Techniques*, Vol. MTT-8, No. 11, November 1960, pp. 633–637.

[3]  Cohn, Seymour, "Thickness Corrections for Capacitive Obstacles and Strip Conductors," *IRE Transactions on Microwave Theory and Techniques*, Vol. MTT-8, No. 11, November 1960, pp. 638–644.

[4]     Horton, M.C., "Loss Calculations for Rectangular Coupled Bars," *IEEE Transactions on Microwave Theory and Techniques*, Vol. MTT-18, October 1970, pp. 736–738.

[5]     Howe, Jr., Harlan, *Stripline Circuit Design*, Artech House, Norwood, MA, 1974.

[6]     Kolker, R.A., "The Amplitude Response of a Coupled Transmission Line, All-Pass Network Having Loss," *IEEE Transactions on Microwave Theory and Techniques*, Vol. MTT-15, August 1967, pp. 438–443.

[7]     Mathaei, G., L. Young, and E.M.T. Jones, *Microwave Filters, Impedance-Matching Networks, and Coupling Structures*, Artech House, Norwood, MA 1980.

[8]     Shelton, P.J., "Impedances and Offset Parallel Coupled Strip Transmission Line," *IEEE Transactions on Microwave Theory and Techniques*, Vol. MTT-14, No. 1, January 1966, pp. 7–15, and "Correction" *IEEE Transactions on Microwave Theory and Techniques*, MTT-14, No. 5, May 1966, p. 249. (Article gives synthesis equations.)

### 4.6.3   Offset Coupled Striplines

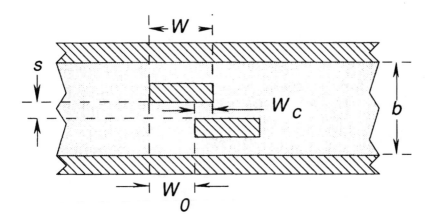

Figure 4.6.3.1:      Offset Coupled Striplines

This configuration allows couplings less than approximately 6 dB to be achieved with practical PCB dimensions.    For $w_c = 0$ use the equations of the last section.  Shelton's [5] equations are normalized to $b = 1.0$.

$$Z_{0,e} = \frac{120.0\,\pi}{\sqrt{\varepsilon_r}}\,\frac{1.0}{C_e}\quad (\Omega)$$
(4.6.3.1)

$$Z_{0,o} = \frac{120.0 \ \pi}{\sqrt{\varepsilon_r}} \frac{1.0}{C_o} \tag{4.6.3.2}$$

where for tight coupling $w_c \geq 0$ (overlapping traces as shown in Figure 4.6.3.1), $w \ / \ (1.0 - s) \geq 0.35$, and $w_c \geq 0.7$:

$$C_o = 2.0 \ w \left( \frac{1.0}{1.0 - s} + \frac{1.0}{s} \right) + C_{fo} = \frac{2.0 \ w}{s \ (1.0 - s)} + C_{fo} \tag{4.6.3.3}$$

$$C_e = \frac{2.0 \ w}{1.0 - s} + C_{fe} \tag{4.6.3.4}$$

$$C_{fo} = \frac{1.0}{\pi} \left\{ - \frac{2.0}{1.0 - s} \log s + \frac{1.0}{s} \log \left[ \frac{p \ r}{(p + s)(1.0 + p)(r - s)(1.0 - r)} \right] \right\} \tag{4.6.3.5}$$

$$C_{fe} = \frac{1.0}{\pi} \left[ \frac{-2.0}{1.0 - s} \log s - 2.0 \log s \right.$$

$$\left. + 4.0 \log (s + p \ r) - \log p \ r \ (p + s)(p + 1.0)(r - s)(1.0 - r) \right] \tag{4.6.3.6}$$

$$r = \frac{p + \dfrac{1.0 + s}{2.0}}{1.0 + p \left( \dfrac{1.0 + s}{2.0 \ s} \right)} \qquad (0 < p < \infty) \tag{4.6.3.7}$$

$$p = \frac{(B - 1.0) \left( \dfrac{1.0 + s}{2.0} \right) + \sqrt{\left( \dfrac{1.0 + s}{2.0} \right)^2 (B - 1.0)^2 + 4.0 \ s \ B}}{2.0} \tag{4.6.3.8}$$

$$B = \frac{p \ r}{s} = \frac{A - 2.0 + \sqrt{A^2 - 4.0 \ A}}{2.0} \tag{4.6.3.9}$$

$$A = \exp \left[ \frac{60 \ \pi^2}{\sqrt{\varepsilon_r} \ Z_0} \left( \frac{1.0 - \rho \ s}{\sqrt{\rho}} \right) \right] \tag{4.6.3.10}$$

$$\rho = \frac{Z_{0,e}}{Z_{0,o}} \tag{4.6.3.11}$$

and for loose coupling $w_c \leq 0$ (traces not overlapping), $w / (1.0 - s) \geq 0.35$, and $2.0 \, w_o / (1.0 + s) \geq 0.85$:

$$C_o = 2.0 \, w \left( \frac{1.0}{1.0 - s} + \frac{1.0}{1.0 + s} \right) + C_{fo} - C_f(a = \infty) \tag{4.6.3.12}$$

$$= \frac{4.0 \, w}{1.0 - s^2} + C_{fo} + C_f(a = \infty)$$

$$C_e = \frac{4.0 \, w}{1.0 - s^2} + C_{fe} + C_f(a = \infty) \tag{4.6.3.13}$$

$$C_{fo} = \frac{2.0}{\pi} \left[ \frac{1.0}{1.0 + s} \log \frac{1.0 + a}{a \, (1.0 - q)} - \frac{1.0}{1.0 - s} \log q \right] \tag{4.6.3.14}$$

$$C_{fe} = \frac{2.0}{\pi \, (1.0 + s)} \log \left( \frac{1.0 + a}{a - a \, q} \right) - \frac{2.0}{1.0 - s} \log q - 2.0 \log \left( \frac{1.0 + a \, q}{a \, q} \right) \tag{4.6.3.15}$$

$$q = \left( \frac{s + 1.0}{2.0} \right) \left( \frac{a + \dfrac{2.0 \, s}{s + 1.0}}{a + \dfrac{s + 1.0}{2.0}} \right) \tag{4.6.3.16}$$

$$a = \sqrt{ \left( \frac{s - k}{s + 1.0} \right)^2 + k } - \frac{s - k}{s + 1.0} \tag{4.6.3.17}$$

$$k = \frac{1.0}{\exp \left( \dfrac{\pi \, \Delta C}{2.0} \right) - 1.0} \tag{4.6.3.18}$$

$$\Delta C = \frac{120 \, \pi \, (\rho - 1.0)}{\sqrt{\varepsilon_r} \, Z_0 \, \sqrt{\rho}} \tag{4.6.3.19}$$

$$\rho = \frac{Z_{0,e}}{Z_{0,o}} \tag{4.6.3.20}$$

$$C_f(a = \infty) = -\frac{2.0}{\pi}\left[\frac{1.0}{1.0 + s} \log\left(\frac{1.0 - s}{2.0}\right) + \frac{1.0}{1.0 - s} \log\left(\frac{1.0 + s}{2.0}\right)\right]$$

(4.6.3.21)

## REFERENCES

[1]   Howe, Jr., Harlan, *Stripline Circuit Design*, Artech House, Norwood, MA, 1974. (Has tables of data that can be used for checking results.)

[2]   King, Charles A., "Math Software Simplifies Power Combiner Design," *Microwaves & RF*, June 1990, pp. 145–151.

[3]   Knighten, James L., "Effect of Conductor Thickness on the Mode Capacitances of Shielded Strip Transmission Lines," *1978 IEEE MTT-S Symposium Digest*, pp. 416–418.

[4]   Mosko, Joseph A., "Coupling Curves for Offset Parallel-Coupled Strip Transmission Lines," *Microwave Journal*, Vol. 10, No. 5, April 1967, pp. 35–37.

[5]   Shelton, P.J., "Impedances and Offset Parallel Coupled Strip Transmission Line," *IEEE Transactions on Microwave Theory and Techniques*, Vol. MTT-14, No. 1, January 1966, pp. 7–15, and "Correction" *IEEE Transactions on Microwave Theory and Techniques*, MTT-14, No. 5, May 1966, p. 249. (Article gives synthesis equations.)

## 4.6.4   Zero-Thickness Edge-Coupled Stripline

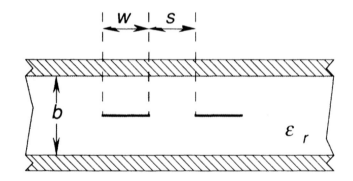

**Figure  4.6.4.1:**      **Edge-Coupled  Stripline**

This zero-thickness structure never occurs in real life; however it is often used to check the accuracy of finite-thickness equations.

$$Z_{0,e} = \frac{30\,\pi}{\sqrt{\varepsilon_r}} \frac{K(k'_e)}{K(k_e)} \quad (\Omega)$$

(4.6.4.1)

$$Z_{0,o} = \frac{30\,\pi}{\sqrt{\varepsilon_r}} \frac{K(k'_o)}{K(k_o)} \quad (\Omega)$$

(4.6.4.2)

where

$$k_e = \tanh\left(\frac{\pi w}{2b}\right) \tanh\left[\frac{\pi(w+s)}{2b}\right]$$

(4.6.4.3)

$$k_o = \tanh\left(\frac{\pi w}{2b}\right) \coth\left[\frac{\pi(w+s)}{2b}\right]$$

(4.6.4.4)

$$k'_e = \sqrt{1 - k_e^2}$$

(4.6.4.5)

$$k'_o = \sqrt{1 - k_o^2}$$

(4.6.4.6)

Equations are exact.

### REFERENCES

[1]   Cohn, Seymour B., "Shielded Coupled-Strip Transmission Line," *IRE Transactions on Microwave Theory and Techniques*, Vol. MTT-3, No. 10, October 1955, pp. 29-38. (This reference is a seminal article giving all the necessary information for design of striplines and edge-coupled striplines.)

[2]   Howard, John, and Wenny C-Lin, "Simple Rules Guide Design of Wideband Stripline Couplers," *Microwaves & RF*, May 1988, pp. 207–211.

[3]   Mathaei, G., L. Young, and E.M.T. Jones, *Microwave Filters, Impedance-Matching Networks, and Coupling Structures*, Artech House, Norwood, MA 1980. (Typo in Eqn 5.05-2 corrected here.)

### 4.6.5    Finite-Thickness Edge-Coupled Striplines

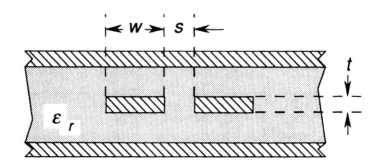

**Figure 4.6.5.1:**    Edge-Coupled Stripline

The equations of Cohn [2] are

$$
Z_{0,e}\left(\frac{w}{b},\frac{t}{b},\frac{s}{b}\right)=\left\{\frac{1.0}{Z_0\left(\frac{w}{b},\frac{t}{b}\right)}\right.
$$

$$
\left.-\frac{C_f'(t/b)}{C_f'(0)}\left[\frac{1.0}{Z_0\left(\frac{w}{b},0\right)}-\frac{1.0}{Z_{0,e}\left(\frac{w}{b},0,\frac{s}{b}\right)}\right]\right\}^{-1}
$$

$$(\Omega)$$

$$(4.6.5.1)$$

for $s/t \geq 5.0$

$$
Z_{0,o}\left(\frac{w}{b},\frac{t}{b},\frac{s}{b}\right)=\left\{\frac{1.0}{Z_0\left(\frac{w}{b},\frac{t}{b}\right)}\right.
$$

$$
\left.-\frac{C_f'(t/b)}{C_f'(0)}\left[\frac{1.0}{Z_{0,o}\left(\frac{w}{b},0,\frac{s}{b}\right)}-\frac{1.0}{Z_0\left(\frac{w}{b},0\right)}\right]\right\}^{-1}
$$

$$(\Omega)$$

$$(4.6.5.2)$$

and for $s/t \le 5.0$

$$Z_{0,o}\left(\frac{w}{b}, \frac{t}{b}, \frac{s}{b}\right) = \left\{ \frac{1.0}{Z_{0,o}\left(\frac{w}{b}, 0, \frac{s}{b}\right)} \right. $$

$$+ \left[ \frac{1.0}{Z_{0,o}\left(\frac{w}{b}, 0, \frac{s}{b}\right)} - \frac{1.0}{Z_0\left(\frac{w}{b}, 0\right)} \right]$$

$$\left. - \frac{2.0}{377}\left(\frac{C_f^{'}(t/b)}{\varepsilon} - \frac{C_f^{'}(0)}{\varepsilon}\right) + \frac{2.0}{377\,s} \right\}^{-1} \quad (\Omega)$$

$$(4.6.5.3)$$

where

$$Z_0\left(\frac{w}{b}, \frac{t}{b}\right) = \text{finite-thickness impedance of an isolated stripline (see Section 3.6.2)}$$

$$Z_0\left(\frac{w}{b}, 0\right) = \text{zero-thickness impedance of an isolated stripline (see Section 3.6.1)}$$

$$Z_{0,e}\left(\frac{w}{b}, 0, \frac{s}{b}\right) = \text{zero-thickness coupled stripline even mode impedance}$$
$$\text{(see Section 4.6.4)}$$

$$Z_{0,o}\left(\frac{w}{b}, 0, \frac{s}{b}\right) = \text{zero-thickness coupled stripline odd mode impedance}$$
$$\text{(see Section 4.6.4)}$$

$$C_f^{'}(t/b) = \frac{0.0885\,\varepsilon_r}{\pi}\left\{ \frac{2.0}{1.0 - t/b}\ln\left(\frac{1.0}{1.0 - t/b} + 1.0\right) \right.$$

$$\left. - \left(\frac{1.0}{1.0 - t/b} - 1.0\right)\ln\left[\frac{1.0}{(1.0 - t/b)^2} - 1.0\right] \right\} \quad (\mu F/cm)[†]$$

$$(4.6.5.4)$$

---

[†] [1]

$$C_f'(0) = \frac{0.0885 \, \varepsilon_r}{\pi} \, 2.0 \ln 2.0 \quad (\mu F / cm)^{\ddagger} \tag{4.6.5.5}$$

The equations [8] for even and odd mode conductor losses (incremental inductance rule) and the dielectric losses:

$$\alpha_{c,e} = \frac{2.0 \, R_s \, \varepsilon_r \, Z0,e}{\eta^2 \, b} \left[ (s/b - 1.0) \, \frac{\partial C_{fe}' / \varepsilon}{\partial s / b} \right.$$

$$\left. + (1.0 + t/b) \left( \frac{\partial C_{fe}' / \varepsilon}{\partial t / b} + \frac{\partial C_f' / \varepsilon}{\partial t / b} \right) + 2.0 \, \frac{w/b + 1.0 - t/b}{(1.0 - t/b)^2} \right] \tag{4.6.5.6}$$

$$\alpha_{c,o} = \frac{2 \, R_s \, \varepsilon_r \, Z0,o}{\eta^2 \, b} \left[ (s/b - 1.0) \, \frac{\partial C_{fo}' / \varepsilon}{\partial s / b} \right.$$

$$\left. + (1.0 + t/b) \left( \frac{\partial C_{fo}' / \varepsilon}{\partial t / b} + \frac{\partial C_f' / \varepsilon}{\partial t / b} \right) + 2.0 \, \frac{w/b + 1.0 - t/b}{(1 - t/b)^2} \right] \tag{4.6.5.7}$$

$$\alpha_{d,e} = \alpha_{d,o} = \frac{\pi \sqrt{\varepsilon_r}}{\lambda} \tan \delta \tag{4.6.5.8}$$

where $\lambda$ is the wavelength. The remaining capacitances are found with the following.
For small $s$

$$C_{fo}'\left(\frac{t}{b}, \frac{s}{b}\right) = C_f'\left(0, \frac{s}{b}\right) + \varepsilon \frac{t}{s} \tag{4.6.5.9}$$

For $s/t \geq 10.0$

$$C_{fo}'\left(\frac{t}{b}, \frac{s}{b}\right) = C_f'\left(\frac{t}{b}\right) \frac{C_{fo}'\left(0, \frac{s}{b}\right)}{C_f'(0)} \tag{4.6.5.10}$$

For $s/t \leq 10.0$

---

$\ddagger$ [1], valid for $w/(b-t) \geq 0.35$ to 1.2% accuracy.

$$C_{f,o} = C_f' \left(0, \frac{s}{b}\right) + \frac{\varepsilon\, Z_0\left(\frac{s}{t}, 0\right)}{377} \qquad (4.6.5.11)$$

where

$$Z_0\left(\frac{s}{t}, 0\right) = \text{the impedance of a single zero-thickness strip having width } s,$$

ground plane spacing $t$, and $\varepsilon_r = 1$

The mode velocities are slightly unequal and are calculated with the even and odd mode effective dielectric constants

$$\varepsilon_{eff,e} = 1.0 \qquad (4.6.5.12)$$

$$\varepsilon_{eff,o} = \left[ \frac{1.0 + 2.0\, Z_{0,o}\left(\frac{w}{b}, 0, \frac{s}{b}\right) Z_0\left(\frac{s}{b}, 0\right) \Big/ \eta_0^2}{1.0 + 2.0\, \varepsilon_r\, Z_{0,o}\left(\frac{w}{b}, 0, \frac{s}{b}\right) Z_0\left(\frac{s}{t}, 0\right) \Big/ \eta_0^2} \right]^{-1} \qquad (4.6.5.13)$$

## REFERENCES

[1]   Cohn, Seymour B., "Characteristic Impedance of the Shielded-Strip Transmission Line," *IRE Transactions on Microwave Theory and Techniques,* Vol. MTT-2, No. 3, July 1954, pp. 52-57.

[2]   Cohn, Seymour B., "Shielded Coupled-Strip Transmission Line," *IRE Transactions on Microwave Theory and Techniques,* Vol. MTT-3, No. 10, October 1955, pp. 29-38. (The preceding reference is a seminal article giving all the necessary information for design of striplines and coupled striplines.)

[3]   Cohn, Seymour, "Thickness Corrections for Capacitive Obstacles and Strip Conductors," *IRE Transactions on Microwave Theory and Techniques,* Vol. MTT-8, No. 11, November 1960, pp. 638–644.

[4]   Getsinger, William J., "Coupled Rectangular Bars Between Parallel Plates," *IRE Transactions on Microwave Theory and Techniques,* Vol. MTT-10, No. 1, pp. 65–72.

[5]   Horton, M.C., "Loss Calculations for Rectangular Coupled Bars," *IEEE Transactions on Microwave Theory and Techniques,* Vol. MTT-18, October 1970, pp. 736–738.

[6]   Kolker, R.A., "The Amplitude Response of a Coupled Transmission Line, All-Pass Network Having Loss," *IEEE Transactions on Microwave Theory and Techniques,* Vol. MTT-15, August 1967, pp. 438–443.

[7]   Kumar, Mahesh, and B.N. Das, "A Directional Coupler Using Asymmetrically Coupled Strip-Lines," *International Journal of Electronics,* Vol. 44, No. 2, 1978, pp. 167–171.

[8]   Tripathi, V.K., "Loss Calculations for Coupled Transmission-Line Structures," *IEEE Transactions on Microwave Theory and Techniques*, Vol. MTT-20, No. 2, February 1972, pp. 178–180. ([7], equation (2) has two apparent typos. $\partial C_{fe}/\varepsilon'$ probably is $\partial C_{fe}'/\varepsilon$ and $\partial C_{f/\varepsilon}'$ is probably $\partial C_f'/\varepsilon$.)

### 4.6.6   Shielded Edge-Coupled Striplines

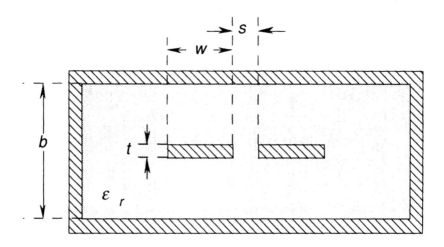

**Figure   4.6.6.1:**      **Shielded   Edge-Coupled   Striplines**

Chao [1] describes a variable coupler using this configuration. By varying the side wall distance as the height, $b$, is varied, the coupling is controlled while keeping the characteristic impedance approximately constant. Chao's equations are

$$Z_{0,e} = \frac{\eta_0}{\sqrt{\varepsilon_r}} \frac{\varepsilon}{C_{0,e}} \qquad\qquad (4.6.6.1)$$

$$Z_{0,o} = \frac{\eta_0}{\sqrt{\varepsilon_r}} \frac{\varepsilon}{C_{0,o}} \qquad\qquad (4.6.6.2)$$

where

$$\frac{C_{0,e}}{\varepsilon} = 2.0 \left( \frac{C_p}{\varepsilon} + \frac{C_{fo}'}{\varepsilon} + \frac{C_{fom}'}{\varepsilon} \right) \qquad\qquad (4.6.6.3)$$

$$\frac{C_{0,o}}{\varepsilon} = 2.0 \left( \frac{C_p}{\varepsilon} + \frac{C_{fe}{'}}{\varepsilon} + \frac{C_{fom}{'}}{\varepsilon} \right) \qquad (4.6.6.4)$$

$$\frac{C_p}{\varepsilon} = \frac{2.0 \, w}{b - t} \qquad (4.6.6.5)$$

$$\frac{C_{fo}{'}}{\varepsilon} \approx \frac{2.0}{\pi} \ln \left[ 1.0 + \coth \frac{\pi}{2.0} \frac{s}{b} \right] + \frac{t}{s} \left( 1.0 - \frac{s}{2.0 \, b} \right) \qquad (4.6.6.6)$$

$$\frac{C_{fe}{'}}{\varepsilon} \approx \frac{2.0}{\pi} \ln \left[ 1.0 + \tanh \frac{\pi}{2.0 \, b} \frac{s}{b} \right] \left( 0.89 + \frac{2.14 \, t}{b} \right) \qquad (4.6.6.7)$$

$$\frac{C_{fom}{'}}{\varepsilon} \approx \frac{2.0}{\pi} \ln \left[ 1.0 + \coth \frac{\pi}{2.0} \frac{m}{b} + \frac{t}{m} \left( 1.0 - \frac{m}{2.0 \, b} \right) \right] \qquad (4.6.6.8)$$

The approximations for $C_{fe}{'}$ and $C_{fo}{'}$ are accurate to within a few percent for

$0.1 \le t / b \le 0.5$

$s / b < 0.2$

## REFERENCES

[1]     Chao, Gene, "A Wide-Band Variable Microwave Coupler," *IEEE Transactions on Microwave Theory and Techniques*, Vol. MTT-18, No. 9, September 1970, pp. 576–583.

## 4.6.7  Asymmetric Edge-Coupled Striplines

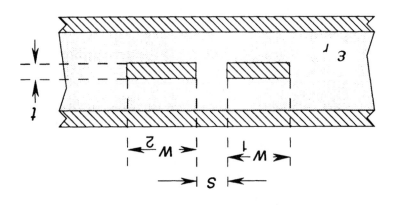

**Figure 4.6.7.1:**   Asymmetric Edge-Coupled Stripline

The general equations for an asymmetric coupler are found in Cristal [3]. For this stucture, we define four impedances:

$$Z_{0,e}^1 = \frac{1.0}{v_p \, C_1} \tag{4.6.7.1}$$

$$Z_{0,e}^2 = \frac{1.0}{v_p \, C_2} \tag{4.6.7.2}$$

$$Z_{0,o}^1 = \frac{1.0}{v_p \, (C_1 + 2.0 \, C_{12})} \tag{4.6.7.3}$$

$$Z_{0,o}^2 = \frac{1.0}{v_p \, (C_2 + 2.0 \, C_{12})} \tag{4.6.7.4}$$

where

$$\frac{C_1}{\varepsilon} = \text{capacitance of line 1 to ground}$$

$$\frac{C_2}{\varepsilon} = \text{capacitance of line 2 to ground}$$

$$\frac{C_{12}}{\varepsilon} = \text{capacitance of line 1 to line 2}$$

These can evaluated with the equations of Cohn [2] and in Section 4.6.5. The coupling and line impedances are defined as:

$$k = \frac{D}{\sqrt{A \; B}} \qquad (4.6.7.5)$$

$$R_a = \sqrt{\frac{A}{B}} \sqrt{A \; B - D^2} \qquad (4.6.7.6)$$

$$R_b = \sqrt{\frac{B}{A}} \sqrt{A \; B - D^2} \qquad (4.6.7.7)$$

where $A$, $B$, and $D$ are defined in terms of the admittances (reciprocal of impedances above):

$$A = \frac{Y_{0,o}{}^1 + Y_{0,e}{}^1}{2.0} \qquad (4.6.7.8)$$

$$B = \frac{Y_{0,o}{}^2 + Y_{0,e}{}^2}{2.0} \qquad (4.6.7.9)$$

$$D = \frac{Y_{0,o}{}^1 - Y_{0,e}{}^1}{2.0} \qquad (4.6.7.10)$$

The coupling and the impedance transformation ratio have the restrictions that

$$\frac{1.0}{k^2} \geq \frac{R_2}{R_1} \qquad (4.6.7.11)$$

$$\frac{R_2}{R_1} \geq 1.0 \qquad (4.6.7.12)$$

### REFERENCES

[1] Bedair, S.S., "Characteristics of Asymmetrical Coupled Shielded Microstrip Lines," *IEE Proceedings*, Vol. 132, Pt. H, No. 5, August 1985, pp. 342–343.

[2] Cohn, Seymour B., "Shielded Coupled-Strip Transmission Line," *IRE Transactions on Microwave Theory and Techniques,* Vol. MTT-3, No. 4, October 1955, pp. 29-38. (A seminal article giving all the necessary information for design of striplines and coupled striplines.)

[3] Cristal, Edward G., "Coupled-Transmission-Line Directional Couplers with Coupled Lines of Unequal Characteristic Impedances," *IEEE Transactions on*

*Microwave Theory and Techniques*, Vol. MTT-14, No. 7, July 1966, pp. 337–346.

[4]    Fouladian, Jamshid, and Protap Pramanick, "Generalized Analytical Equations for Strip Transmission Lines," *Microwave Journal*, November 1989, pp. 165–168.

[5]    Kumar, Mahesh, and B.N. Das, "A Directional Coupler Using Asymmetrically Coupled Strip-Lines," *International Journal of Electronics*, Vol. 44, No. 2, 1978, pp. 167–171.

[6]    Kumar, Mahesh, and B.N. Das, "A Uniformly Coupled Asymmetric TEM-mode Transmission-Line Directional Coupler," *International Journal of Electronics*, Vol. 41, No. 6, 1976, pp. 565–572.

[7]    Mathaei, G., L. Young, and E.M.T. Jones, *Microwave Filters, Impedance-Matching Networks, and Coupling Structures*, Artech House, Norwood, MA, 1980.

## 4.7 COUPLED SLABLINES

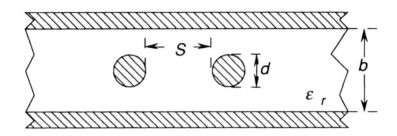

**Figure 4.7.1: Coupled Slablines**

This structure is used when analyzing coupled wires or coupled rods as used in interdigital filters.

$$Z_{0,e} = \frac{59.952}{\sqrt{\varepsilon_r}} \ln\left[ \frac{0.523962}{f_1(x)\,f_2(x,\,y)\,f_3(x,\,y)} \right] \quad (\Omega) \tag{4.7.1.1}$$

$$Z_{0,o} = \frac{59.952}{\sqrt{\varepsilon_r}} \ln\left[ \frac{0.523962\,f_3(x,\,y)}{f_1(x)\,f_4(x,\,y)} \right] \quad (\Omega) \tag{4.7.1.2}$$

where

$$f_1(x) = \frac{x\,a(x)}{b(x)} \tag{4.7.1.3}$$

$$f_2(x) = \begin{cases} c(y) - x\,d(y) + e(x)\,g(y) & y < 0.9 \\ 1.0 + .004\,e^{(0.9\,-\,y)} & y \geq 0.9 \end{cases} \tag{4.7.1.4}$$

$$f_3(x) = \tanh\left[ \frac{\pi(x\,+\,y)}{w} \right] \tag{4.7.1.5}$$

$$f_4(x) = \begin{cases} k(y) - x\,l(y) + m(x)\,n(y) & y < 0.9 \\ 1.0 & y \geq 0.9 \end{cases} \tag{4.7.1.6}$$

$$a(x) = 1.0 + e^{(16.0\,x\,-\,18.272)} \tag{4.7.1.7}$$

$$b(x) = \sqrt{5.905 - x^4} \tag{4.7.1.8}$$

$$c(y) = -0.8107 \, y^3 + 1.3401 y^2 - 0.6929 \, y + 1.0892 + \frac{0.014002}{y} - \frac{0.000636}{y^2}$$

(4.7.1.9)

$$d(y) = 0.11 - 0.83 \, y + 1.64 \, y^2 - y^3$$

(4.7.1.10)

$$e(x) = -0.15 \, e^{(-13.0 \, x)}$$

(4.7.1.11)

$$g(y) = 2.23 \, e^{(-7.01 \, y + 10.24 \, y^2 - 27.58 \, y^3)}$$

(4.7.1.12)

$$k(y) = 1.0 + 0.01 \left( -0.0726 - \frac{0.2145}{y} + \frac{0.222573}{y^2} - \frac{0.012823}{y^3} \right)$$

(4.7.1.13)

$$l(y) = 0.01 \left( -0.26 + \frac{0.6866}{y} + \frac{0.0831}{y^2} - \frac{0.0076}{y} \right)$$

(4.7.1.14)

$$m(x) = -0.1098 + 1.2138 \, x - 2.2535 \, x^2 + 1.1313 \, x^3$$

(4.7.1.15)

$$n(y) = -0.019 - \frac{0.016}{y} + \frac{0.0362}{y^2} - \frac{0.00243}{y^3}$$

(4.7.1.16)

where

$$x = d / b$$

(4.7.1.17)

$$y = s / b$$

(4.7.1.18)

The ground is required to be $\geq 5 \, (2 \, d + s)$ wide.

The equations for even and odd mode losses [6]:

$$\alpha_{c,e} = \frac{R_s \, \varepsilon_r \, Z_{0,e}}{\eta^2 \, b} \left[ (s/b - 1.0) \, \frac{\partial C_{oe}/\varepsilon}{\partial s/b} + (1.0 + d/b) \left( \frac{\partial C_{oe}/\varepsilon}{\partial d/b} \right) \right]$$

(4.7.1.19)

$$\alpha_{c,o} = \frac{R_s \, \varepsilon_r \, Z_{0,o}}{\eta^2 \, b} \left[ (s/b - 1.0) \, \frac{\partial C_{oo}/\varepsilon}{\partial s/b} + (1.0 + d/b) \left( \frac{\partial C_{oe}/\varepsilon}{\partial d/b} \right) \right]$$

(4.7.1.20)

$$\alpha_{d,e} = \alpha_{d,o} = \frac{\pi \sqrt{\varepsilon_r}}{\lambda} \tan \delta \tag{4.7.1.21}$$

Rosloniec [5] suggests we set the initial guesses for synthesis to:

$$x^{(0)} = \frac{4.0}{\pi \exp \left( \dfrac{-Z_0}{59.952\sqrt{0.987 - 0.171\ k - 1.723\ k^2}} \right)} \tag{4.7.1.22}$$

$$y^{(0)} = \frac{1.0}{\pi \ln \left( \dfrac{r + 1.0}{r - 1.0} \right)} - x^{(0)} \tag{4.7.1.23}$$

Newton's method is then used by the program given in [5] to solve for $x$ and $y$ based on the designer's required $Z_{0,o}$ and $Z_{0,e}$.

## REFERENCES

[1]    Agarwal, A., et al., "Coupled Bars in Rectangular Coaxial," Electronics Letters, Vol. 25, No. 1, January 5, 1989, pp. 66–67.

[2]    Cristal, Edward G., "Coupled Circular Cylindrical Rods Between Parallel Ground Planes," IEEE Transactions on Microwave Theory and Techniques, Vol. MTT-12, No. 4, July 1964, pp. 428–439.

[3]    Jones, E.M.T., and J.T. Bolljahn, "Coupled Strip-Transmission-Line Filters and Directional Couplers, IRE-MTT, April 1956, Vol. MTT-4, No. 2, pp. 75–81.

[4]    Levy, Ralph, "Conformal Transformations Combined with Numerical Techniques, with Applications to Coupled-Bar Problems," IEEE Transactions on Microwave Theory and Techniques, Vol. MTT-28, No. 4, April 1980, pp. 369–375.

[5]    Rosloniec, Stanislaw, "An Improved Algorithm for the Computer–Aided Design of Coupled Slab Lines," IEEE Transactions on Microwave Theory and Techniques, Vol. MTT-37, No. 1, January 1989, pp. 258–261.

[6]    Tripathi, V.K., "Loss Calculations for Coupled Transmission-Line Structures," IEEE Transactions on Microwave Theory and Techniques, Vol. MTT-20, No. 2, February 1972, pp. 178–180.

## 4.8    COUPLED SUSPENDED SUBSTRATE OR COUPLED INVERTED MICROSTRIP LINES

The unshielded and offset versions of these structures are not available, however the following structures can be used when the line is centered in a shield.

### 4.8.1    Edge-Coupled Suspended Substrate Lines

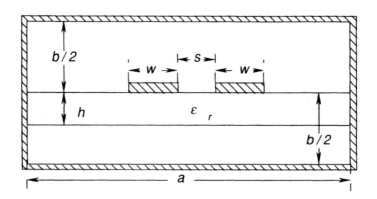

**Figure 4.8.1.1:        Edge-Coupled Suspended Substrate Lines**

In Wang *et al.* [4] edge-coupled suspended substrate is analyzed. As for single lines, the suspended substrate configuration has lower loss and less sensitivity to physical dimensions than an equivalent microstrip or stripline.

The equations were derived by curve-fitting results of the method of lines. They are

$$Z_{0,e} = \frac{\eta_0\,\pi}{2.0\,\sqrt{\varepsilon_{r.e}}} \frac{1.0}{\ln\left(2.0\,\dfrac{1.0 + \sqrt{k_1'}}{1.0 - \sqrt{k_1'}}\right)} \tag{4.8.1.1}$$

$$Z_{0,o} = \frac{\eta_0\,\pi}{2.0\,\sqrt{\varepsilon_{r.o}}} \frac{1.0}{\ln\left(2.0\,\dfrac{1.0 + \sqrt{k_2'}}{1.0 - \sqrt{k_2'}}\right)} \tag{4.8.1.2}$$

and

$$\varepsilon_{r.e} = \left\{\left[1.0 - D\,\frac{s}{a} + H\,\ln\frac{w}{a} + R\right](\ln \varepsilon_r)\,\varepsilon_r^{\,0.06}\left(\frac{\varepsilon_r}{2.8}\right)^{(b/a - 0.4)/3.0}\right\}^{-2}$$

$$(4.8.1.3)$$

$$\varepsilon_{r,o} = \left[ 1.0 - E \; (\ln \tfrac{s}{a} - F) \; (\ln \tfrac{w}{a} - G)(\ln \varepsilon_r) \; \varepsilon_r^{\; 0.08} \right]^{-2} \qquad (4.8.1.4)$$

where

$$k_1' = \sqrt{1.0 - k_1^2} \qquad (4.8.1.5)$$

$$k_1 = \frac{\text{tg } \vartheta_2'}{\tan \vartheta_1'} \qquad (4.8.1.6)$$

$$k_2' = \sqrt{1.0 - k_2^2} \qquad (4.8.1.7)$$

$$k_2 = \sqrt{\frac{1.0 + \tan^2 \vartheta_2'}{1.0 + \tan^2 \vartheta_1'}} \qquad (4.8.1.8)$$

$$\vartheta_1' = Q_1 \tanh \left[ \frac{Q_2 \; (2.0 \; w + s)}{b} \right] + Q_3 \qquad (4.8.1.9)$$

$$\vartheta_2' = Q_1 \tanh \left( \frac{Q_2 \; s}{b} \right) + Q_3 \qquad (4.8.1.10)$$

$$Q_1 = 0.0361168 \; e^{\; 5.12 \; k} + 1.4404 \qquad (4.8.1.11)$$

$$Q_2 = -0.0316177 \; e^{\; 3.947 \; k} + 1.07319 \qquad (4.8.1.12)$$

$$Q_3 = 0.15988 \; k^3 - 0.0895 \; k^2 + 0.02535 \; k \qquad (4.8.1.13)$$

$$k = \sqrt{1.0 - \left( \frac{e^{\pi \; a \; / \; b} - 2.0}{e^{\pi \; a \; / \; b} + 2.0} \right)^4} \qquad (4.8.1.14)$$

$$D = k_{11} \left( \frac{h}{a} \right)^3 + k_{12} \left( \frac{h}{a} \right)^2 + k_{13} \left( \frac{h}{a} \right) + k_{14} \qquad (4.8.1.15)$$

$$E = 0.0001 \left( 7.7291 \; \left\{ \frac{b}{a} + 6.9924 \left[ 9.5 \left( \frac{h}{a} \right)^2 - 1.0 \right] \right\} \right.$$

$$\times\ (\ln \frac{h}{a} + 2.1657) + 561.62 \left(\frac{h}{a}\right)^2 - 6.7772 \left[\frac{h}{a}\right]\ ) \tag{4.8.1.16}$$

$$F = 60.4762 \left\{\left(\frac{b}{a} - \frac{6.0\ h}{a}\right)^2 \left(1.0 + \frac{2.0\ h}{a}\right) - 1.6012 \left[1.0 - 8.0 \left(\frac{h}{a}\right)^2\right]\right\}$$

$$\times (e^{h/a} - 1.04684) + 1721.61 \left(\frac{h}{a}\right)^2 + 19.027 \left(\frac{h}{a}\right) \tag{4.8.1.17}$$

$$G = 0.05154 \left\{-\left(\frac{b}{a} - 0.8\right)^2 + 0.3653 \left[1.0 - 8.0 \left(\frac{h}{a}\right)^2\right]\right\}$$

$$\times (e^{h/a} + 54.554) + 6551.53 \left(\frac{h}{a}\right)^3 + 34.76 \left(\frac{h}{a}\right) \tag{4.8.1.18}$$

$$H = k_{21} \left(\frac{h}{a}\right)^3 + k_{22} \left(\frac{h}{a}\right)^2 + k_{23} \left(\frac{h}{a}\right) + k_{24} \tag{4.8.1.19}$$

$$R = k_{31} \left(\frac{h}{a}\right)^2 + k_{32} \left(\frac{h}{a}\right) + k_{33} \tag{4.8.1.20}$$

$$k_{11} = -28.794 \left(\frac{b}{a}\right)^2 + 29.958 \left(\frac{b}{a}\right) + 52.861 \tag{4.8.1.21}$$

$$k_{12} = 0.92658 \left(\frac{b}{a}\right)^2 + 3.2713 \left(\frac{b}{a}\right) - 20.653 \tag{4.8.1.22}$$

$$k_{13} = -1.5075 \left(\frac{b}{a}\right)^2 + 1.8914 \left(\frac{b}{a}\right) + 0.75778 \tag{4.8.1.23}$$

$$k_{14} = -0.25477 \left(\frac{b}{a}\right)^3 + 0.60480 \left(\frac{b}{a}\right)^2 - 0.48538 \left(\frac{b}{a}\right) + 0.16713 \tag{4.8.1.24}$$

$$k_{21} = -11.459 \left(\frac{b}{a}\right)^2 + 18.744 \left(\frac{b}{a}\right) - 17.109 \tag{4.8.1.25}$$

$$k_{22} = 7.4499 \left(\frac{b}{a}\right)^2 - 11.927 \left(\frac{b}{a}\right) + 7.4606 \tag{4.8.1.26}$$

$$k_{23} = -0.33432 \left(\frac{b}{a}\right)^2 + 0.552 \left(\frac{b}{a}\right) - 0.37211 \qquad (4.8.1.27)$$

$$k_{24} = 0.014317 \left(\frac{b}{a}\right)^2 + 0.030418 \left(\frac{b}{a}\right) - 0.0314 \qquad (4.8.1.28)$$

$$k_{31} = -48.729 \left(\frac{b}{a}\right)^3 + 115.45 \left(\frac{b}{a}\right)^2 - 92.939 \left(\frac{b}{a}\right) + 22.5953 \qquad (4.8.1.29)$$

$$k_{32} = 1.4237 \ln \left(\frac{b}{a} + 3.2411\right) - 0.18519 \qquad (4.8.1.30)$$

$$k_{33} = 0.014478 \ln \left(\frac{b}{a} + 0.23568\right) - 0.024308 \qquad (4.8.1.31)$$

Equations are valid for

$2.22 \leq \varepsilon_r \leq 3.8$

$0.05 \leq w / a \leq 0.2$

$0.054 \leq s / a \leq 0.26$

$0.4 \leq b / a \leq 0.85$

$0.04 \leq h / a \leq 0.12$

The equations were compared with the results of SuperCompact™ and found to agree to within 5%.

## REFERENCES

[1] Bhartia, Prakash, and Protap Pramanick, "Computer-Aided Design Models for Broadside-Coupled Striplines and Millimeter-Wave Suspended Substrate Microstrip Lines," *IEEE Transactions on Microwave Theory and Techniques*, Vol. MTT-36, No. 11, November 1988, pp. 1476–1481 and errata, Vol. MTT-37, No. 10, p. 1658.

[2] Bochtler, U. and F. Endress, "CAD Program Designs Stripline Couplers," *Microwaves & RF*, December 1986, pp. 91–95.

[3] Smith, J. I., "The Even- and Odd-Mode Capacitance Parameters for Coupled Lines in Suspended Substrate," *IEEE Transactions on Microwave Theory and Techniques*, Vol. MTT-19, No. 5, May 1971, pp. 424–430.

[4]   Wang, Yi Yun, *et al.*, "Analysis and Synthesis Equations for Edge-Coupled Suspended Substrate Microstrip Line," *1989 IEEE MTT-S Symposium Digest,* pp. 1123–1126.

### 4.8.2   Shielded Suspended Substrate Broadside Coupled Lines

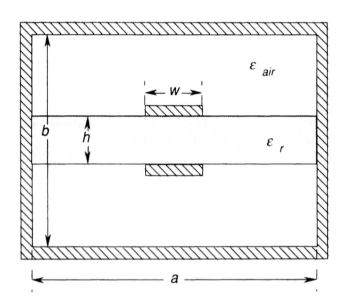

**Figure 4.8.2.1:**      **Broadside Coupled Shielded Suspended Stripline**

$$Z_{0,o} = \frac{\eta_0}{2\sqrt{\varepsilon_{eff,o}}} \left\{ S_o + T_o \ln \left[ \frac{0.2}{w/b} + \sqrt{1.0 + \frac{0.23}{(w/b)^2}} \right] \right\} \quad (\Omega) \quad (4.8.2.1)$$

$$Z_{0,e} = \frac{\eta_0}{2\pi\sqrt{\varepsilon_{eff,e}}} \left\{ S_e + T_e \ln \left[ \frac{12.0}{w/b} + \sqrt{1.0 + \frac{16.0}{(w/b)^2}} \right] \right\} \quad (\Omega)$$

$$(4.8.2.2)$$

$$\varepsilon_{eff,o} = \left[ 1.0 + 0.5 \left\{ H_o - P_o \ln \left[ \frac{w}{b} + \sqrt{(w/b)^2 + 1.0} \right] \right\} \right.$$

$$\times \left( \ln \frac{1.0}{\sqrt{\varepsilon_r}} + \frac{1.0}{\sqrt{\varepsilon_r}} - 1.0 \right) \Bigg]^{-1} \tag{4.8.2.3}$$

$$\varepsilon_{eff,e} = \frac{1.0}{\left[ 1.0 + (H_e - P_e \ln \frac{w}{b}) \ln \frac{1.0}{\sqrt{\varepsilon_r}} \right]^2} \tag{4.8.2.4}$$

where

$$S_o = -0.1073 + 1.67080 \frac{h}{b} + 0.007484 \frac{a}{b} \tag{4.8.2.5}$$

$$T_o = 0.4768 + 2.1295 \frac{h}{b} - 0.01278 \frac{a}{b} \tag{4.8.2.6}$$

$$H_o = 0.7210 - 0.3568 \frac{h}{b} + 0.02132 \frac{a}{b} \tag{4.8.2.7}$$

$$P_o = -0.3035 + 0.3743 \frac{h}{b} + 0.07274 \frac{a}{b} \tag{4.8.2.8}$$

$$S_e = -2.6528 + 0.9452 \frac{h}{b} - 0.4531 \frac{a}{b} \tag{4.8.2.9}$$

$$T_e = 1.4793 - 1.1903 \frac{h}{b} - 0.04511 \frac{a}{b} \tag{4.8.2.10}$$

$$H_e = 0.2245 + 0.7192 \frac{h}{b} - 0.1022 \frac{a}{b} \tag{4.8.2.11}$$

$$P_e = 0.001356 + 0.06590 \frac{h}{b} + 0.01951 \frac{a}{b} \tag{4.8.2.12}$$

### REFERENCES

[1]    Bhartia, Brakash, and Protap Pramanick, "Computer-Aided Design Models for Broadside-Coupled Striplines and Millimeter-Wave Suspended Substrate Microstrip Lines," *IEEE Transactions on Microwave Theory and Techniques*, Vol. MTT-36, No. 11, November 1988, pp. 1476–1481, and "Corrections to 'Computer-Aided Design Models for Broadside-Coupled Striplines and Millimeter-Wave Suspended Substrate Microstrip Lines,'" Vol. MTT-37, No. 10, October 1989, p. 1658.

[2]    Shu, Yong-hui, Xiao-xia Qi, and Yun-ji Wang, "Analysis Equations for Shielded Suspended Substrate Microstrip Line and Broadside-Coupled Stripline," *1980 IEEE MTT-S International Microwave Symposium Digest*, pp. 693–696.

## 4.9   OTHER COUPLED LINES

### 4.9.1   Coupled Slotlines

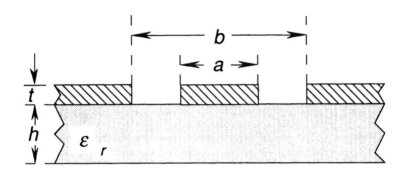

**Figure  4.9.1.1:**       Coupled  Slotlines

As can be seen from the figure, coupled slotlines are strongly related to the coplanar waveguide structure and the slotline structure.  Knorr and Kuchler calculate the impedances and effective dielectric constant of this structure and present curves in [1].  Although closed-form equations are not yet available, some observations can be made about the structure at the extremes.

For a narrow center conductor or tight coupling ($a / h \ll 1.0$), the structure resembles a single slotline with slot width $b - a$ whose impedance is $Z_{0,sl}$ and

$$Z_{0,e} = \frac{Z_{0,sl}}{2.0} \tag{4.9.1.1}$$

For very wide center conductor ($a / h \gg 1.0$) or loose coupling between the slots, the structure becomes two uncoupled slotlines or CPW.  In this situation, if we calculate the characteristic impedance of the coplanar waveguide structure, $Z_{0,cpw}$, the symmetric coupled slotline has

$$Z_{0,o} = 2.0 \, Z_{0,cpw} \tag{4.9.1.2}$$

REFERENCE

[1]     Knorr, Jeffrey B., and Klaus-Dieter Kuchler, "Analysis of Coupled Slots and Coplanar Strips on Dielectric Substrate," *IEEE Transactions on Microwave Theory and Techniques*, Vol. MTT-23, No. 7, July 1975, pp. 541–548.

# Chapter 5

# Transmission Line
# Components and
# Discontinuities

## 5.1   DISCONTINUITY MODELS

A discontinuity occurs whenever our assumption that the line can be modeled as a uniform line having length is violated. This can be a bend, a step in width, a T junction, a change in dielectric constant, a wall in close proximity, or a transition to another type of line. We need to calculate discontinuity effects to minimize them and in other cases to use them to create a desired circuit (*e.g.*, a filter).

Often the discontinuity is small relative to the signal being propagated and can be accurately approximated by lumped-element equivalent circuits. Lumped-element models are best for understanding the effects of and remedies for discontinuities. For more accuracy over frequency the model may include transmission lines. As shown in Chapter 1, for small size relative to a wavelength, a transmission line can often be replaced by an equivalent lumped element. Several equivalent models are possible for each discontinuity.

In this chapter the equations for calculating the discontinuity models are given. These models are made up of the usual lumped elements and lengths of transmission lines. Reference planes (denoted $T_n$ and a dotted line on each figure) indicate the physical location of the discontinuity model. The physical structure between the reference planes is replaced with the model.

## 5.2   COAXIAL DISCONTINUITIES

The original work in coaxial discontinuities was published in 1944 in two articles, by Whinnery *et al.*, [7] and [8]. Somlo followed these up with corrections and closed-form equations [5, 6] based on numerical calculations including higher order modes.

### 5.2.1   Coaxial Step in Inner Conductor

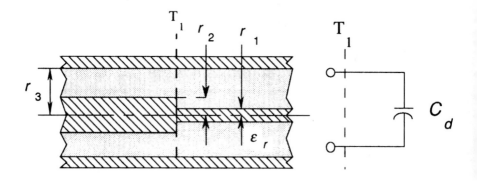

**Figure 5.2.1.1:**      **Coaxial Step in Inner Conductor**

The discontinuity capacitance for a step on the inner conductor is [6, 8]

$$C_d = 2.0 \, \varepsilon_r \, \pi \, r_2 C'_{d_1} \tag{5.2.1.1}$$

where

$$C'_{d_1} = \frac{\varepsilon}{100.0 \, \pi} \left[ \frac{\alpha^2 + 1.0}{\alpha} \ln \left( \frac{1.0 + \alpha}{1.0 - \alpha} \right) - 2.0 \ln \left( \frac{4.0 \, \alpha}{1.0 - \alpha^2} \right) \right]$$

$$+ 1.11 \times 10^{-15} (1.0 - \alpha)(\tau - 1.0) \quad (\text{F / cm}) \tag{5.2.1.2}$$

$$\alpha = \frac{r_3 - r_2}{r_3 - r_1} \tag{5.2.1.3}$$

$$\tau = \frac{r_3}{r_1} \tag{5.2.1.4}$$

which is accurate to ±0.3 fF / cm for

$$0.01 \leq \alpha \leq 1.0$$

$$1.0 \leq \tau \leq 6.0$$

In Young [9], Green notes that that the effect of the discontinuity on the fields is minimal at a distance greater than 2 $c$ away from the step for this structure. The effect of frequency can be approximately corrected [8] with the formula

$$\frac{C(f)}{C_d} = \frac{1.0}{\sqrt{1.0 - (2.0 \, r_2 / \lambda)^2}} \qquad (5.2.1.5)$$

which is valid for

$$0.0 \leq r_2 \leq \lambda / 2.0$$

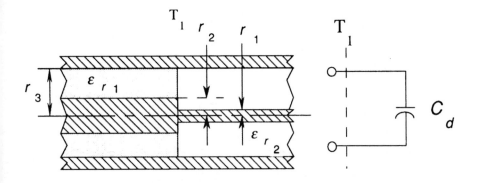

**Figure 5.2.1.2:** **Coaxial Step on Inner Conductor with Dielectric Change**

For a dielectric discontinuity in the same plane as the inner conductor step (Figure 5.2.1.2), the approximating equations are

$$C_d \cong 2.0 \, \pi \, r_3 \, \varepsilon_{r2} \, C_{d_1}' \qquad (5.2.1.6)$$

where

$$C_{d_1}' = \frac{\varepsilon}{100.0 \, \pi} \left[ \frac{\alpha^2 + 1.0}{\alpha} \ln \left( \frac{1.0 + \alpha}{1.0 - \alpha} \right) - 2.0 \ln \left( \frac{4.0 \, \alpha}{1.0 - \alpha^2} \right) \right]$$

$$+ 1.11 \times 10^{-15} (1.0 - \alpha) (\tau - 1.0) \quad (F / cm) \qquad (5.2.1.7)$$

$$\alpha = \frac{r_3 - r_2}{r_3 - r_1} \qquad (5.2.1.8)$$

$$\tau = \frac{r_3}{r_1} \tag{5.2.1.9}$$

This structure is encountered in coaxial air lines that are suppported by dielectric beads. The step on the inner conductor is introduced to compensate for the dielectric discontinuity.

In this situation and referring to the figure above, $\varepsilon_{r1} = \varepsilon_0$ and $\varepsilon_{r2}$ is the bead dielectric. Two of these discontinuities occur as shown in the figure below.

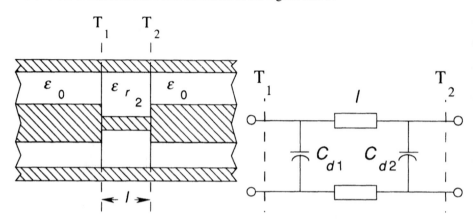

**Figure 5.2.1.3:** Compensated Coaxial Dielectric Bead Support

REFERENCES

[1]  Gwarek, Wojciech K., "Computer-Aided Analysis of Arbitrarily Shaped Coaxial Discontinuities," *IEEE Transactions on Microwave Theory and Techniques*, Vol. MTT-36, No. 2, February 1988, pp. 337–342.

[2]  Marcuvitz, N.,*Waveguide Handbook*, McGraw-Hill, New York, 1955. Reprint with errata and preface by N. Marcuvitz, Peter Peregrinus Ltd., 1986.

[3]  Saad, T., ed., *The Microwave Engineer's Handbook*, Vol. 1, Artech House, Norwood, MA, 1971.

[4]  Silvester, P., and Ivan A. Cermak, "Analysis of Coaxial Line Discontinuities by Boundary Relaxation," *IEEE Transactions on Microwave Theory and Techniques*, Vol. MTT-17, No. 8, August 1969, pp. 489–495.

[5]  Somlo, P.I., "Calculating Coaxial Transmission-Line Step Capacitance," *IEEE Transactions on Microwave Theory and Techniques*, Vol. MTT-11, September 1963, p. 454. (Gives curves fit to charts of Whinnery *et al..*)

[6]    Somlo, P.I., "The Computation of Coaxial Line Step Capacitances," *IEEE Transactions on Microwave Theory and Techniques*, Vol. MTT-15, No. 1, January 1967, pp. 48–53. (Corrections to Whinnery *et al.* and Somlo 1963.)

[7]    Whinnery, J.R., *et al.*, "Coaxial-Line Discontinuities," *Proceedings of the IRE*, November 1944, pp. 695–709.

[8]    Whinnery, J.R., *et al.*, "Equivalent Circuits for Discontinuities in Transmission Lines," *Proceedings of the IRE*, February 1944, pp. 99–114.

[9]    Young, Leo, ed., *Advances in Microwaves*, Volume 2, "The Numerical Solution of Transmission Line Problems," by Harry E. Green, Academic Press, New York, 1967. (Improves on [7] and [8] for $b/a > 2$ and gives results in equation and table form rather than graphically.)

### 5.2.2   Coaxial Step in Outer Conductor

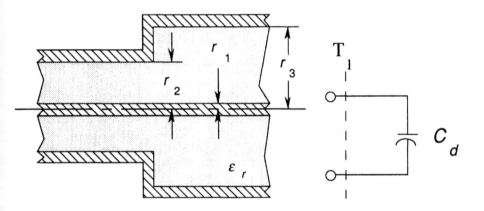

**Figure 5.2.2.1:     Coaxial Step in Outer Conductor**

A step on the outer conductor is modeled by a capacitance

$$C_d = 2.0 \, \varepsilon_r \, \pi \, r_1 \, C'_{d_2} \qquad (5.2.2.1)$$

where

$$C'_{d_2} = \frac{\varepsilon}{100.0 \, \pi} \left( \frac{\alpha^2 + 1.0}{\alpha} \ln \frac{1.0 + \alpha}{1.0 - \alpha} - 2.0 \ln \frac{4.0 \, \alpha}{1.0 - \alpha^2} \right)$$

$$+ \, 4.12 \times 10^{-15} \, (0.8 - \alpha) \, (\tau - 1.4) \quad \text{(F / cm)} \qquad (5.2.2.2)$$

$$\alpha = \frac{r_2 - r_1}{r_3 - r_1} \qquad (5.2.2.3)$$

$$\tau = \frac{r_3}{r_1} \qquad (5.2.2.4)$$

The equation is accurate to $\pm 0.6$ fF / cm for

$$0.01 \leq \alpha \leq 0.7$$

$$1.5 \leq \tau \leq 6.0$$

The effect of frequency can be approximately corrected [6] with the formula

$$\frac{C(f)}{C_d} = \frac{1.0}{\sqrt{1.0 - (2.0 \, r_3 / \lambda)^2}} \qquad (5.2.2.5)$$

which is valid for

$$0.0 \leq r_3 \leq \lambda / 2.0$$

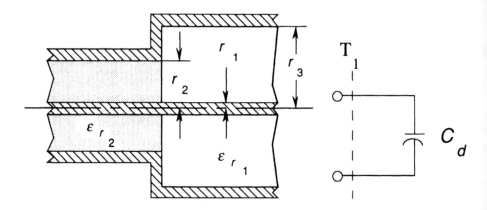

**Figure 5.2.2.2:** **Coaxial Step in Outer Conductor with Dielectric Change**

The capacitance of the step in the outer conductor with a corresponding dielectric step is approximated with

$$C_d = 2.0 \, \varepsilon_{r2} \, \pi \, r_1 \, C_{d_2}' \qquad (5.2.2.6)$$

where

$$C'_{d_2} = \frac{\varepsilon}{100.0\ \pi} \left( \frac{\alpha^2 + 1.0}{\alpha} \ln \frac{1.0 + \alpha}{1.0 - \alpha} - 2.0 \ln \frac{4.0\ \alpha}{1.0 - \alpha^2} \right)$$

$$+ 4.12 \times 10^{-15}\ (0.8 - \alpha)\ (\tau - 1.4) \quad (F/cm) \tag{5.2.2.7}$$

$$\alpha = \frac{r_2 - r_1}{r_3 - r_1} \tag{5.2.2.8}$$

$$\tau = \frac{r_3}{r_1} \tag{5.2.2.9}$$

REFERENCES

[1]  Gwarek, Wojciech K., "Computer-Aided Analysis of Arbitrarily Shaped Coaxial Discontinuities," *IEEE Transactions on Microwave Theory and Techniques*, Vol. MTT-36, No. 2, February 1988, pp. 337–342.

[2]  Marcuvitz, N., *Waveguide Handbook*, McGraw-Hill, New York, 1955. Reprint with errata and preface by N. Marcuvitz, Peter Peregrinus Ltd., 1986.

[3]  Saad, T., ed., *The Microwave Engineer's Handbook*, Vol. 1, Artech House, Norwood, MA, 1971.

[4]  Silvester, P., and Ivan A. Cermak, "Analysis of Coaxial Line Discontinuities by Boundary Relaxation," *IEEE Transactions on Microwave Theory and Techniques*, Vol. MTT-17, No. 8, August 1969, pp. 489–495.

[5]  Somlo, P.I., "The Computation of Coaxial Line Step Capacitances," *IEEE Transactions on Microwave Theory and Techniques*, Vol. MTT-15, No. 1 January 1967, pp. 48–53. (Corrections to Whinnery, *et al.* and Somlo 1963.)

[6]  Whinnery, J.R., *et al.*, "Equivalent Circuits for Discontinuities in Transmission Lines," *Proceedings of the IRE*, February 1944, pp. 99–114.

### 5.2.3   Coaxial Simultaneous Step in Inner and Outer Conductors

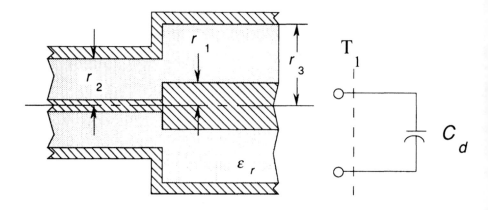

**Figure 5.2.3.1:**      **Coaxial Step in Inner and Outer Conductors**

The discontinuity capacitance of this structure is modeled by the sum of two discontinuity capacitances:

$$C_d = 2.0 \, \varepsilon_r \, \pi \, c \, C'_{d_1} + 2.0 \, \varepsilon_r \, \pi \, a \, C'_{d_2} \tag{5.2.3.1}$$

where

$$C'_{d_1} = \frac{\varepsilon}{100.0 \, \pi} \left( \frac{\alpha^2 + 1.0}{\alpha} \ln \frac{1.0 + \alpha}{1.0 - \alpha} - 2.0 \ln \frac{4.0 \, \alpha}{1.0 - \alpha^2} \right)$$

$$+ \, 1.11 \times 10^{-15} \, (1.0 - \alpha) \, (\tau - 1.0) \quad \text{(F / cm)} \tag{5.2.3.2}$$

$$C'_{d_2} = \frac{\varepsilon}{100.0 \, \pi} \left( \frac{\alpha^2 + 1.0}{\alpha} \ln \frac{1.0 + \alpha}{1.0 - \alpha} - 2.0 \ln \frac{4.0 \, \alpha}{1.0 - \alpha^2} \right)$$

$$+ \, 4.12 \times 10^{-15} \, (0.8 - \alpha) \, (\tau - 1.4) \quad \text{(F / cm)} \tag{5.2.3.3}$$

The effect of frequency can be approximately corrected [6] with the formula

$$\frac{C(f)}{C_d} = \frac{1.0}{\sqrt{1.0 - (2.0 \, b \, / \, \lambda)^2}} \tag{5.2.3.4}$$

which is valid for

$$0.0 \leq b \leq \lambda / 2.0$$

Some simple equations that give good results for the discontinuity capacitance of this structure over frequency are given in [1]:

$$C(f) \cong a + \frac{b}{\sqrt{1.0 - \left(\frac{f}{f_0}\right)^2}} \qquad (5.2.3.5)$$

where

$a$, $b$ are functions of the dimensions and dielectric constants $\varepsilon_{r1}$ and $\varepsilon_{r2}$.

$f_0$ is the cutoff frequency of the first TM -mode.

The values of $a$, $b$, and $f_0$ from [1] are given in the table below.

| $\varepsilon_{r1}$ | $\varepsilon_{r2}$ | $f_0$ | $a$ | $b$ |
|---|---|---|---|---|
| 1 | 1 | 49.42 GHz | 2.125 | 12.88 |
| 1 | 2 | 34.94 GHz | 4.339 | 25.69 |
| 1 | 10 | 15.63 GHz | 20.93 | 128.9 |

**Table 5.2.3.1:** **Step Capacitance Frequency Dependent Parameters**

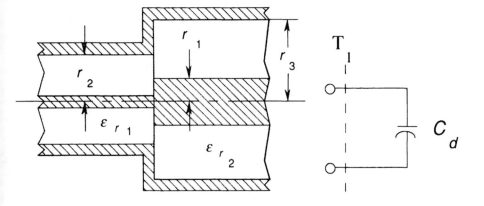

**Figure 5.2.3.2:** **Coaxial Step in Inner and Outer Conductor with Dielectric Change**

For lines with a discontinuity in the dielectric as well

$$C_d \cong 2.0 \, \varepsilon_{r1} \, \pi \, r_1 \, C_{d_2}' + 2.0 \, \varepsilon_{r2} \, \pi \, r_2 \, C_{d_1}' \qquad (5.2.3.6)$$

where

$$C'_{d_1} = \frac{\varepsilon}{100.0\ \pi} \left[ \frac{\alpha_1{}^2 + 1.0}{\alpha_1} \ln \left( \frac{1.0 + \alpha_1}{1.0 - \alpha_1} \right) - 2.0 \ln \left( \frac{4.0\ \alpha_1}{1.0 - \alpha_1{}^2} \right) \right]$$

$$+ 1.11 \times 10^{-15} (1.0 - \alpha_1)(\tau_1 - 1.0) \quad (\text{F}/\text{cm}) \tag{5.2.3.7}$$

$$C'_{d_2} = \frac{\varepsilon}{100.0\ \pi} \left[ \frac{\alpha_2{}^2 + 1.0}{\alpha_2} \ln \frac{1.0 + \alpha_2}{1.0 - \alpha_2} - 2.0 \ln \frac{4.0\ \alpha_2}{1.0 - \alpha_2{}^2} \right]$$

$$+ 4.12 \times 10^{-15} (0.8 - \alpha_2)(\tau_2 - 1.4) \quad (\text{F}/\text{cm}) \tag{5.2.3.8}$$

and

$$\alpha_1 = \frac{r_2 - r_1}{r_2 - r_0} \tag{5.2.3.9}$$

$$\alpha_2 = \frac{r_2 - r_1}{r_3 - r_1} \tag{5.2.3.10}$$

$$\tau_1 = \frac{r_2}{r_0} \tag{5.2.3.11}$$

$$\tau_2 = \frac{r_3}{r_1} \tag{5.2.3.12}$$

Another type of coaxial step (shown below) is modeled by Young [7].

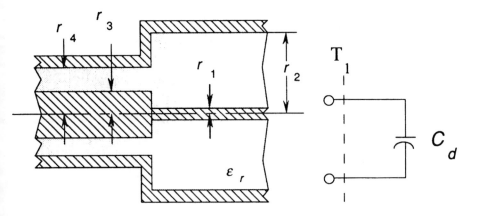

**Figure 5.2.3.3:**         Another Coaxial Step in Inner and Outer Conductors

Young suggests modeling this with the series combination of two equivalent steps (similar to above). The equivalent steps are a step on the inner conductor and a step on the outer conductor.

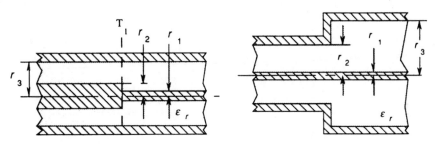

**Figure 5.2.3.4:**         Double Step is Calculated from Equivalent Discontinuities

First we calculate from the original discontinuity's dimensions (subscripted $o$)

$$r = 0.5 \exp\left( \frac{\log r_1 \log \frac{r_4}{r_3} - \log r_3 \log \frac{r_2}{r_1}}{\log \frac{r_4}{r_3} - \log \frac{r_2}{r_1}} \right) \tag{5.2.3.13}$$

The step on the inner conductor portion of the equivalent model is now calculated with

$$r_1 = r_{1,o}$$

$$r_2 = r_{3,o}$$

$r_3 = r$

and the step on the outer conductor portion of the equivalent model is calculated with

$r_1 = r$

$r_2 = r_{4,o}$

$r_3 = r_{2,o}$

The accuracy of this approximation is within 10% of numeric calculations.

### REFERENCES

[1]   Gogioso, L., *et al.*, "A Variational Approach to Compute the Equivalent Capacitance of Coaxial Line Discontinuities," *1979 IEEE MTT-S International Microwave Symposium Digest*, pp. 580–582.

[2]   Gwarek, Wojciech K., "Computer-Aided Analysis of Arbitrarily Shaped Coaxial Discontinuities," *IEEE Transactions on Microwave Theory and Techniques*, Vol. MTT-36, No. 2, February 1988, pp. 337–342.

[3]   Saad, T., ed., *The Microwave Engineer's Handbook*, Vol. 1, Artech House, Norwood, MA, 1971.

[4]   Silvester, P., and Ivan A. Cermak, "Analysis of Coaxial Line Discontinuities by Boundary Relaxation," *IEEE Transactions on Microwave Theory and Techniques*, Vol. MTT-17, No. 8, August 1969, pp. 489–495.

[5]   Somlo, P.I., "The Computation of Coaxial Line Step Capacitances," *IEEE Transactions on Microwave Theory and Techniques*, Vol. MTT-15, No. 1, January 1967, pp. 48–53. (Corrections to Whinnery *et al.* and Somlo 1963.)

[6]   Whinnery, J.R., *et al.*, "Equivalent Circuits for Discontinuities in Transmission Lines," *Proceedings of the I.R.E.*, February 1944, pp. 99–114.

[7]   Young, Leo, *ed.*, *Advances in Microwaves*, Volume 2, "The Numerical Solution of Transmission Line Problems," by Harry E. Green, Academic Press, New York, 1967. (Improves on [82] and [83] for $b/a > 2$ and gives results in equation form rather than graphically)

### 5.2.4   Gap in Coaxial Center Conductor

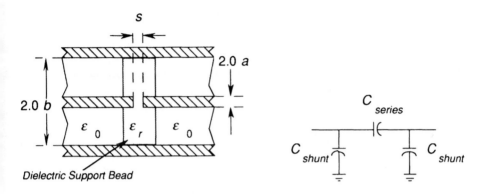

**Figure  5.2.4.1:**      Gap in Coaxial Center Conductor

Dawirs [1] shows that the coaxial gap can be modeled as a $\pi$ capacitor circuit. By comparison of various source's equations with measurement data it is determined that the best model is to use the $C_{series}$ of Young [5], and the $C_{shunt}$ of Whinnery et al. [4]. The value of $C_{shunt}$ is most easily calculated with the closed-form equations of Somlo [3].

For the purposes of calculation the field is assumed to concentrate in the dielectric support (higher $\varepsilon_r$ than the surrounding dielectric). The $\varepsilon_r$ of the support bead is used for the discontinuity calcuations.

$$C_{series} = \frac{0.7056 \, \varepsilon_r \, a^2}{s} \text{ (pF)} \tag{5.2.4.1}$$

where $a$ and $s$ are in inches.  Equation (5.2.4.1) is valid for

$$b - a < \lambda$$

$$s \ll \lambda$$

$$s \ll b - a$$

where $\lambda$ is the free-space wavelength.

$$C_{shunt} = 2.0 \, \varepsilon_r \, \pi \, b \, C_{d_1}' \tag{5.2.4.2}$$

where

$$C'_{d_1} = \frac{2.0\ \varepsilon}{100.0\ \pi} \left( 1.477 - \ln 4.0\ \alpha \right) \tag{5.2.4.3}$$

$$\alpha = 1.0 - \frac{a}{b} \tag{5.2.4.4}$$

Also see the coaxial open end discontinuity, which is the above structure with a short inserted in the gap. The open end, therefore, has half the value of $C_{series}$ and the same value for $C_{shunt}$ (however, there is only one $C_{shunt}$ in the model).

REFERENCES

[1]   Dawirs, Harvel N., "Equivalent Circuit of a Series Gap in the Center Conductor of a Coaxial Transmission Line," *IEEE Transactions on Microwave Theory and Techniques*, MTT-17, No. 2, February 1969, pp. 127–129.

[2]   Gwarek, Wojciech K., "Computer-Aided Analysis of Arbitrarily Shaped Coaxial Discontinuities," *IEEE Transactions on Microwave Theory and Techniques*, Vol. MTT-36, No. 2, February 1988, pp. 337–342.

[3]   Somlo, P.I., "The Computation of Coaxial Line Step Capacitances," *IEEE Transactions on Microwave Theory and Techniques*, Vol. MTT-15, No. 1, January 1967, pp. 48–53. (Corrections to Whinnery, *et al.* and Somlo 1963.)

[4]   Whinnery, J.R., *et al.*, "Coaxial-Line Discontinuities," *Proceedings of the IRE*, Vol. 32, November 1944, pp. 695–709.

[5]   Young, Leo, "The Practical Realization of Series-Capacitive Couplings for Microwave Filters," *Microwave Journal*, December 1962, pp. 79–81.

### 5.2.5   Coaxial Open Circuit

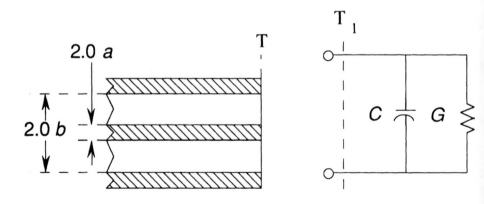

**Figure 5.2.5.1:**   Coaxial Open Circuit

The coaxial open circuit can be part of a filter or other circuit. The fringing fields of an open circuit can also be used to probe a dielectric sample and measure its permittivity.

The equivalent circuit's conductance is calculated with:

$$\frac{G}{Y_0} = \frac{1.0}{\ln \frac{a}{b}} \int_0^{\pi / 2.0} \frac{d\theta}{\sin \theta} \left[ J_0(k \, a \, \sin \, \theta) - J_0(k \, b \, \sin \, \theta) \right]^2 \tag{5.2.5.1}$$

$$\cong \frac{2}{3 \ln \frac{a}{b}} \left[ \frac{\pi^2 \, (b^2 - a^2)}{\lambda^2} \right]^2 \tag{5.2.5.2}$$

The susceptance is:

$$\frac{B}{Y_0} = \frac{1.0}{\pi \ln \frac{a}{b}} \int_0^\pi \left[ 2.0 \, Si\left( k \, \sqrt{a^2 + b^2 - 2.0 \, a \, b \, \cos \, \phi} \, \right) \right.$$

$$\left. - Si\left( 2.0 \, k \, a \, \sin \frac{\phi}{2.0} \right) - Si\left( 2.0 \, k \, b \, \sin \frac{\phi}{2.0} \right) d\phi \right] \tag{5.2.5.3}$$

$$\cong \frac{8.0 \, (a + b)}{\lambda_0 \ln \frac{a}{b}} \left[ E\left( \frac{2.0 \, \sqrt{a \, b}}{a + b} \right) - 1.0 \right] \tag{5.2.5.4}$$

where

$E(x) =$ complete elliptic integral of the second kind

$Si(x) =$ sine integral function

$J_0(x) =$ Bessel function of the first kind and order 0

$\lambda_0 =$ free-space wavelength

$$k = \frac{2.0 \, \pi}{\lambda_0}$$

The approximations are valid for

$$a \ll \lambda_0$$

$$b \ll \lambda$$

Stated accuracy of the exact equations is better than 10%. The approximations agree to better than 15% for

$$\frac{a - b}{\lambda} < 0.10$$

$$\frac{a}{b} \geq 2.0$$

with increasing accuracy for increasing $a / b$ (higher impedances).[1]

REFERENCES

[1]   Gwarek, Wojciech K., "Computer-Aided Analysis of Arbitrarily Shaped Coaxial Discontinuities," *IEEE Transactions on Microwave Theory and Techniques*, Vol. MTT-36, No. 2, February 1988, pp. 337–342.

[2]   Marcuvitz, N., *Waveguide Handbook*, McGraw-Hill, New York, 1955. Reprint with errata and preface by N. Marcuvitz, Peter Peregrinus Ltd., 1986.

## 5.2.6   Enclosed Coaxial Open Circuit

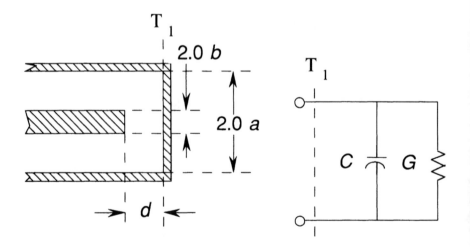

**Figure  5.2.6.1:**      **Enclosed  Coaxial  Open  Circuit**

---

[1]A 50 $\Omega$ line has $a / b$ = 2.3 for an air dielectric; $a / b$ increases for other dielectrics.

The equivalent circuit of the coaxial open into its shield is a capacitance from the center conductor to the shield at the reference plane $T_1$. The normalized susceptance is [2]

$$\frac{B}{Y_0} = \left(\frac{4.0\,b}{\lambda_0} \ln \frac{a}{b}\right)\left[\frac{\pi\,b}{4.0\,d} + \ln\left(\frac{a-b}{d}\right)\right] \qquad (5.2.6.1)$$

where

$\lambda_0$ = free-space wavelength

This model is valid for:

$2\,\pi\,d \ll \lambda_0$

$d \ll a - b$

$\lambda_0 > 2\,(a - b)$

## REFERENCES

[1] Gwarek, Wojciech K., "Computer-Aided Analysis of Arbitrarily Shaped Coaxial Discontinuities," *IEEE Transactions on Microwave Theory and Techniques*, Vol. MTT-36, No. 2, February 1988, pp. 337–342.

[2] Marcuvitz, N., *Waveguide Handbook*, McGraw-Hill, New York, 1955. Reprint with errata and preface by N. Marcuvitz, Peter Peregrinus Ltd., 1986.

[3] Young, Leo, ed., *Advances in Microwaves*, Volume 2, Academic Press, New York, 1967.

## 5.3   PAIRED-LINE DISCONTINUITIES

Analysis of paired-line transmission line discontinuities are not generally included in available literature. They are generally not used in filters and circuits. The designer of these structures may be able to make use of the equations of a similar structure for approximate design. The design's success would then rely on further experimentation.

### 5.3.1   Open End

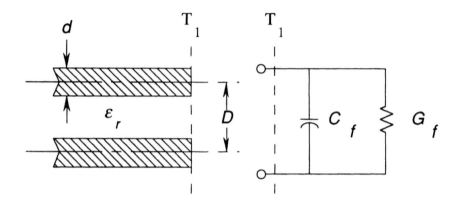

Figure  5.3.1.1:        Paired-Line Open End

In [1] the equations for the excess capacitance of the open end of a two-wire transmission line were experimentally determined due to the intractibility of an exact solution to be

$$\frac{C_f}{C_0} = \frac{D}{-3.954 + \sqrt{\left[2.564 \cosh^{-1}\left(\frac{D}{d}\right)\right]^2 + (3.954)^2}} \quad \text{(unitless)} \quad (5.3.1.1)$$

where

$C_0$ is the capacitance per unit length of the transmission line

$d$ and $D$ are defined in Section 3.3.1 and have the same dimensions

which is valid for

$$\frac{C_f}{C_0} \ll \lambda$$

and results from curve-fitting data taken over the range of $1.67 \leq \cosh^{-1}\left(\dfrac{D}{d}\right) \leq 5.0$. A parallel conductance, $G_f$, is also present and can be calculated with

$$G_f = \frac{\eta\,(\beta\,D)^4 \left[\dfrac{(C_f/C_0)}{D}\right]^2}{12.0\,\pi} \quad \text{(Siemens)} \tag{5.3.1.2}$$

where $\eta$ is defined in Section 1.1 and the other terms have already been defined.

Green [1] notes that the fringing field lines from the open extend approximately $D\,/\,4$ back from the open end and to infinity ($\gg D$) on the open end.

REFERENCES

[1]    Green, H.E., and J.D. Cashman, "End Effect in Open-Circuited Two-Wire Transmission Lines," *IEEE Transactions on Microwave Theory and Techniques*, Vol. MTT-34, No. 1, January 1986, pp. 180–182.

## 5.4    COPLANAR WAVEGUIDE DISCONTINUITIES

The coplanar waveguide structure is growing in popularity because of its single-sided nature; however closed-form design equations for discontinuities are not yet readily available. Design curves are available in some references. Houdart [1] catalogs many of the structures shown here and others.

REFERENCES

[1]    Houdart, M., "Coplanar Lines: Application to Broadband Microwave Integrated Circuits," *6th European Microwave Conference Proceedings*, 1976, pp. 49–53.

### 5.4.1    Coplanar Waveguide Shunt Capacitance

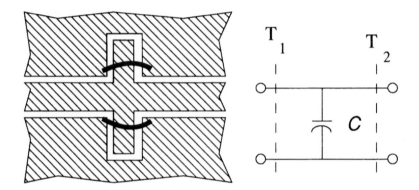

**Figure  5.4.1.1:      Shunt  Capacitance  in  CPW**

This line can be modeled as ideal transmission lines in the absence of a model for the cross junction. The open end effects can be modeled with the aid of curves in [2] and [3] as described later. Multiple lines can be repeated to build up higher capacitances. Bond wires (shown as dark lines in Figure 5.4.1.1) between the grounds are used to suppress undesired modes.

REFERENCES

[1]    Houdart, M., "Coplanar Lines: Application to Broadband Microwave Integrated Circuits," *6th European Microwave Conference Proceedings*, 1976, pp. 49–53.

[2]    Naghed, Mohsen, and Ingo Wolff, "A Three-Dimensional Finite-Difference Calculation of Equivalent Capacitances of Coplanar Waveguide Discontinuities," *1980 IEEE MTT-S International Microwave Symposium Digest*, pp. 1143–1146.

[3]    Simons, Rainee N., and George E. Ponchak, "Modeling of Some Coplanar Waveguide Discontinuities," *IEEE Transactions on Microwave Theory and Techniques,* Vol. MTT-36, No. 12, December 1988, pp. 1796–1803.

## 5.4.2   Coplanar Waveguide Series Capacitance

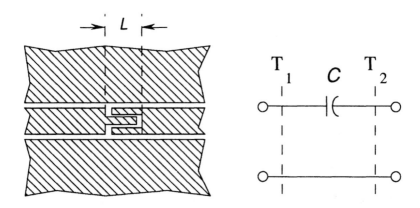

**Figure  5.4.2.1:**      Series Capacitance in CPW

Again, no general closed-form equations are available. Multiple fingers may be interleaved to increase the capacitance. Williams and Schwarz [5] use the above configuration in a bandpass filter design and note that the radiation losses of the above structure are lower than the simpler end-coupled CPW series capacitance. Dib *et al.* [2] give a graph for one set of dimensions. They found that the structure resonated when $L$ was slightly less than $\lambda / 4$. In Dib *et al.* [3] closed-form discontinuity model equations are given for a single set of physical dimensions as a function of the finger length $L$.

In lieu of design equations a starting point may be the relations for coupled CPW as described in Chapter 4 together with a technique after Bates [1].

REFERENCES

[1]    Bates, R.N., "Design of Microstrip Spur-Line Band-Stop Filters," *Microwaves, Optics and Acoustics,* Vol. 1, No. 6, November 1977, pp. 209–214.

[2]    Dib, N.I., *et al.*, "Coplanar Waveguide Discontinuities for P-I-N Diode Switches and Filter Applications," *1990 IEEE MTT-S International Microwave Symposium Digest,* pp. 399–402.

[3]    Dib, N.I., *et al.*, "Theoretical and Experimental Characterization of Coplanar Waveguide Discontinuities for Filter Applications," *IEEE Transactions on Microwave Theory and Techniques,* Vol. MTT-39, No. 5, May 1991, pp. 873–882.

[4]     Houdart, M., "Coplanar Lines:  Application to Broadband Microwave Integrated Circuits," *6th European Microwave Conference Proceedings*, 1976, pp. 49–53.

[5]     Williams, Dylan F., and S.E. Schwarz, "Design and Performance of Coplanar Waveguide Bandpass Filters," *IEEE Transactions on Microwave Theory and Techniques*, Vol. MTT-31, No. 7, July 1983, pp. 558–566.

### 5.4.3   Coplanar Waveguide Shunt Inductance

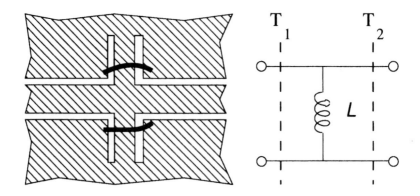

**Figure   5.4.3.1:**     **Shunt Inductance in CPW**

This structure is cataloged in Houdart [1].  Kibuuka *et al.* [2] use this in a snaked or bent form as part of a band-pass filter.  No cross model is available, so ideal equations for the short-circuited lines must be used.  The bond wires (shown as dark lines in Figure 5.4.3.1) are used to suppress higher order modes.

REFERENCES

[1]     Houdart, M., "Coplanar Lines:  Application to Broadband Microwave Integrated Circuits," *6th European Microwave Conference Proceedings*, 1976, Rome, Italy, pp. 49–53.

[2]     Kibuuka, Godfrey, *et al.*, "Coplanar Lumped Elements and their Application in Filters on Ceramic and Gallium Arsenide Substrates," *19th European Microwave Conference Proceedings*, September 4–7, 1989, London, England, pp. 656–661.

## 5.4.4   Coplanar Waveguide Series Inductance

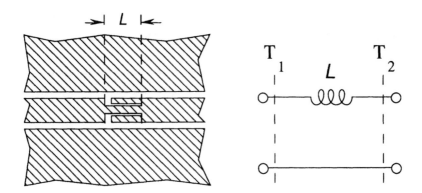

**Figure 5.4.4.1:**     **Series Inductance in CPW**

In Dib *et al.* [2], this structure is shown to resonate somewhat below the frequency where $L = \lambda / 4.0$. In Dib *et al.* [3] closed-form discontinuity model equations are given for a single set of physical dimensions as a function of the finger length $L$. In lieu of a full set of design equations a starting point may be the relations for coupled CPW as described in Chapter 4 together with a technique after Bates [1].

REFERENCES

[1]   Bates, R.N., "Design of Microstrip Spur-Line Band-Stop Filters," *Microwaves, Optics and Acoustics,* Vol. 1, No. 6, November 1977, pp. 209–214.

[2]   Dib, N.I., *et al.*, "Coplanar Waveguide Discontinuities for P-I-N Diode Switches and Filter Applications," *1990 IEEE MTT-S International Microwave Symposium Digest*, pp. 399–402.

[3]   Dib, N.I., *et al.*, "Theoretical and Experimental Characterization of Coplanar Waveguide Discontinuities for Filter Applications," *IEEE Transactions on Microwave Theory and Techniques*, Vol. MTT-39, No. 5, May 1991, pp. 873–882.

[4]   Houdart, M., "Coplanar Lines: Application to Broadband Microwave Integrated Circuits," *6th European Microwave Conference Proceedings*, 1976, Rome, Italy, pp. 49–53.

[5]   Houdart, M., *et al.*, "Coplanar Lines: Application to Lumped and Semi-Lumped Microwave Integrated Circuits," *7th European Microwave Conference Proceedings*, 1977, Copenhagen, Denmark, pp. 450–454.

[6]    Simons, Rainee N., and George E. Ponchak, "Modeling of Some Coplanar Waveguide Discontinuities," *IEEE Transactions on Microwave Theory and Techniques*, Vol. MTT-36, No. 12, December 1988, pp. 1796–1803.

### 5.4.5    Coplanar Waveguide Open End with Connected Grounds

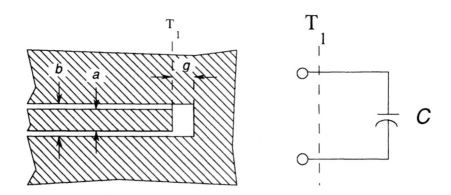

**Figure  5.4.5.1:**        **CPW  Open  End  with  Connected  Grounds**

The coplanar waveguide open end occurs in filters and matching circuits.  Curves of the capacitance are found in Naghed and Wolff [1] and Simons and Ponchak [3] for a limited number of cases.  For $g / h \geq 0.3$ the capacitance is approximately independent of $g$. Curve-fitting the results of Naghed and Wolff [2] gives the following equations for design. For $a / b = 0.75$:

$$C = 13.49807 + \frac{0.4487826}{g / h} - \frac{0.0047486308}{(g / h)^2} \quad \text{(fF)} \tag{5.4.5.1}$$

For $a / b = 0.50$:

$$C = 9.7179749 + \frac{0.34173829}{g / h} - \frac{0.0042329111}{(g / h)^2} \quad \text{(fF)} \tag{5.4.5.2}$$

For $a / b = 0.33$:

$$C = 7.3706512 + \frac{0.20880575}{g / h} - \frac{0.002343378}{(g / h)^2} \quad \text{(fF)} \tag{5.4.5.3}$$

For $a / b = 0.25$:

$$C = 5.9622775 + \frac{0.16905027}{g / h} - \frac{0.0019673178}{(g / h)^2} \quad \text{(fF)} \tag{5.4.5.4}$$

Equations are valid to approximately ±2.5% over the range of the graphs

$$0.03 \leq g / h \leq 1.0$$

Interpolation between values of $a / b$ can be used elsewhere.

REFERENCES

[1]  Naghed, Mohsen, and Ingo Wolff, "A Three-Dimensional Finite-Difference Calculation of Equivalent Capacitances of Coplanar Waveguide Discontinuities," *1980 IEEE MTT-S International Microwave Symposium Digest*, pp. 1143–1146.

[2]  Naghed, Mohsen, and Ingo Wolff, "Equivalent Capacitances of Coplanar Waveguide Discontinuities and Interdigitated Capacitors Using a Three-Dimensional Finite Difference Method," *IEEE Transactions on Microwave Theory and Techniques*, Vol. MTT-38, No. 12, December 1990, pp. 1808–1815.

[3]  Simons, Rainee N., and George E. Ponchak, "Modeling of Some Coplanar Waveguide Discontinuities," *IEEE Transactions on Microwave Theory and Techniques,* December 1988, pp. 1796–1803.

## 5.4.6   CPW Open End

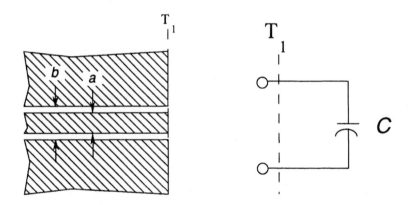

**Figure 5.4.6.1:**     Coplanar Waveguide Open End

Analysis techniques are described in [1]. Design curves are available in Naghed and Wolff [2]. For the case of $\varepsilon_r = 9.8$ and $h = 0.635$ mm, curve fits to their curves are given below. These curves are valid to within approximately ±5% over the range

$$0.1 \leq w / d \leq 0.9$$

For $h / b = 1.0$:

$$C = -5.3001726 + 33.062752 \sqrt{a / b} \quad \text{(fF)} \qquad (5.4.6.1)$$

For $h / b = 0.2$:

$$C = -1.4051152 + 22.07531 \sqrt{a / b} \quad \text{(fF)} \qquad (5.4.6.2)$$

For $h / b = 0.1$:

$$C = -0.99986182 + 17.285616 \sqrt{a / b} \quad \text{(fF)} \qquad (5.4.6.3)$$

REFERENCES

[1]    Jansen, R.H., "Hybrid Mode Analysis of End Effects of Planar Microwave and Millimetrewave Transmission Lines," *IEE Proceedings,* Vol. 128, Pt. H, No. 2, April 1981, pp. 77–86.

[2]    Naghed, Mohsen, and Ingo Wolff, "A Three-Dimensional Finite-Difference Calculation of Equivalent Capacitances of Coplanar Waveguide Discontinuities," *1980 IEEE MTT-S International Microwave Symposium Digest,* pp. 1143–1146.

[3]    Naghed, Mohsen, and Ingo Wolff, "Equivalent Capacitances of Coplanar Waveguide Discontinuities and Interdigitated Capacitors Using a Three-Dimensional Finite Difference Method," *IEEE Transactions on Microwave Theory and Techniques,* Vol. MTT-38, No. 12, December 1990, pp. 1808–1815.

### 5.4.7   Coplanar Waveguide Series Gap

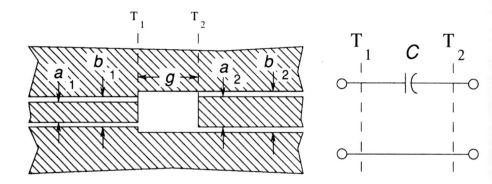

**Figure 5.4.7.1:     CPW Series Gap**

Williams and Schwarz [4] compare this structure to simple interdigital CPW capacitors (as in Figure 5.4.2.1). They also experimented with overlaying this structure with dielectric material and dielectric material with a conductive mode suppressing ribbon.

Some design curves are also given for the equivalent circuit. Simons and Ponchak [3] include shunt capacitances in their model and also present curves.

Naghed [1] gives curves for the case of $\varepsilon_r = 9.8$, $h = 0.635$, $a_1 / h = 0.2$, and $a_1 / b_1 = a_2 / b_2 = 0.56$. Curve fits are given below.

For $a_2 / a_1 = 1.0$:

$$C_{p1} = C_{p2} = \frac{-0.47822096 + 57.893523 \frac{g}{h}}{1.0 + 5.8451737 \frac{g}{h}} \quad \text{(fF)} \tag{5.4.7.1}$$

$$C_g = \frac{14.099684 - 11.843073 \frac{g}{h} + 1.0433114 \left(\frac{g}{h}\right)^2}{1.0 + 10.502696 \frac{g}{h} + 18.503307 \left(\frac{g}{h}\right)^2} \quad \text{(fF)} \tag{5.4.7.2}$$

For $a_2 / b_1 = 2.0$:

$$C_{p1} = \frac{-4.0235493 + 35.576495 \frac{g}{h}}{1.0 + 2.7541323 \frac{g}{h}} \quad \text{(fF)} \tag{5.4.7.3}$$

$$C_{p2} = \frac{1.0 - 7.9340513 \frac{g}{h} + 37.923732 \left(\frac{g}{h}\right)^2}{0.087137596 - 0.59125367 \frac{g}{h} + 2.941263 \left(\frac{g}{h}\right)^2} \quad \text{(fF)} \tag{5.4.7.4}$$

$$C_g = \frac{18.99756 - 1.8997741 \frac{g}{h}}{1.0 + 9.9022677 \frac{g}{h} + 6.8034362 \left(\frac{g}{h}\right)^2} \quad \text{(fF)} \tag{5.4.7.5}$$

For $a_2 / b_1 = 3.0$:

$$C_{p1} = \frac{-6.2682507 + 32.193198 \frac{g}{h}}{1.0 + 3.6295239 \frac{g}{h}} \quad \text{(fF)} \tag{5.4.7.6}$$

$$C_{p2} = \sqrt{315.94893 + 9.6539534 / x} \tag{5.4.7.7}$$

$$C_g = \frac{20.102184 - 2.00947 \frac{g}{h}}{1.0 + 7.7736146 \frac{g}{h}} \quad \text{(fF)} \qquad (5.4.7.8)$$

It was attempted to make these equations correct asymptotically; however, the accuracy and range of the data made this impossible to do in every case. These CAE equations are valid in general to within < 5% over the range

$$0.02 \le g / h \le 0.3$$

### REFERENCES

[1]    Drissi, M'hamed, *et al.*, "Analysis of Coplanar Waveguide Radiating End Effects Using the Integral Equation Technique," *IEEE Transactions on Microwave Theory and Techniques*, Vol. MTT-39, No. 1, January 1991.

[2]    Naghed, Mohsen, and Ingo Wolff, "A Three-Dimensional Finite-Difference Calculation of Equivalent Capacitances of Coplanar Waveguide Discontinuities," *1980 IEEE MTT-S International Microwave Symposium Digest*, pp. 1143–1146.

[3]    Naghed, Mohsen, and Ingo Wolff, "Equivalent Capacitances of Coplanar Waveguide Discontinuities and Interdigitated Capacitors Using a Three-Dimensional Finite Difference Method," *IEEE Transactions on Microwave Theory and Techniques*, Vol. MTT-38, No. 12, December 1990, pp. 1808–1815.

[4]    Simons, Rainee N., and George E. Ponchak, "Modeling of Some Coplanar Waveguide Discontinuities," *IEEE Transactions on Microwave Theory and Techniques,* Vol. MTT-36, No. 12, December 1988, pp. 1796–1803.

[5]    Williams, Dylan F., and S.E. Schwarz, "Design and Performance of Coplanar Waveguide Bandpass Filters," *IEEE Transactions on Microwave Theory and Techniques*, Vol. MTT-31, No. 7, July 1983, pp. 558–566.

### 5.4.8   Coplanar Waveguide Abrupt Width Change

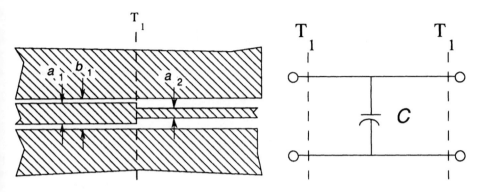

**Figure 5.4.8.1:**     **CPW Abrupt Width Change**

A limited number of curves for this discontinuity are given in Simons and Ponchak [1]. Curve fits to this data are given here.

For $\varepsilon_r = 2.1$, $a_1 = 10.5$ mil, $a_1 / b_1 = 0.86$, $h = 62$ mil, and $1.1 < S_1 / S_2 \leq 2.0$:

$$C_s = 0.01335 + 0.00772 \left(\frac{S_1}{S_2}\right) \tag{5.4.8.1}$$

For $\varepsilon_r = 2.2$, $a_1 = 11$ mil, $a_1 / b_1 = 0.9$, $h = 125$ mil, and $1.1 < S_1 / S_2 \leq 3.0$:

$$C_s = \sqrt{0.0062268592 - \frac{0.0048471339}{(S_1 / S_2)^2}} \tag{5.4.8.2}$$

For $\varepsilon_r = 2.2$, $a_1 = 9$ mil, $a_1 / b_1 = 0.9$, $h = 125$ mil, and $1.1 < S_1 / S_2 \leq 3.0$:

$$C_s = \sqrt{0.0024722655 - \frac{0.0023832659}{(S_1 / S_2)^2}} \tag{5.4.8.3}$$

These equations fit the curves of [1] over the specified range to better than ±6%.

#### REFERENCES

[1]   Simons, Rainee N., and George E. Ponchak, "Modeling of Some Coplanar Waveguide Discontinuities," *IEEE Transactions on Microwave Theory and Techniques,* Vol. MTT-36, No. 12, December 1988, pp. 1796–1803.

### 5.4.9   CPW Abrupt Ground Plane Width Change

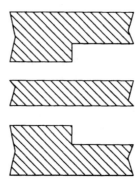

**Figure 5.4.9.1:**   Coplanar Waveguide Abrupt Ground Plane Step

There is very little information on this structure in the literature. Kuo and Itoh [1] give plots of *s*-parameters for a range of step sizes.

REFERENCE

[1]   Kuo, Chih-Wen, and Tatsuo Itoh, "Characterization of the Coplanar Waveguide Step Discontinuity Using the Transverse Resonance Method," *19th European Microwave Conference Proceedings*, September 4–7, 1989, London, England, pp. 662–665.

### 5.4.10   Abrupt Coplanar Waveguide Bend

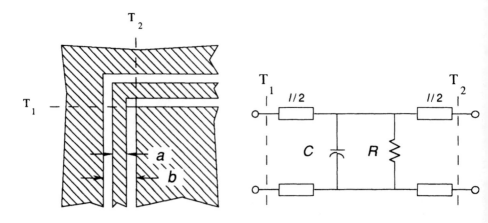

**Figure 5.4.10.1:**   CPW Abrupt Bend

The capacitive discontinuity at a bend in CPW is caused by the difference in path length between the inner and outer metallization slots. Simons and Ponchak [1] report very good results compensating this difference with a small piece of dielectric on the inner slot of the bend to slow down this part of the wave. The model is described and was measured for one set of parameters in [2]. The resistive component is due to radiation losses at the corner.

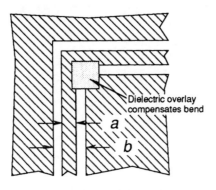

**Figure 5.4.10.2:** **Compensated Abrupt Bend in CPW**

REFERENCES

[1]    Simons, Rainee N., and George E. Ponchak, "Modeling of Some Coplanar Waveguide Discontinuities," *IEEE Transactions on Microwave Theory and Techniques,* Vol. MTT-36, No. 12, December 1988, pp. 1796–1803.

[2]    Simons, Rainee N., *et al.,* "Channelized Coplanar Waveguide: Discontinuities, Junctions, and Propagation Characteristics," *1989 IEEE MTT-S Symposium Digest,* pp. 915–918.

## 5.4.11 CPW Short Circuit

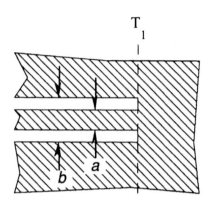

**Figure 5.4.11.1:    CPW Shorted End**

The CPW short is not well covered in the literature. It can be modeled as an ideal line without discontinuities. A similar structure, coupled slots, is analyzed in [3].

REFERENCES

[1]    Bartolucci, G., and J. Piotrowski, "Full-Wave Analysis of Shielded Coplanar Waveguide Short-End," *Electronics Letters*, Vol. 26, No. 19, September 13, 1990, pp. 1615–1616.

[2]    Drissi, M'hamed, *et al.*, "Analysis of Coplanar Waveguide Radiating End Effects Using the Integral Equation Technique," *IEEE Transactions on Microwave Theory and Techniques*, Vol. MTT-39, No. 1, January 1991.

[3]    Jansen, R.H., "Hybrid Mode Analysis of End Effects of Planar Microwave and Millimetrewave Transmission Lines," *IEE Proceedings,* Vol. 128, Pt. H, No. 2, April 1981, pp. 77–86.

## 5.4.12 CPW Air Bridge

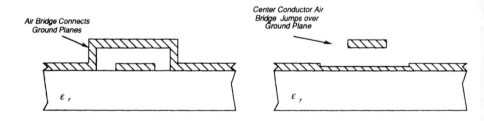

**Figure 5.4.12.1:    CPW Air Bridges**

The air bridge is used in MMIC CPW to bridge signal conductors and also to break up modes by shorting the ground planes together. Because of the small dimensions involved in MMICs and multiple crossings, the capacitance between the lines becomes significant and must be corrected. Koster [1] measured the two types of bridges and various compensation schemes to reduce the capacitance's effects.

The second configuration of Figure 5.4.12.1 was shown to have significantly lower losses. Koster also recommends a similar MMIC coplanar waveguide T-junction as having fewer problems with higher order modes.

REFERENCE

[1]     Koster, N.H.L., *et al.*, *19th European Microwave Conference Proceedings*, September 4–7, 1989, London, England, pp. 666–671.

## 5.4.13 CPW Interdigital Capacitor

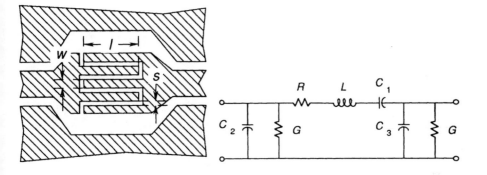

Figure  5.4.13.1:    CPW  Interdigital  Capacitor

Design equations are not available for this configuration. Kibuuka *et al.* [1] found that the $\varepsilon_{eff}$ of the modes of this structure are nearly equal and gives the model shown in Figure 5.4.13.1 for one set of dimensions. Naghed and Wolff [2] calculated this structure using the finite difference technique. The inductance element of the model, $L$, becomes significant for long fingers.

The interdigital capacitor can be analyzed by coupled lines with the correct conditions set at the ends. As an alternative, the interdigital capacitor of Chapter 7, could be used.

REFERENCES

[1]     Kibuuka, Godfrey, *et al.*, "Coplanar Lumped Elements and their Application in Filters on Ceramic and Gallium Arsenide Substrates," *19th European Microwave Conference Proceedings*, September 4–7, 1989, London, England, pp. 656–661.

[2]     Naghed, Mohsen, and Ingo Wolff, "A Three-Dimensional Finite-Difference Calculation of Equivalent Capacitances of Coplanar Waveguide Discontinuities," *1980 IEEE MTT-S International Microwave Symposium Digest*, pp. 1143–1146.

[3]     Naghed, Mohsen, and Ingo Wolff, "Equivalent Capacitances of Coplanar Waveguide Discontinuities and Interdigitated Capacitors Using a Three-Dimensional Finite Difference Method," *IEEE Transactions on Microwave Theory and Techniques*, Vol. MTT-38, No. 12, December 1990, pp. 1808–1815.

## 5.5    MICROSTRIP LINE DISCONTINUITIES

### 5.5.1    Microstrip Line Bend

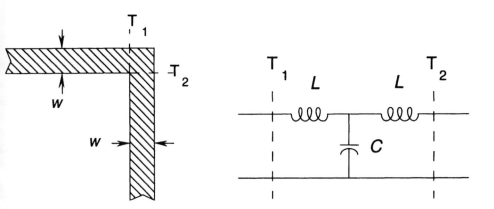

Figure 5.5.1.1:  Microstrip Line Abrupt Bend

The discontinuity created by a bend in a microstrip line is given by the model above [2] where:

$$L/h = \frac{100.0}{2.0}\left(4.0\sqrt{w/h} - 4.21\right) \text{ (nH / m)} \tag{5.5.1.1}$$

and for $w/h < 1$:

$$C/w = \frac{(14.0\,\varepsilon_r + 12.5)(w/h) - (1.83\,\varepsilon_r - 2.25)}{\sqrt{w/h}} + \frac{0.02\,\varepsilon_r}{w/h} \quad \text{(pF / m)}$$

$$\tag{5.5.1.2}$$

or for $w/h \geq 1$:

$$C/w = (9.5\,\varepsilon_r + 1.25)(w/h) + 5.2\,\varepsilon_r + 7.0 \quad \text{(pF/m)} \tag{5.5.1.3}$$

These equations are accurate to within 5% for

$$2.5 \leq \varepsilon_r \leq 15.0$$

$$0.1 \leq w/h \leq 5.0$$

Kirschning [7] gives equations for $C$ and $L$ as

289

$$C = 0.001\,h\left[(10.35\,\varepsilon_r + 2.5)\,(w/h)^2 + (2.6\,\varepsilon_r + 5.64)\,(w/h)\right]\quad(\text{pF})$$

(5.5.1.4)

$$L = 0.22\,h\left[1.0 - 1.35\,e^{-0.18(w/h)^{1.39}}\right]\quad(\text{nH})$$

(5.5.1.5)

where $w$ and $h$ are in mm. Equations are valid for

$$2.0 \le \varepsilon_r \le 13.0$$

$$0.2 \le w/h \le 6.0$$

Agreement of calculated and measured resonant frequencies using these equations was to 0.3%.

REFERENCES

[1]    Douville, Rene J.P., and David S. James, "Experimental Study of Microstrip Bends and Their Compensation," *IEEE Transactions on Microwave Theory and Techniques*, Vol. MTT-26, No. 3, March 1978, pp. 175–181.

[2]    Garg, Ramesh, and I.J. Bahl, "Microstrip discontinuities," *International Journal of Electronics*, Vol. 45, No. 1, 1978, pp. 81–87.

[3]    Hammerstad, E., "Computer-Aided Design of Microstrip Couplers with Accurate Discontinuity Models," *1981 IEEE MTT-S International Microwave Symposium Digest*, pp. 54–56.

[4]    Hammerstad, E.O., and F. Bekkadal, *Microstrip Handbook*, The University of Trondheim, ELAB report STF44 A74169, February 1975.

[5]    Hill, A., and V.K. Tripathi, "Analysis and Modeling of Coupled Right Angle Microstrip Bend Discontinuities," *1989 IEEE MTT-S Symposium Digest*, pp. 1143–1146.

[6]    Hill, Achim, "An Efficient Algorithm of the Analysis of Passive Microstrip Discontinuities for Microwave and Millimeter Wave Integrated Circuits in a Shielding Box," *Compact Software*, 1990, pp. 9–10.

[7]    Kirschning, M., *et al.*, "Measurement and Computer-Aided Modeling of Microstrip Discontinuities by an Improved Resonator Method," *IEEE Microwave Theory and Techniques Symposium Digest 1983*, pp. 495–497.

[8]    Mehran, Reza, "The Frequency-Dependent Scattering Matrix of Microstrip Right-Angle Bends, T-Junctions and Crossings," *AEÜ*, Band 29, Heft 11, 1975, pp. 454–460.

[9]    Mehran, Reza, "Frequency Dependent Equivalent Circuits for Microstrip Right-Angle Bends, T-Junctions and Crossings," *AEÜ*, Band 30, Heft 2, 1976, pp. 80–82.

[10]   Menzel, Wolfgang and Ingo Wolff, "A Method for Calculating the Frequency-Dependent Properties of Microstrip Discontinuities," *IEEE Transactions on Microwave Theory and Techniques*, Vol. MTT-25, No. 2, February 1977, pp. 107–112.

[11]   Silvester, P., and Peter Benedek, "Microstrip Discontinuity Capacitances for Right-Angle Bends, T Junctions, and Crossings," *IEEE Transactions on Microwave Theory and Techniques*, Vol. MTT-21, No. 5, May 1973, pp. 341–346, and "Correction to 'Microstrip Discontinuity Capacitances for Right-Angle Bends, T Junctions, and Crossings,'" Vol. MTT-23, No. 5, May 1975, p. 456.

[12]   Thomson, Alistair F., and Anand Gopinath, "Calculation of Microstrip Discontinuity Inductances," *IEEE Transactions on Microwave Theory and Techniques*, Vol. MTT-23, No. 8, August 1975, pp. 648–655.

### 5.5.2   Optimal Right-Angle Mitered Bend

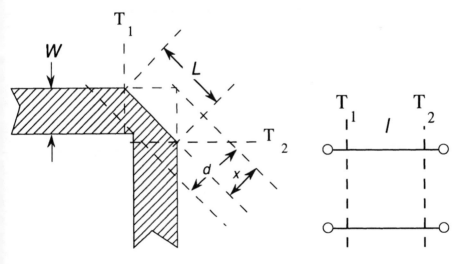

**Figure 5.5.2.1:**     **Optimal Microstrip Line Mitered Bend**

The optimal mitre occurs for:

$$M = 52 + 65 \ e^{-1.35 \ w/h} = \frac{100 \ x}{d} \quad (\%) \tag{5.5.2.1}$$

or, more conveniently,

$$\frac{L}{w} = \sqrt{2} \ [1.04 + 1.3 \ e^{-1.35 \ w/h}] \tag{5.5.2.2}$$

$$d - x = \sqrt{2} \ w - \frac{L}{2} \tag{5.5.2.3}$$

These equations are valid for $w/h \geq .25$, $\varepsilon_r \leq 25$, $\pm 4$ (sic, %?) accuracy. In Chadha [2], the optimal mitre is found to be $w/h = 2.895$. Chadha also recommends that for bends made of lines of unequal widths that more material be removed from the wider line. Figure 5.5.3.2 compares the references.

The equivalent electrical length of the mitre, $l$, is calculable from arbitrary bend equations of [1] by setting the angle $\alpha$ to 90° resulting in

$$l = \frac{L}{\sqrt{2}} \tag{5.5.2.4}$$

Other techniques of equalizing the bend are shown in Figure 5.5.2.2 below. Analytic equations for these structures are not available. Douville [3] shows that the 45° double bend is always better than a single right-angle bend.

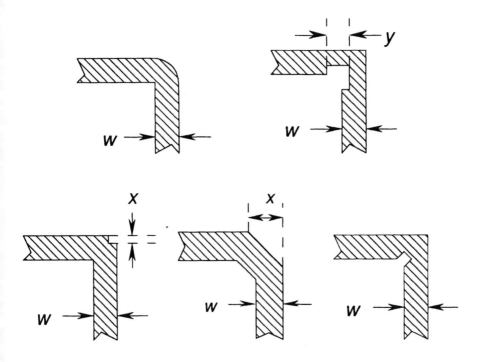

**Figure 5.5.2.2:**     **Various Techniques of Bend Compensation**

REFERENCES

[1]     Anders, Peter, and Fritz Arndt, "Microstrip Discontinuity Capacitances and
        Inductances for Double Steps, Mitred Bends with Arbitrary Angle, and Asymmetric
        Right-Angle Bends," *IEEE Transactions on Microwave Theory and Techniques*,
        Vol. MTT-28, No. 11, November 1988, pp. 1213–1217.

[2]     Chadha, Rakesh, and K.C. Gupta, "Compensation of Discontinuities in Planar
        Transmission Lines," *IEEE Transactions on Microwave Theory and Techniques*,
        Vol. MTT-30, No. 12, December 1982, pp. 2151–2156.

[3]     Douville, Rene J.P., and David S. James, "Experimental Study of Microstrip
        Bends and Their Compensation," *IEEE Transactions on Microwave Theory and
        Techniques*, Vol. MTT-26, No. 3, March 1978, pp. 175–181.

[4]     Hill, Achim, "An Efficient Algorithm of the Analysis of Passive Microstrip
        Discontinuities for Microwave and Millimeter Wave Integrated Circuits in a
        Shielding Box," *Compact Software*, 1990, pp. 9–10.

[5]     Menzel, Wolfgang and Ingo Wolff, "A Method for Calculating the Frequency-
        Dependent Properties of Microstrip Discontinuities," *IEEE Transactions on*

*Microwave Theory and Techniques*, Vol. MTT-25, No. 2, February 1977, pp. 107–112.

[6]    Moore, John, and Hao Ling, "Characterization of a 90° Microstrip Bend with Arbitrary Miter via the Time-Domain Finite Difference Method," *IEEE Transactions on Microwave Theory and Techniques*, Vol. MTT-38, No. 4, April 1990, pp. 404–410.

[7]    Rizzoli, Vittorio, "A General Approach to the Resonance Measurement of Asymmetric Microstrip Discontinuities," *1980 IEEE MTT-S International Microwave Symposium Digest*, pp. 422–424.

[8]    Zheng, J.X., and David C. Chang, "Numerical Modelling of Chamfered Bends and Other Microstrip Junctions of General Shape in MMICs," *1990 IEEE MTT-S Symposium Digest*, pp. 709–712.

### 5.5.3   Optimal Arbitrary Angle Mitered Bend

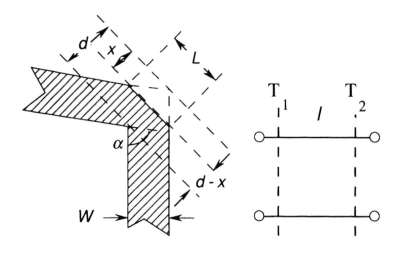

**Figure 5.5.3.1:**     **Arbitrary Mitered Bend**

The optimal mitre dimensions are shown in Figure 5.5.3.2 for a range of angles and $w/h$ ratios. This figure combines the results of a number of references for comparison and convenience.

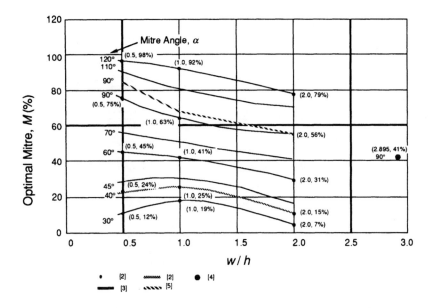

**Figure 5.5.3.2:** **Optimum Mitres for Arbitrary Bend Angles**

$$M = \frac{100 \, x}{d} \quad (\%) \tag{5.5.3.1}$$

$$d - x = \sqrt{2} \, w \left(1.0 - \frac{M}{100}\right) \tag{5.5.3.2}$$

The arbitrary angle bend can be compensated by taking material off of the outside of the corner. This can be done by mitering. Figure 5.5.3.2 shows the optimal mitre percentage to achieve a matched impedance as a function of the line's $w \, / \, h$ ratio. The accuracy is expected to be within a few percent [3]. For comparison the equation of [5] is plotted for the optimal 90° mitred bend. The other references are also shown.

The electrical length of the bend is calculated:

$$\frac{l_k}{w} = \frac{2 \, M}{100 \sin \alpha} \tag{5.5.3.3}$$

REFERENCES

[1]     Anders, Peter, and Fritz Arndt, "Beliebig abgeknickte Mikrostrip-Leitungen mit bogenförmigen Übergang," *Archiv Für Elektronik und Übertragungstechnik*, Band 33, Heft 3, March 1979, pp. 93–99.

[2]    Anders, Peter und Fritz Arndt, "Kompensierte Mikrostrip-Leitungsknicke mit beliebigem Knickwinkel," *Archiv Für Elektronik und Übertragungstechnik*, Vol. 33, No. 9, September 1979, pp. 371–375.

[3]    Anders, Peter, and Fritz Arndt, "Microstrip Discontinuity Capacitances and Inductances for Double Steps, Mitred Bends with Arbitrary Angle, and Asymmetric Right-Angle Bends," *IEEE Transactions on Microwave Theory and Techniques*, Vol. MTT-28, No. 11, November 1980, pp. 1213–1217.

[4]    Chadha, Rakesh, and K.C. Gupta, "Compensation of Discontinuities in Planar Transmission Lines," *IEEE Transactions on Microwave Theory and Techniques*, Vol. MTT-30, No. 12, December 1982, pp. 2151–2156.

[5]    Douville, Rene J.P., and David S. James, "Experimental Study of Microstrip Bends and Their Compensation," *IEEE Transactions on Microwave Theory and Techniques*, Vol. MTT-26, No. 3, March 1978, pp. 175–181.

[6]    Hammerstad, E., "Computer-Aided Design of Microstrip Couplers with Accurate Discontinuity Models,"*1981 IEEE MTT-S International Microwave Symposium Digest*, pp. 54–56.

[7]    Hammerstad, E.O., and F. Bekkadal, *Microstrip Handbook*, The University of Trondheim, ELAB report STF44 A74169, February 1975.

[8]    Kirschning, M., *et al.*, "Measurement and Computer-Aided Modeling of Microstrip Discontinuities by an Improved Resonator Method," *IEEE Microwave Theory and Techniques Symposium Digest 1983*, pp. 495–497. (Equation for lumped equivalent of a 50% mitre on a right-angle bend.)

[9]    Zheng, J.X., and David C. Chang, "Numerical Modelling of Chamfered Bends and Other Microstrip Junctions of General Shape in MMICs," *1990 IEEE MTT-S International Microwave Symposium Digest*, pp. 709–712.

## 5.5.4  Microstrip Line Rounded Bend

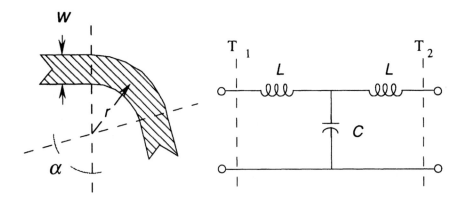

**Figure 5.5.4.1:**     **Microstrip Line Rounded Bend**

Curves of model parameters are available in Weisshar *et al.* [2]; however, no closed-form equations have been derived to date. The dispersion relative to a straight line has been derived by curve-fitting experimental results in Roy *et al.* [1] and are:

$$\varepsilon_{eff,bend}(f) = \varepsilon_r - \frac{\varepsilon_r - \varepsilon_{eff}(0)\,\exp\left(\frac{\sqrt[4]{h/r}}{12.0}\right)}{1.0 + \left(\frac{f}{f_c}\right)^2} \tag{5.5.4.1}$$

where

$$f_c = \frac{c}{4.0\,h\,\sqrt{\varepsilon_r - 1.0}} \tag{5.5.4.2}$$

$\varepsilon_{eff}(0)$ is defined in (3.5.1.3)

$r$ is the mean radius of the curve

which is valid for

$2.0 \le \varepsilon_r \le 12.0$

$0.8 \le w/h \le 5.0$

$$\frac{\lambda_g}{4.0\ \pi\ h} < r/h \leq \infty$$

$0.0 \leq f \leq 12$ GHz

This equation showed good agreement with dispersion equations of a straight line when $r$ is infinity.

REFERENCES

[1]     Roy, Jibendu Sekhar, *et al.*, "Dispersion Characteristics of Curved Microstrip Transmission Lines," *IEEE Transactions on Microwave Theory and Techniques*, Vol. MTT-38, No. 8, August 1990, pp. 1366–1370.

[2]     Weisshar, A., *et al.*, "Modeling of Radial Microstrip Bends," *1990 IEEE MTT-S Symposium Digest*, pp. 1051–1054.

### 5.5.5   Optimal Microstrip Line Rounded Bend

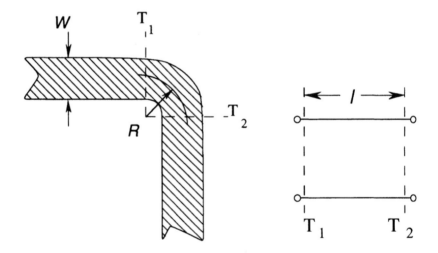

**Figure  5.5.5.1:**      **Optimal Microstrip Line Rounded Bend**

As an alternative to the mitred corners of the previous sections, a microstrip line may be gradually bent to make a corner.  In [2] it is stated that $R \geq 4.0\ h$ will give VSWR < 1.05. Generally, this is not a space efficient technique to use.  U-bends can be made in a similar fashion as shown in Figure 5.5.5.2 below.

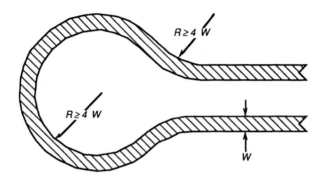

**Figure 5.5.5.2:** **Microstrip Line U-bend**

The line can be modeled as a section of line having length equal to the center line length of the curved section. This model does not include the changes in $\varepsilon_{eff}$ and the effects of coupling between parts of the curve.

## REFERENCES

[1]  Anders, Peter, and Fritz Arndt, "Beliebig abgeknickte Mikrostrip-Leitungen mit bogenförmigen Übergang," *AEÜ*, Band 33, Heft 3, March 1979, pp. 93–99.

[2]  Niehenke, Edward C., "Additional Notes Microwave Transmission Media Microstrip Discontinuities."

[3]  Roy, Jibendu Sekhar, *et al.*, "Dispersion Characteristics of Curved Microstrip Transmission Lines," *IEEE Transactions on Microwave Theory and Techniques*, Vol. MTT-38, No. 8, August 1990, pp. 1366–1370.

[4]  Weisshaar, A., *et al.*, "Modeling of Radial Microstrip Bends," *1990 IEEE MTT-S International Microwave Symposium Digest*, pp. 1051–1054.

## 5.5.6 Unequal Width Bend

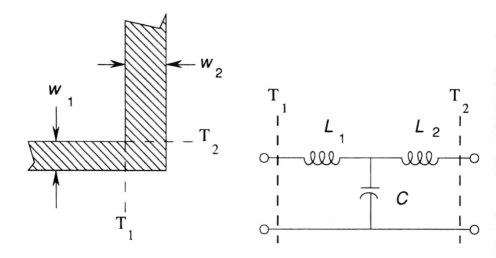

**Figure 5.5.6.1:** **Unequal Width Bend**

Graphs of $s$ parameters of this junction are available in [1] and [2]. The curves of [2] have been fit with closed-form equations. These can be interpolated for other values of $w_1 / h$. For this model (low frequency) $L_1 = L_2$.

For $w_1 / h = 2.0$,

$$\frac{C}{C'h} = 0.23073798 + 1.5514797 \left(\frac{w_2}{w_1}\right) - 0.052050577 \left(\frac{w_2}{w_1}\right)^2 \tag{5.5.6.1}$$

$$\frac{2.0\,L}{L'\,h} = \frac{0.84199442 - 1.4452286 \left(\frac{w_2}{w_1}\right) + 2.3264141 \left(\frac{w_2}{w_1}\right)^2}{1.0 - 0.79920542 \left(\frac{w_2}{w_1}\right) + 2.7511988 \left(\frac{w_2}{w_1}\right)^2} \tag{5.5.6.2}$$

For $w_1 / h = 1.0$,

$$\frac{C}{C'h} = 0.23550754 + 0.41160143 \left(\frac{w_2}{w_1}\right) + 0.21758643 \left(\frac{w_2}{w_1}\right)^2 \tag{5.5.6.3}$$

$$\frac{2.0\,L}{L'\,h} = \frac{0.27972652 - 1.0950536 \left(\frac{w_2}{w_1}\right) + 1.4761673 \left(\frac{w_2}{w_1}\right)^2}{1.0 + 1.0158358 \left(\frac{w_2}{w_1}\right) + 4.9673273 \left(\frac{w_2}{w_1}\right)^2} \tag{5.5.6.4}$$

For $w_1 / h = 0.5$,

$$\frac{C}{C'\,h} = 0.15274236 - 0.033719111 \left(\frac{w_2}{w_1}\right) - 0.23887456 \left(\frac{w_2}{w_1}\right)^2 \qquad (5.5.6.5)$$

$$\frac{2.0\,L}{L'\,h} = \frac{0.4550564 - 3.1343404 \left(\frac{w_2}{w_1}\right) + 2.3297251 \left(\frac{w_2}{w_1}\right)^2}{1.0 + 8.2371314 \left(\frac{w_2}{w_1}\right) + 4.1304198 \left(\frac{w_2}{w_1}\right)^2} \qquad (5.5.6.6)$$

The variables $C'$ and $L'$ are the incremental capacitance and inductance of the single strip. These equations fit the curves to better than approximately 2.5% for the valid region of

$$0.1 \leq w_2 / w_1 \leq 0.9$$

The original curve accuracy is a few percent. Frequency dependency of the parameters is also available in Mehran [3].

REFERENCES

[1]    Anders, Peter, and Fritz Arndt, "Microstrip Discontinuity Capacitances and Inductances for Double Steps, Mitred Bends with Arbitrary Angle, and Asymmetric Right-Angle Bends," *IEEE Transactions on Microwave Theory and Techniques*, Vol. MTT-28, No. 11, November 1988, pp. 1213–1217.

[2]    Mehran, Reza, "The Frequency-Dependent Scattering Matrix of Microstrip Right-Angle Bends, T-Junctions and Crossings," *Archiv Für Elektronik und Übertragungstechnik*, Band 29, Heft 11, 1975, pp. 454–460.

[3]    Mehran, Reza, "Frequency Dependent Equivalent Circuits for Microstrip Right-Angle Bends, T-Junctions and Crossings," *Archiv Für Elektronik und Übertragungstechnik*, , Band 30, Heft 2, 1976, pp. 80–82.

### 5.5.7    Microstrip Line Radial Stub

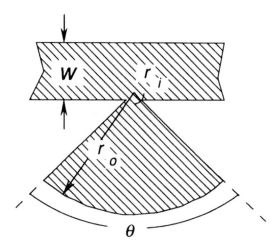

Figure  5.5.7.1:        Radial  Stub

This element presents a low impedance to ground at a precisely located point. It is useful because it is physically shorter than the equivalent transmission line. The two basic types are series and shunt as depicted below:

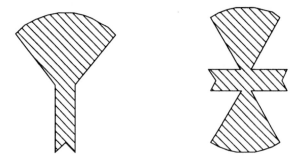

Figure  5.5.7.2:        Radial  Stubs:    (a)  Series    (b)  Shunt

Multiple stubs can be present on either the series or shunt configuration. In the shunt configuration, the added symmetry of multiple stubs may improve bandwidth.

The relations of March [8] include the effect of losses:

$$Z_{in} = -j \frac{120 \, \pi \, h}{r_i \, \theta \, \sqrt{\varepsilon_{eff}}} \text{Ct}(\beta \, r_i, \, \beta \, r_o) \quad (\Omega)$$

(5.5.7.1)

or

$$C = \frac{1.0}{\omega \dfrac{120 \, \pi \, h}{r_i \, \theta \, \sqrt{\varepsilon_{eff}}} \, Ct(\beta \, r_i, \, \beta \, r_o)} \quad (F) \qquad (5.5.7.2)$$

where,

$\varepsilon_{eff} = \varepsilon_{eff}$ of microstripline having width, $w_{eq,stub} = (r_i + r_o) \sin\left(\dfrac{\theta}{2}\right)$ (5.5.7.3)

$$Ct(x, y) \equiv \frac{Y_0(x) \, J_1(y) - J_0(x) \, Y_1(y)}{J_1(x) \, Y_1(y) - Y_1(x) \, J_1(y)} = \text{large radial cotangent function [9]}$$

$$(5.5.7.4)$$

For small conductor losses we define

$$k = \beta - j \, \alpha \qquad (5.5.7.5)$$

$$\beta = \frac{2 \, \pi \sqrt{\varepsilon_{eff}}}{\lambda_0} \qquad (5.5.7.6)$$

$$\alpha \cong \frac{R(r)}{\lambda_0} \qquad (5.5.7.7)$$

and the Bessel functions of complex argument can be written as

$$J_0(k \, r) = J_0(\beta \, r) + j \, \alpha \, r \, J_1(\beta \, r) \qquad (5.5.7.8)$$

$$J_1(k \, r) = J_1(\beta \, r) - j \, \alpha \, r \left[ J_0(\beta \, r) - \frac{J_1(\beta \, r)}{\beta} \right] \qquad (5.5.7.9)$$

$$Y_0(k \, r) = Y_0(\beta \, r) + j \, \alpha \, r \, Y_1(\beta \, r) \qquad (5.5.7.10)$$

$$Y_1(k \, r) = Y_1(\beta \, r) - j \, \alpha \, r \left[ Y_0(\beta \, r) - \frac{Y_1(\beta \, r)}{\beta} \right] \qquad (5.5.7.11)$$

$J_i(\beta \, r) = i$th order Bessel function of the first kind having real argument

$Y_i(\beta \, r) = i$th order Bessel function of the second kind having real argument

Atwater [3] reports good results modeling the radial stub as a series of short transmission lines similar to the delta stub model depicted in Figure 5.5.8.2. A complete model should include a microstrip line open end as described elsewhere in this chapter.

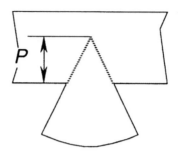

**Figure 5.5.7.3:**      **Definition of Radial Stub Insertion Depth, *P***

Giannini [5] notes that for shunt stubs, high frequencies, high dielectric constants, and $r_i / w$ large the above relations may need to be corrected with effective dimensions. The effective radii $r_i$ and $r_o$ replace the actual radii in the calculations above.

$$r_{i,e} = r + \frac{2.0\ \Delta}{\sin \dfrac{\theta}{2.0}} \tag{5.5.7.12}$$

$$r_{o,e} = r_o \sqrt{1.0 + \frac{2.0\ h}{\pi\ r_o} \ln \left(\frac{\pi\ r_o}{2.0\ h}\right) + 1.7726} \tag{5.5.7.13}$$

where

$$r = \frac{2.0\ P + w_{eff} - w}{2.0\ \cos \dfrac{\theta}{2.0}} \tag{5.5.7.14}$$

$$\Delta = \frac{w_{eff,2} - \alpha\ r}{2.0} \tag{5.5.7.15}$$

and

$w_{eff}$ = the effective width of a microstrip line having width $w$

$w_{eff,2}$ = the effective width of a microstrip line having width $\alpha\ r$

$P$ = depth of insertion, defined in Figure 5.5.7.3 above

Angles greater than 180° have the disadvantage of requiring an additional section of line to access the radial stub, and the advantage of greater bandwidth. Atwater [2] has shown data that such stubs also have a wider bandwidth than the corresponding quarter-wave line. The loss can be calculated by analogy with the standard transmission line [8]:

$$\alpha_c = \frac{R_s \sqrt{\varepsilon_{eff}}}{120 \; \pi \; h} \tag{5.5.7.16}$$

Giannini [4] reported measurement data comparing 180° stubs to three 60° stubs. The three 60° stub configuration yielded approximately 20% wider bandwidths.

**Figure 5.5.7.4:**    **(a) Three 60° Radial Stubs   (b) A Single 180° Radial Stub**

## REFERENCES

[1]    Allen, George W., "Radial Line Stub Design," *Ham Radio*, February 1988, pp. 65–66.

[2]    Atwater, Harry A., "Microstrip Reactive Circuit Elements," *IEEE Transactions on Microwave Theory and Techniques*, Vol. MTT-31, No. 6, June 1983, pp. 488–491.

[3]    Atwater, Harry A., "The Design of the Radial Line Stub: A Useful Microstrip Circuit Element," *Microwave Journal*, November 1985, pp. 149–156.

[4]    Giannini, F., *et al.*, "A Very Broadband Matched Termination Utilizing Non-Grounded Radial Lines," *17th European Microwave Conference Proceedings*, 1987, pp. 1027–1031.

[5]  Giannini, F., *et al.*, "Shunt-Connected Microstrip Radial Stubs," *IEEE Transactions on Microwave Theory and Techniques*, Vol. MTT-34, No. 3, March 1986, pp. 363–366.

[6]  Giannini, Franco, and Claudio Paolini, "Modelling of Shunt Connected Single Radial Stub for CAD Applications," *Alta Frequenza*, Vol. LVII, No. 5, June 1988, pp. 227–232.

[7]  Giannini, F., *et al.*, "CAD Oriented Lossy Modelling of Radial Lines," *1987 International Microwave Symposium Proceedings,* pp. 859–864.

[8]  March, Steven L., "Analyzing Lossy Radial-Line Stubs," *IEEE Transactions on Microwave Theory and Techniques*, Vol. MTT-33, No. 3, pp. 269–271, March 1985.

[9]  Marcuvitz, N.,*Waveguide Handbook*, McGraw-Hill, 1955.  Reprint with errata and preface by N. Marcuvitz.  Peter Peregrinus Ltd., 1986.  (Large radial cotangent function defined, p. 33)

[10]  Syrett, B.A., "A Broad-Band Element for Microstrip Bias or Tuning Circuits," *IEEE Transactions on Microwave Theory and Techniques*, Vol. MTT-28, No. 8, August 1980, pp. 925–927.

### 5.5.8  Delta Stub

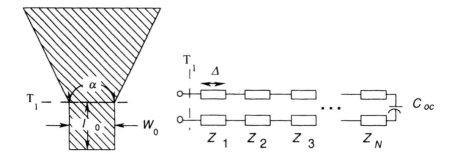

**Figure  5.5.8.1:**     **Delta  Stub**

The delta stub is used as a circuit element where a short is desired.  Because of the straight sides it may be easier to lay out than the radial stub and it is smaller than an open-circuited transmission line.

The open end of the delta stub is transferred by the equivalent line length to a short circuit at the reference plane T.  Coimbra [1, 2] recommends modeling the delta stub by a series of $n$ transmission line segments of length $\Delta$ as shown below.  The number, $n$, is chosen so each segment is $\ll \lambda_g$ in length.  Each segment's characteristic impedance is calculated with the standard microstrip line equations with the exception of the final section, which includes the open end discontinuity.

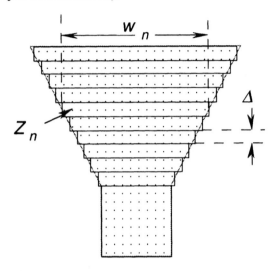

Figure 5.5.8.2:  Modeling the Delta Stub

Coimbra used an iterative technique for synthesis. The impedance of the cascaded lines is calculated as each section is added until the desired open-circuit impedance is achieved.

REFERENCES

[1] Coimbra, Mauro de Lima, "A New Kind of Radial Stub and Some Applications," *14th European Microwave Conference Proceedings*, 1984, Liège, Belgium, pp. 516–521.

[2] Coimbra, Mauro L., "The Generalized Delta Stubs," *1987 International Microwave Symposium Proceedings*, pp. 1071–1075.

### 5.5.9   Microstrip Line Cross

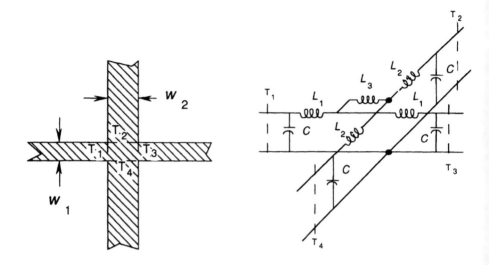

**Figure 5.5.9.1:**        **Microstrip Line Cross**

The cross junction is used when we wish to parallel two stubs to get a lower impedance. The design equations [3] are:

$$\frac{C}{w_1} = \left\{ \left[ \frac{86.6\, w_2}{h} - 30.9 \left( \frac{w_2}{h} \right)^{1/2} + 367.0 \right] \log \left( \frac{w_1}{h} \right) + \left( \frac{w_2}{h} \right)^3 \right.$$

$$\left. + \frac{74.0\, w_2}{h} + 130.0 \right\} \left( \frac{w_1}{h} \right)^{-1/3} - 240.0 + \frac{2.0}{w_2/h} - \frac{1.5\, w_1}{h} \left( 1.0 - \frac{w_2}{h} \right) \ (\text{pF}/\text{m})$$

$$\text{(5.5.9.1)}$$

$$L_1/h = \left\{ \left[ \frac{165.6\, w_2}{h} + 31.2 \sqrt{\frac{w_2}{h}} - 11.8 \left( \frac{w_2}{h} \right)^2 \right] \frac{w_1}{h} - \frac{32.0\, w_2}{h} \right.$$

$$\left. + 3.0 \right\} \left( \frac{w_1}{h} \right)^{-3/2} \ (\text{nH}/\text{m}) \qquad \text{(5.5.9.2)}$$

$$L_2/h = \left\{ \left[ \frac{165.6\, w_1}{h} + 31.2 \sqrt{\frac{w_1}{h}} - 11.8 \left( \frac{w_1}{h} \right)^2 \right] \frac{w_2}{h} - \frac{32.0\, w_1}{h} \right.$$

$$+ \ 3.0 \ \Bigg\} \left(\frac{w_2}{h}\right)^{-3/2} \ (\text{nH / m}) \tag{5.5.9.3}$$

$$L_3 / h = 337.5 + \left(1.0 + \frac{7}{w_1 / h}\right)\frac{1.0}{w_2 / h} - \frac{5.0 \ w_2}{h} \cos\left[\frac{\pi \ (1.5 \ h - w_1)}{2 \ h}\right]$$

$$(\text{nH / m})$$

$$\tag{5.5.9.4}$$

The equation for $C$ is valid for

$$\varepsilon_r = 9.9$$

$$0.3 \le w_1 / h \le 3.0$$

$$0.1 \le w_2 / h \le 3.0$$

Equations for $L_1$, $L_2$, and $L_3$ are valid for

$$0.5 \le w_1 / h \le 2.0$$

$$0.5 \le w_2 / h \le 2.0$$

The equations are accurate to within 5%.

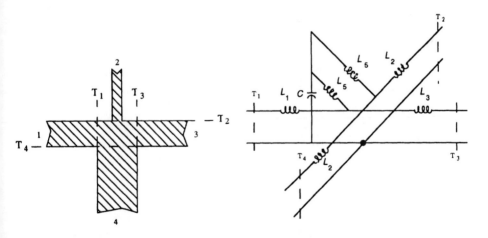

**Figure 5.5.9.2:**     **Asymmetric Cross Junction**

Akello [1] reports on measurements of the asymmetric cross junction. For $Z_2 \gg Z_1$, $Z_3$, $Z_4$ the affect of line 2 is shown to be small. This allows the discontinuity at the junction to be approximated with the equations for a microstrip line T.

REFERENCES

[1]    Akello, R.J., *et al.*, "Equivalent Circuit of the Asymmetric Crossover Junction," *Electronics Letters*, Vol. 13, No. 4, February 17, 1977, pp. 117–118.

[2]    Easter, Brian, "The Equivalent Circuit of Some Microstrip Discontinuities," *IEEE Transactions on Microwave Theory and Techniques*, Vol. MTT-23, No. 8, August 1975, pp. 655–660.

[3]    Garg, Ramesh, and I.J. Bahl, "Microstrip Discontinuities," *International Journal of Electronics*, Vol. 45, No. 1, 1978, pp. 81–87.

[4]    Giannini, F., *et al.*, "An Improved Equivalent Model for Microstrip Cross-Junction," *19th European Microwave Conference Proceedings*, September 4–7, 1990, London, England, pp. 1226–1231.

[5]    Gopinath, A., *et al.*, "Equivalent Circuit Parameters of Microstrip Step Change in Width and Cross Junctions," *IEEE Transactions on Microwave Theory and Techniques*, Vol. MTT-24, No. 3, March 1976, pp. 142–144.

[6]    Hammerstad, E.O., and F. Bekkadal, *Microstrip Handbook*, The University of Trondheim, ELAB report STF44 A74169, February 1975.

[7]    Lau, Wai Yuen, "Network Analysis Verifies Models in CAD Packages," *Microwaves & RF*, November 1989, pp. 99–110.

[8]    Mehran, Reza, "The Frequency-Dependent Scattering Matrix of Microstrip Right-Angle Bends, T-Junctions and Crossings," *AEÜ*, Band 29, Heft 11, 1975, pp. 454–460.

[9]    Mehran, Reza, "Frequency Dependent Equivalent Circuits for Microstrip Right-Angle Bends, T-Junctions and Crossings," *AEÜ*, Band 30, Heft 2, 1976, pp. 80–82.

[10]   Menzel, Wolfgang and Ingo Wolff, "A Method for Calculating the Frequency-Dependent Properties of Microstrip Discontinuities," *IEEE Transactions on Microwave Theory and Techniques*, Vol. MTT-25, No. 2, February 1977, pp. 107–112.

[11]   Silvester, P., and Peter Benedek, "Microstrip Discontinuity Capacitances for Right-Angle Bends, T Junctions, and Crossings," *IEEE Transactions on Microwave Theory and Techniques*, Vol. MTT-21, No. 5, May 1973, pp. 341–346 and "Correction to 'Microstrip Discontinuity Capacitances for Right-Angle Bends, T Junctions, and Crossings,'" Vol. MTT-23, No. 5, May 1975, p. 456.

[12]   Wu, Shih-Chang, *et al.*, "A Rigorous Dispersive Characterization of Microstrip Cross and Tee Junctions," *1990 IEEE MTT-S International Microwave Symposium Digest*, pp. 1151–1154.

### 5.5.10  Microstrip Line T Junction

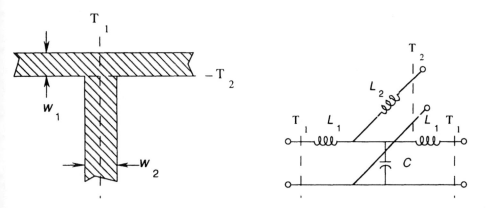

**Figure  5.5.10.1:**     **Microstrip  Line  T  Junction**

The T junction is used when we wish to split a signal into two paths.  This is used in various combiner/dividers.  A number of references give design curves for the discontinuity.  Design equations are available in [3]

$$\frac{C_T}{w_1} = \frac{100.0}{\tanh(0.0072\,Z_0)} + 0.64\,Z_0 - 261.0 \quad (\text{pF}/\text{m})$$

For $0.5 \le w_1/h \le 2.0$ and $0.5 \le w_2/h \le 2.0$,

$$\frac{\Delta L_1}{h} = \frac{-w_2}{h}\left[\frac{w_2}{h}\left(-0.016\,\frac{w_1}{h} + 0.064\right) + \frac{0.016}{\frac{w_1}{h}}\right]L_{w,1}$$

For $1.0 \le w_1/h \le 2.0$ and $0.5 \le w_2/h \le 2.0$,

$$\frac{\Delta L_2}{h} = \left[\left(\frac{0.12\,w_1}{h} - 0.47\right)\frac{w_2}{h} + 0.195\,\frac{w_1}{h} - 0.357\right.$$

$$\left. + 0.0283\,\sin\left(\frac{\pi\,w_1}{h} - 0.75\,\pi\right)\right]L_{w,2}$$

For each line width we calculate

$$L_{w,n} = \frac{Z_0(w_n) \sqrt{\varepsilon_{eff}}}{c}$$

The equations are valid to an accuracy of better than 5% for $\varepsilon_r = 9.9$ and

$$25.0 \leq Z_0 \leq 100.0$$

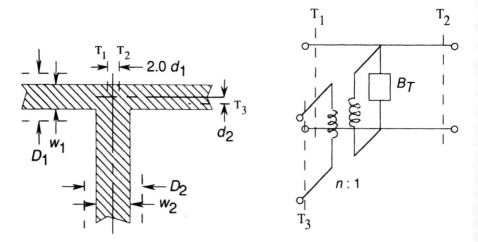

**Figure 5.5.10.2:** **Microstrip Line T Junction**

An alternative model, shown in Figure 5.5.10.2, is given in Hammerstad [4]. The design equations are

$$d_1 = \frac{0.05 \, D_2 \, Z_1 \, n^2}{Z_2}$$

$$d_2 = \frac{D_1}{2.0} - 0.16 \frac{D_1 \, Z_1}{Z_2} \left[ 1.0 + \left( \frac{2.0 \, D_1}{\lambda} \right)^2 - 2.0 \ln \frac{Z_1}{Z_2} \right]$$

$$n = \frac{\sin\left( \frac{2.0 \, \pi \, D_1 \, Z_1}{2.0 \, \lambda \, Z_2} \right)}{\frac{2.0 \, \pi \, D_1 \, Z_1}{2.0 \, \lambda \, Z_2}}$$

for $\frac{Z_1}{Z_2} \leq 0.5$

$$B_T = \frac{-D_1 \left( 1.0 - 2.0 \, D_1 / \lambda \right)}{\lambda \, Z_2}$$

and for $\frac{Z_1}{Z_2} \geq 0.5$

$$B_T = \frac{-D_1 \left( \frac{3.0 \, Z_1}{Z_2} - 2.0 \right) \left( 1.0 - 2.0 \, D_1 / \lambda \right)}{\lambda \, Z_2}$$

where

$$D_i = \frac{\eta_0 \, h}{\sqrt{\varepsilon_{eff,i}} \, Z_i}$$

## REFERENCES

[1]    Dydyk, Michael, "Master the T-Junction and Sharpen Your MIC Designs," *Microwaves*, May 1977, pp. 184–186.

[2]    Easter, Brian, "The Equivalent Circuit of Some Microstrip Discontinuities," *IEEE Transactions on Microwave Theory and Techniques*, Vol. MTT-23, No. 8, August 1975, pp. 655–660.

[3]    Garg, Ramesh, and I.J. Bahl, "Microstrip discontinuities," *International Journal of Electronics*, Vol. 45, No. 1, 1978, pp. 81–87.

[4]    Hammerstad, Erik O., "Equations for Microstrip Circuit Design," *5th European Microwave Conference 1975*, Hamburg, pp. 268–272.

[5]    Hammerstad, Erik O., "Computer-Aided Design of Microstrip Couplers with Accurate Discontinuity Models,"*1981 IEEE MTT-S International Microwave Symposium Digest*, pp. 54–56.

[6]    Hammerstad, E.O., and F. Bekkadal, *Microstrip Handbook*, The University of Trondheim, ELAB report STF44 A74169, February 1975.

[7]    Mehran, Reza, "The Frequency-Dependent Scattering Matrix of Microstrip Right-Angle Bends, T-Junctions and Crossings," *Archiv Für Elektronik und Übertragungstechnik*, , Band 29, Heft 11, 1975, pp. 454–460.

[8]    Mehran, Reza, "Frequency Dependent Equivalent Circuits for Microstrip Right-Angle Bends, T-Junctions and Crossings," *Archiv Für Elektronik und Übertragungstechnik*, , Band 30, Heft 2, 1976, pp. 80–82.

[9]    Silvester, P., and Peter Benedek, "Microstrip Discontinuity Capacitances for Right-Angle Bends, T Junctions, and Crossings," *IEEE Transactions on Microwave*

*Theory and Techniques*, Vol. MTT-21, No. 5, May 1973, pp. 341–346, and "Correction to 'Microstrip Discontinuity Capacitances for Right-Angle Bends, T Junctions, and Crossings,'" May 1975, p. 456.

[10] Thomson, Alistair F., and Anand Gopinath, "Calculation of Microstrip Discontinuity Inductances," *IEEE Transactions on Microwave Theory and Techniques*, Vol. MTT-23, No. 8, August 1975, pp. 648–655.

[11] Vogel, R.W., "Effects of the T-Junction Discontinuity on the Design of Microstrip Directional Couplers," *IEEE Transactions on Microwave Theory and Techniques*, Vol. MTT-21, No. 3, March 1973, pp. 145–146.

[12] Wu, Shih-Chang, *et al.*, "A Rigorous Dispersive Characterization of Microstrip Cross and Tee Junctions," *1990 IEEE MTT-S International Microwave Symposium Digest*, pp. 1151–1154.

## 5.5.11 Asymmetrical Microstrip Line T Junction

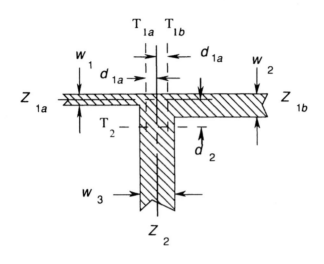

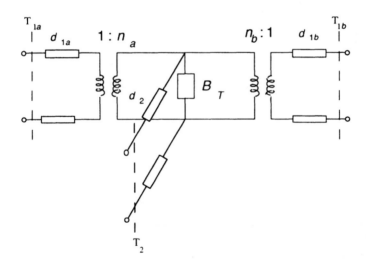

**Figure 5.5.11.1:** **Asymmetric Microstrip Line T Junction and Model**

Hammerstad [1] recommends that we modify the equations for a symmetric T junction to model the asymmetric junction. The equations are

$$\frac{d_{1a}}{D_2} = 0.055 \left[ 1.0 - 1.0 \frac{Z_{1a}}{Z_2} \left( \frac{f}{f_{p1}} \right)^2 \right] \frac{Z_{1a}}{Z_2} \qquad (5.5.11.1)$$

$$\frac{d_{1b}}{D_2} = 0.055 \left[ 1.0 - 1.0 \frac{Z_{1b}}{Z_2} \left(\frac{f}{f_{p1}}\right)^2 \right] \frac{Z_{1b}}{Z_2} \tag{5.5.11.2}$$

$$\frac{d_2}{D_1} = 0.5 - \left[ 0.05 + 0.7 e^{-1.6 \sqrt{Z_{1a} Z_{1b}} \; / \; Z_2} \right.$$

$$\left. + 0.25 \frac{\sqrt{Z_{1a} Z_{1b}}}{Z_2} \left(\frac{f}{f_{p1}}\right)^2 - 0.17 \ln \frac{\sqrt{Z_{1a} Z_{1b}}}{Z_2} \right] \frac{\sqrt{Z_{1a} Z_{1b}}}{Z_2} \tag{5.5.11.3}$$

$$n_a = \sqrt{1.0 - \pi \left(\frac{f}{f_{p1}}\right)^2 \left[ \frac{1.0}{12.0} \frac{Z_{1a}^2}{Z_2} + \left(0.5 - \frac{d_2}{D_1}\right)^2 \right]} \tag{5.5.11.4}$$

$$n_b = \sqrt{1.0 - \pi \left(\frac{f}{f_{p1}}\right)^2 \left[ \frac{1.0}{12.0} \frac{Z_{1b}^2}{Z_2} + \left(0.5 - \frac{d_2}{D_1}\right)^2 \right]} \tag{5.5.11.5}$$

$$\frac{B_T}{Y_2} \frac{\lambda_1}{D_1} = 5.5 \frac{\varepsilon_r + 2.0}{\varepsilon_r} \left[ 1.0 + 0.9 \ln \frac{\sqrt{Z_{1a} Z_{1b}}}{Z_2} + 4.5 \frac{\sqrt{Z_{1a} Z_{1b}}}{Z_2} \left(\frac{f}{f_{p1}}\right)^2 \right.$$

$$\left. - 4.4 \exp \left(\frac{-1.3 \sqrt{Z_{1a} Z_{1b}}}{Z_2}\right) - 20.0 \left(\frac{Z_2}{\eta_0}\right)^2 \right] n - 2.0 \frac{d_1}{D_2} \tag{5.5.11.6}$$

where

$$f_p = 0.4 \, Z \, / \, h$$

The variable $f_p$ is in GHz, $Z$ in $\Omega$, and $h$ in mm. The $Z$'s are calculated with the usual microstrip line equations.

## REFERENCES

[1]    Hammerstad, E., "Computer-Aided Design of Microstrip Couplers with Accurate Discontinuity Models,"*1981 IEEE MTT-S International Microwave Symposium Digest*, pp. 54–56.

[2]    Hammerstad, E.O., and F. Bekkadal, *Microstrip Handbook*, The University of Trondheim, ELAB report STF44 A74169, February 1975.

[3]    Menzel, Wolfgang and Ingo Wolff, "A Method for Calculating the Frequency-Dependent Properties of Microstrip Discontinuities," *IEEE Transactions on Microwave Theory and Techniques*, Vol. MTT-25, No. 2, February 1977, pp. 107–112.

[4]     Thompson, James H., "Simplified Microstrip Discontinuity Modeling Using the Transmission Line Matrix Method Interfaced to Microwave CAD," *Microwave Journal*, July 1990, pp. 79–88.

### 5.5.12 Optimal Microstrip Line T

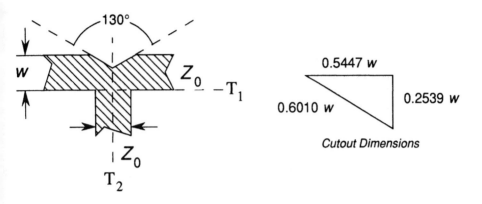

**Figure 5.5.12.1:**     Optimal Microstrip Line T and Notch Dimensions

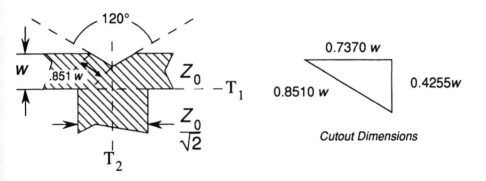

**Figure 5.5.12.2:**     Optimal Microstrip Line T and Notch Dimensions

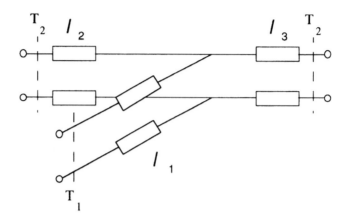

**Figure 5.5.12.3:**   Compensated Microstrip Line T Model

The microstrip line T junction can be compensated by removing a triangular section of the line. For equal impedance lines the optimal compensating notch has an angle of >120° (approximately 130° from Figure 8 [1]) and the optimal length of a side, $a$, is $0.601w$ for 130°. For a main line impedance of $Z_0 / \sqrt{2}$ the optimal angle is 120° and the optimal side is $0.851w$. The dimension $a$ is independent of frequency, dielectric, and other parameters.

The compensated equal impedance T junction reference plane extensions are approximately independent of frequency [1]. For the case shown in Figure 5.5.12.1, the low frequency value is

$$\Delta l_1 / h \cong 0.45$$

$$\Delta l_2 / h = \Delta l_3 / h \cong 0.7$$

For the case shown in Figure 5.5.12.2 the low frequency value is

$$\Delta l_1 / h \cong 0.2$$

$$\Delta l_2 / h = \Delta l_3 / h \cong 0.2$$

Exact values and curves showing the frequency dependency are available in [1], Figures 7 and 9.

In Wu *et al.* [2], another technique of compensating a T is proposed and analyzed by the method of moments. Good results were achieved; however, no design equations are available.

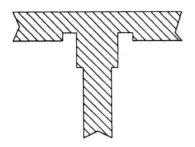

**Figure 5.5.12.4:** Shaped T junction

REFERENCES

[1]     Chadha, Rakesh, and K.C. Gupta, "Compensation of Discontinuities in Planar Transmission Lines," *IEEE Transactions on Microwave Theory and Techniques*, Vol. MTT-30, No. 12, December 1982, pp. 2151–2156.

[2]     Wu, Shih-Chang, *et al.*, "A Rigorous Dispersive Characterization of Microstrip Cross and Tee Junctions," *1990 IEEE MTT-S International Microwave Symposium Digest*, pp. 1151–1154.

## 5.5.13 Microstrip Line Step (Abrupt Change in Width)

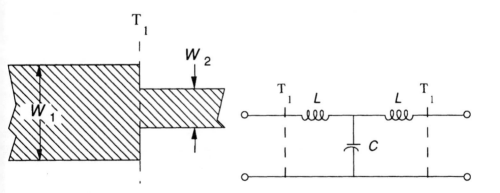

**Figure 5.5.13.1:** Microstrip Line Abrupt Width Change

The impedance step is modeled by the equivalent circuit of Figure 5.5.13.1 above. A number of references [1–5, 8, 13] give design curves. The design equations of [3] are given below.

For $1.5 \leq \dfrac{w_1}{w_2} \leq 3.5$,

$$C_s = \sqrt{w_1 w_2}\left[(10.1\ \log\ \varepsilon_r + 2.33)\ \frac{w_1}{w_2} - 12.6\ \log\ \varepsilon_r - 3.17\right]\quad (\text{pF}/\text{m})$$

(5.5.13.1)

The stated accuracy is better than 10% for

$w_1 > w_2$

$\varepsilon_r \leq 10.0$

For $3.5 \leq \dfrac{w_1}{w_2} \leq 10.0$ and $\varepsilon_r = 9.6$,

$$C_s = \sqrt{w_1 w_2}\left[130.0\ \log\left(\frac{w_1}{w_2}\right) - 44.0\right]\quad (\text{pF}/\text{m})$$

(5.5.13.2)

which is accurate to within 0.5%. The inductance is calculated with

$$L_s = h\left[40.5\left(\frac{w_1}{w_2} - 1.0\right) - 75.0\ \log\left(\frac{w_1}{w_2}\right) + 0.2\left(\frac{w_1}{w_2} - 1.0\right)\right]\quad (\text{nH}/\text{m})$$

(5.5.13.3)

This equation is accurate to within 5.0% for

$\dfrac{w_1}{w_2} \leq 5.0$

$\dfrac{w_2}{h} = 1.0$

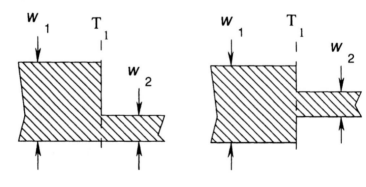

**Asymmetric Step**          **Symmetric Step**

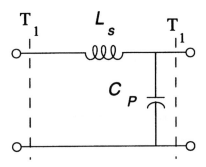

**Figure 5.5.13.2:**          **Microstrip Line Steps and Equivalent Circuit**

A frequency-dependent model of the asymmetric step is presented in Hoffman [7]. The
equivalent circuit is shown in Figure 5.5.13.2 and the model parameters are

$$L_s = \frac{Z_{L1}}{2.0\,\pi f}\,\frac{4.0\,w_{eff,1}}{\Delta\,\lambda_1}\left\{\ln\left[\left(\frac{1.0-a^2}{4.0\,a}\right)\left(\frac{1.0+a}{1.0-a}\right)^{(a+1/a)/2.0}\right]\right.$$

$$+\,2.0\,\frac{A+B+2.0\,D}{A\,B-D^2}+\left(\frac{w_{eff,1}}{2.0\,\Delta\,\lambda_1}\right)^2\left(\frac{1.0-a}{1.0+a}\right)^{4.0\,a}$$

$$\left.\left(\frac{5.0\,a^2-1.0}{1.0-a^2}+\frac{4.0\,a^2\,D}{3.0\,A}\right)^2\right\}$$

$$(5.5.13.4)$$

$$C_P \cong (0.012 + 0.0039\,\varepsilon_r)\,(w_1 - w_2)\quad\text{(pF)}\qquad\qquad(5.5.13.5)$$

where

$$A = \left(\frac{1.0+a}{1.0-a}\right)^{2.0\,a}\left[\frac{1.0+\sqrt{1.0-\left(\frac{2.0\,w_{eff,1}}{\Delta\,\lambda_1}\right)^2}}{1.0-\sqrt{1.0-\left(\frac{2.0\,w_{eff,1}}{\Delta\,\lambda_1}\right)^2}}\right]-\frac{1.0+3.0\,a^2}{1.0-a^2}$$

$$(5.5.13.6)$$

$$B = \left(\frac{1.0+a}{1.0-a}\right)^{a/2.0}\left[\frac{1.0+\sqrt{1.0-\left(\frac{2.0\,w_{eff,2}}{\Delta\,\lambda_2}\right)^2}}{1.0-\sqrt{1.0-\left(\frac{2.0\,w_{eff,2}}{\Delta\,\lambda_2}\right)^2}}\right]+\frac{3.0+a^2}{1.0-a^2}$$

$$(5.5.13.7)$$

$$D = \left(\frac{4.0\, a}{1.0 - a^2}\right)^2 \tag{5.5.13.8}$$

$$a = \frac{w_{eff,2}}{w_{eff,1}} \tag{5.5.13.9}$$

$$\lambda_n = \frac{c_0}{f\sqrt{\varepsilon_{eff,n}}} \tag{5.5.13.10}$$

$$\Delta = \begin{cases} 1 & \text{for asymmetrical steps} \\ 2 & \text{for symmetrical steps} \end{cases} \tag{5.5.13.11}$$

and

$$w_{eff,n} = w_n + \frac{w_{eff,static} - w_n}{1.0 + f/f_{p5}} \tag{5.5.13.12}$$

$$w_{eff,static} = \frac{\eta_0\, h}{Z_{L0}} \tag{5.5.13.13}$$

$Z_{L0}$ = the impedance of the line calculated with $\varepsilon_r$ set to 1.0 $\qquad(5.5.13.14)$

$$f_{p5} = \frac{c}{2.0\, w\, \sqrt{\varepsilon_r}} \tag{5.5.13.15}$$

Equations are accurate to within a few percent for

$$w_1 > w_2$$

$$0 < Z_{L1} < Z_{L2}$$

$$0 < f < f_{g,HE2}$$

where

$$f_{g,HE2} = \frac{c}{w_{eff,1}\sqrt{\varepsilon_{eff}}} \tag{5.5.13.16}$$

REFERENCES

[1]     Anders, Peter, and Fritz Arndt, "Microstrip Discontinuity Capacitances and Inductances for Double Steps, Mitred Bends with Arbitrary Angle, and Asymmetric Right-Angle Bends," *IEEE Transactions on Microwave Theory and Techniques*, Vol. MTT-28, No. 11, November 1988, pp. 1213–1217.

[2]     Benedek, Peter, and P. Silvester, "Equivalent Capacitances for Microstrip Gaps and Steps," *IEEE Transactions on Microwave Theory and Techniques*, Vol. MTT-20, No. 11, November 1972, pp. 729–733.

[3]     Garg, Ramesh, and I.J. Bahl, "Microstrip discontinuities," *International Journal of Electronics*, Vol. 45, No. 1, 1978, pp. 81–87.

[4]     Gopinath, A., *et al.*, "Equivalent Circuit Parameters of Microstrip Step Change in Width and Cross Junctions," *IEEE Transactions on Microwave Theory and Techniques*, Vol. MTT-24, No. 3, March 1976, pp. 142–144.

[5]     Gupta, Chandra, and Anand Gopinath, "Equivalent Circuit Capacitance of Microstrip Step Change in Width," *IEEE Transactions on Microwave Theory and Techniques*, Vol. MTT-25, No. 10, October 1977, pp. 819–822.

[6]     Hammerstad, E.O., and F. Bekkadal, *Microstrip Handbook*, The University of Trondheim, ELAB report STF44 A74169, February 1975.

[7]     Hoffman, Reinmut, *Handbook of Microwave Integrated Circuits,* Artech House, Norwood, MA, 1987.

[8]     Horton, R., "Equivalent Representation of an Abrupt Impedance Step in Microstrip Line," *IEEE Transactions on Microwave Theory and Techniques*, Vol. MTT-21, No. 8, August 1973, pp. 562–564.

[9]     Kumar, D., *et al.*, "Study of an Impedance Step in a Microstrip Line," *International Journal of Electronics*, Vol. 43, No. 1, 1977, pp. 73–76.

[10]    Menzel, Wolfgang and Ingo Wolff, "A Method for Calculating the Frequency-Dependent Properties of Microstrip Discontinuities," *IEEE Transactions on Microwave Theory and Techniques*, Vol. MTT-25, No. 2, February 1977, pp. 107–112.

[11]    Rautio, James C., "An Experimental Investigation of the Microstrip Step Discontinuity," *IEEE Transactions on Microwave Theory and Techniques*, Vol. MTT-37, No. 11, November 1989, pp. 1816–1818.

[12]    Rizzoli, Vittorio, "A General Approach to the Resonance Measurement of Asymmetric Microstrip Discontinuities," *1980 IEEE MTT-S International Microwave Symposium Digest*, pp. 422–424.

[13]    Thomson, Alistair F., and Anand Gopinath, "Calculation of Microstrip Discontinuity Inductances," *IEEE Transactions on Microwave Theory and Techniques*, Vol. MTT-23, No. 8, August 1975, pp. 648–655.

## 5.5.14 Compensated Microstrip Line Step

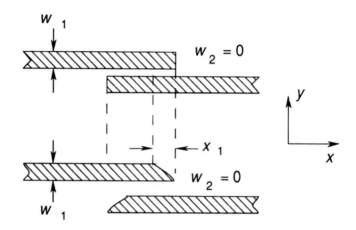

**Figure  5.5.14.1:**      Compensated  Open  End  Discontinuities

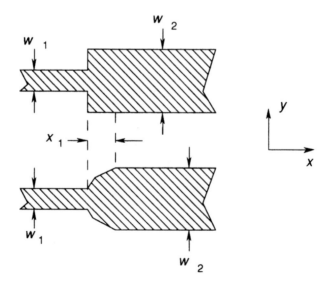

**Figure  5.5.14.2:**      Compensated  Step  Discontinuities

Step discontinuities occur in impedance transformers, filters, and resonators, when interfacing to a device pin and in many other situations. Although the impedance step is desired, the discontinuity creates unwanted parasitics. The technique of this section removes this parasitic.

The presence of a step discontinuity in any TEM line can be compensated by the addition of a taper near the discontinuity as shown in the figures above (upper figures are uncompensated, lower are compensated). The cross-section is varied continuously to move the transmission line's inherent capacitance gradually into fringing capacitance, thus keeping the impedance constant until the step is complete. This does not remove the impedance change, just the parasitics associated with the step.

For microstrip line, this procedure of removing line capacitance and replacing it with fringing capacitance is used. Hoefer [1] gives

$$\frac{dy}{dx} = \pm \sqrt{\left[\frac{\dfrac{\eta_0 h}{2.0\, y\, \varepsilon_r}\, \dfrac{\varepsilon_{eff}(w_2)}{Z_0(\varepsilon_r = 1.0,\, w_2)} - 1.0}{\dfrac{\eta_0 h}{2.0\, y\, \varepsilon_r}\, \dfrac{\varepsilon_{eff}(2.0\, y)}{Z_0(\varepsilon_r = 1.0,\, 2.0\, y)} - 1.0}\right]^2 - 1.0} \qquad (5.5.14.1)$$

where

$\varepsilon_{eff}(w)$ = the effective relative dielectric constant of a microstrip line having width $w$ (see Chapter 3)

$Z_0(\varepsilon_r = 1.0, w)$ = the impedance of microstrip line having width $w$ and the dielectric replaced with air (see Chapter 3)

$y$ = distance from centerline of the microstrip line

To use this equation, Hoefer recommends starting with $y = w_1 / 2.0$ and using approximately 50 steps. The change in width $dy / dx$ is calculated and added to the previous step width until the desired final width $w_2$ is reached.

Fewer steps can be used initially and then increased until the contour dimensions converge. This process and the equation above can be used for impedance steps and open end corrections. Asymmetric compensations are also described in Hoefer.

### REFERENCES

[1]     Hoefer, Wolfgang J.R., "A Contour Formula for Compensated Microstrip Steps and Open Ends," *1983 IEEE MTT-S Symposium Digest,* pp. 524–526.

[2]     Hutchings, J.L., *et al.*, "Contour Program Smoothes 'Strip' Discontinuities," *Microwaves & RF*, November 1987, pp. 129–139.

[3]   Malherbe, J.A.G., and Andre F. Steyn, "The Compensation of Step Discontinuities in TEM-Mode Transmission Lines," *IEEE Transactions on Microwave Theory and Techniques*, Vol. MTT-26, No. 11, November 1978, pp. 883–885.

[4]   STRIP, "Compensated discontinuities Stripline and Microstripline," Design Program.

## 5.5.15 Optimal Microstrip Line Linear Taper

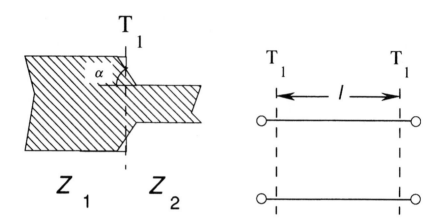

Figure   5.5.15.1:      Optimal Linear Tapers

A simple approximate compensation for the strays of the impedance step is the linear taper. The optimal linear taper for 1:2 and 1:$\sqrt{2}$ steps is found by Chadha and Gupta [1] to be 60° centered at the original location of the step. Chadha and Gupta also gives plots of reference plane shift amount (shifts towards low $Z$ line).

REFERENCES

[1]   Chadha, Rakesh, and K.C. Gupta, "Compensation of Discontinuities in Planar Transmission Lines," *IEEE Transactions on Microwave Theory and Techniques*, Vol. MTT-30, No. 12, December 1982, pp. 2151–2156.

## 5.5.16 Microstrip Line Double Step

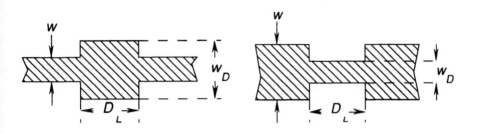

Figure 5.5.16.1: Microstrip Line Double Step

Impedance steps can occur repetitively in filters. Usually these are modeled as a series of transmission lines connected with step discontinuities. In cases where they are spaced closely, the steps will interact. In Hill [3], it is shown with a full wave analysis that two steps separated by $D_L = w_D$ interact significantly and cannot be modeled by cascading the lines and step discontinuities. Hill also found that for $D_L = 3.0 \times w_D$ and $D_L = 8.0 \times w_D$ the isolated models were valid. In Anders and Arndt [1], curves are shown for $D_L / w_D$ less than 2.0. For the cases shown, uncoupled models appear to work well for $D_L > 2.0 \times w_D$.

Giannini *et al.* [2] describe an advanced model that can be used with existing CAE tools to model interacting steps.

### REFERENCES

[1]   Anders, Peter, and Fritz Arndt, "Microstrip Discontinuity Capacitances and Inductances for Double Steps, Mitred Bends with Arbitrary Angle, and Asymmetric Right-Angle Bends," *IEEE Transactions on Microwave Theory and Techniques*, Vol. MTT-28, No. 11, November 1980, pp. 1213–1217.

[2]   Giannini, F., *et al.*, "Enhanced Model for Interacting Step-Discontinuities," *1989 IEEE MTT-S Symposium Digest,* pp. 251–254.

[3]   Hill, Achim, "An Efficient Algorithm of the Analysis of Passive Microstrip Discontinuities for Microwave and Millimeter Wave Integrated Circuits in a Shielding Box," *Compact Software*, 1990, pp. 9–10.

[4]   Parker, T.W., and T.C. Cisco, "Proximity Effects of Transmission Line Discontinuities," *1974 IEEE MTT-S Symposium Digest,* pp. 81–83.

## 5.5.17 Microstrip Line Open

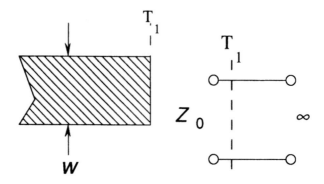

**Figure 5.5.17.1:**    Microstrip Line Open

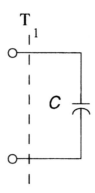

**Figure 5.5.17.2:**    Alternative Microstrip Line Open Model

The open end adds a capacitance that effectively lengthens the transmission line. The additional length can be calculated ($h$ is as shown in Figure 3.5.1.1) [10]:

$$\frac{\Delta l}{h} = \left( \frac{\xi_1 \xi_3 \xi_5}{\xi_4} \right) \tag{5.5.17.1}$$

$$\xi_1 = 0.434907 \left[ \frac{\varepsilon_{eff}^{\,0.81} - 0.26}{\varepsilon_{eff}^{\,0.81} - 0.189} \right] \left[ \frac{(w/h)^{\,0.8544} + 0.236}{(w/h)^{\,0.8544} + 0.87} \right] \tag{5.5.17.2}$$

$$\xi_2 = 1 + \frac{(w/h)^{\,0.371}}{2.358 \, \varepsilon_r + 1} \tag{5.5.17.3}$$

$$\xi_3 = 1.0 + \frac{0.5247 \tan^{-1}\left[0.084 \ (w \ / \ h)^{1.9413} \ / \ \xi_2\right]}{\varepsilon_{eff}^{0.9236}} \tag{5.5.17.4}$$

$$\xi_4 = 1.0 + 0.0377 \tan^{-1}\left[0.067 \ (w \ / \ h)^{1.456}\right]\left[6.0 - 5.0 \ e^{0.036(1.0 \ - \ \varepsilon_r)}\right] \tag{5.5.17.5}$$

$$\xi_5 = 1.0 - 0.218 \ e^{-7.5 \ w \ / \ h} \tag{5.5.17.6}$$

When compared with numerical results at 1.0 GHz, accuracy is better than 2.5% for

$$0.01 \le \frac{w}{h} \le 100.0$$

$$\varepsilon_r \le 50.0$$

## REFERENCES

[1]  Bhat, Bharathi, and Shiban Kishen Koul, "Lumped Capacitance, Open-Circuit End Effects, and Edge-Capacitance of Microstrip-Like Transmission Lines for Microwave and Millimeter-Wave Applications," *IEEE Transactions on Microwave Theory and Techniques*, Vol. MTT-32, No. 4, April 1984, pp. 433–439.

[2]  Boix, Rafael R., and Manuel Horno, "Lumped Capacitance and Open End Effects of Striplike Structures in Multilayered and Anisotropic Substrates," *IEEE Transactions on Microwave Theory and Techniques*, Vol. MTT-37, No. 10, October 1989, pp. 1523–1527.

[3]  Farrar, Andrew, and A.T. Adams, "Computation of Lumped Microstrip Capacities by Matrix Methods—Rectangular Sections and End Effect," *IEEE Transactions on Microwave Theory and Techniques*, Vol. MTT-19, No. 5, May 1971, pp. 495–496.

[4]  Garg, Ramesh, and I.J. Bahl, "Microstrip Discontinuities," *International Journal of Electronics*, Vol. 45, No. 1, 1978, pp. 81–87.

[5]  Hammerstad, E., "Computer-Aided Design of Microstrip Couplers with Accurate Discontinuity Models,"*1981 IEEE MTT-S International Microwave Symposium Digest*, pp. 54–56. (Improves on Silvester and Benedek [13] for $w \ / \ h \ge$ 3.0.)

[6]  Hammerstad, Erik O., "Equations for Microstrip Circuit Design," *5th European Microwave Conference 1975*, Hamburg, pp. 268–272.

[7]  Hammerstad, E.O., and F. Bekkadal, *Microstrip Handbook*, The University of Trondheim, ELAB report STF44 A74169, February 1975. (Page 87, Figure III.3.2, compares 6 references)

[8]   Hoffman, Reinmut, *Handbook of Microwave Integrated Circuits,* Artech House, Norwood, MA, 1987.

[9]   Jansen, Rolf H., and Norbert H.L. Koster, "Accurate Results on the End Effect of Single and Coupled Microstrip Lines for Use in Microwave Circuit Design," *Archiv Für Elektronik und Übertragungstechnik,* Band 34, Heft 11, 1980, pp. 453–459.

[10]   Kirschning, M., *et al.,* "Accurate Model for Open End Effect of Microstrip Lines," *Electronics Letters,* Vol. 17, No. 3, February 5, 1981, pp. 123–126.

[11]   Lau, Wai Yuen, "Network Analysis Verifies Models in CAD Packages," *Microwaves & RF,* November 1989, pp. 99–110.

[12]   Railton, C.J., and J.P. McGeehan, "Characterisation of Microstrip Open-End with Rectangular and Trapezoidal Cross-Section," *Electronics Letters,* Vol. 26, No. 11, May 24, 1990, pp. 685–686.

[13]   Silvester, P., and Peter Benedek, "Equivalent Capacitances of Microstrip Open Circuits," *IEEE Transactions on Microwave Theory and Techniques,* Vol. MTT-20, No. 8, August 1972, pp. 511–516.

[14]   Yang, Hung-Yu, *et al.,* "Microstrip Open-End and Gap Discontinuities in a Substrate-Superstrate Structure," *IEEE Transactions on Microwave Theory and Techniques,* Vol. 37, No. 10, October 1989, pp. 1542–1546.

### 5.5.18  Coupled Microstrip Line Open End

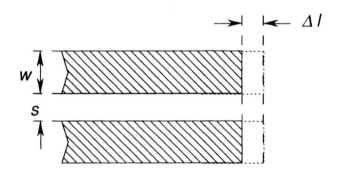

**Figure 5.5.18.1:**   **Coupled Microstrip Line Open End**

Coupled microstrip open ends occur in couplers and coupled line filters. The effects of open ends in these circuits are modeled by an electrical lengthening of the physical strips after Kirschning and Jansen [3].

$$\Delta l_e = \left[ \Delta l(2u, \, \varepsilon_r) - \Delta l(u, \, \varepsilon_r) + 0.0198 \, h \, g^{R_1} \right] e^{-0.328 \, g^{2.244}} + \Delta l(u, \, \varepsilon_r)$$

$$(5.5.18.1)$$

$$\Delta l_o = \left[ \Delta l(u, \varepsilon_r) - h R_3 \right] \left( 1.0 - e^{-R_4} \right) + h R_3 \qquad (5.5.18.2)$$

where

$\Delta l(u, \varepsilon_r)$ = length correction for a single line of width $w$

$\Delta l(2u, \varepsilon_r)$ = length correction for a single line of width $2 w$

$$R_1 = 1.187 \left( 1.0 - e^{-0.069 \, u^{2.1}} \right) \qquad (5.5.18.3)$$

$$R_2 = 0.343 \, u^{0.6187} + \frac{0.45 \, \varepsilon_r}{1.0 + \varepsilon_r} \, u^{\left( 1.357 + \frac{1.65}{1.0 + 0.7 \, \varepsilon_r} \right)} \qquad (5.5.18.4)$$

$$R_3 = 0.2974 \left( 1.0 - e^{-R_2} \right) \qquad (5.5.18.5)$$

$$R_4 = (0.271 + 0.0281 \, \varepsilon_r) \, g^{\left( \frac{1.167 \, \varepsilon_r}{0.66 + \varepsilon_r} \right)} + \left( \frac{1.025 \, \varepsilon_r}{0.687 + \varepsilon_r} \right) g^{\left( \frac{0.958 \, \varepsilon_r}{0.706 + \varepsilon_r} \right)}$$

$$(5.5.18.6)$$

The values of $\Delta l(u, \varepsilon_r)$ and $\Delta l(2u, \varepsilon_r)$ are calculated with the relationships in this book for the microstrip line open discontinuity. The stated accuracy to the numerical analysis for the length corrections is 5%.

Lau [4] compared simulations with network analysis data for coupled microstrip lines. The open end effect was found significant for frequencies over about 24 GHz. The effect of the open ends coupling to the opposite line was also modeled by the addition of 2 fF at each of the lines.

### REFERENCES

[1]  Hammerstad, E., and Ø. Jensen, "Accurate Models for Microstrip Computer-Aided Design," *1980 IEEE MTT-S International Microwave Symposium Digest*, pp. 407–409.

[2]  Jansen, Rolf H., and Norbert H.L. Koster, "Accurate Results on the End Effect of Single and Coupled Microstrip Lines for Use in Microwave Circuit Design," *AEÜ*, Band 34, Heft 11, 1980, pp. 453–459.

[3]     Kirschning, Manford, and Rolf H. Jansen, "Accurate Wide-Range Design Equations for the Frequency-Dependent Characteristic of Parallel Coupled Microstrip Lines," *IEEE Transactions on Microwave Theory and Techniques*, Vol. MTT-32, No. 1, January 1984, pp. 83–90.

[4]     Lau, Wai Yuen, "Network Analysis Verifies Models in CAD Packages," *Microwaves & RF*, November 1989, pp. 99–110.

[5]     Uwano, Tomoki, "Characterization of Microstrip Open End in the Structure of a Parallel-Coupled Stripline Resonator Filter," *IEEE Transactions on Microwave Theory and Techniques*, Vol. MTT-39, No. 3, March 1991, pp. 595–600.

## 5.5.19 Microstrip Line Gap

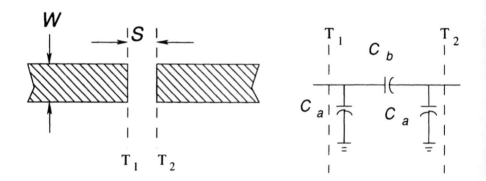

**Figure 5.5.19.1:    Microstrip Line Gap**

The microstrip gap is used in filters and other circuits as a coupling element. It also occurs as a stray circuit element across surface mount components.

Hammerstad [5] fits the curves presented in [1] with the following simple relations:

$$B_g = 2.4 \frac{Y h}{\lambda} \frac{u + 0.1}{u + 1.0} \sqrt{\frac{\varepsilon_r + 2.0}{\varepsilon_r + 1.0}} \ln \left[ \frac{\varepsilon_r s}{h (\varepsilon_r + 2.0)} \right] \tag{5.5.19.1}$$

The capacitance of $C_a$ can be modeled by an equivalent length of transmission line, $\Delta_p$:

$$\Delta_p = \Delta_e \tanh^2 \sqrt{0.5 \ s \ / \ \Delta_e} \tag{5.5.19.2}$$

$$\Delta_g = \Delta_p + \frac{B_g \lambda_g}{2.0 \ \pi \ Y} \tag{5.5.19.3}$$

where

$Y$ = line admittance (S)                     (5.5.19.4)

$u = w / h$                     (5.5.19.5)

$$\Delta_e = 0.412\, h\, \frac{\varepsilon_{eff} + 0.3}{\varepsilon_{eff} - 0.258}\, \frac{u + 0.262}{u + 0.813} = \text{length extension of an open end}$$

(5.5.19.6)

which are valid for

$\varepsilon_r$ = 9.8 and 25.0 $\leq Z_0 \leq$ 70.0

$\varepsilon_r$ = 2.2 and 35.0 $\leq Z_0 \leq$ 100.0

The accuracy is 0.05 for $B_g / Y$ and 0.1 $h$ for $\Delta_g$.

For references such as [1] that give equations or charts of $C_{even}$ and $C_{odd}$ the relations to the model of Figure 5.5.14.1 are:

$$C_a = C_{even} / 2.0$$                     (5.5.19.7)

$$C_b = C_{odd} / 2.0 - C_{even} / 4.0$$                     (5.5.19.8)

The effect of $C_a$ can also be modeled by replacing $C_a$ with the equivalent line extension. The relation is

$$\Delta l = C_a\, Z_0\, v_p$$                     (5.5.19.9)

Özmehmet [9] improves the above model by adding frequency dependence and the effects of radiation losses (a resistor $R_r$ in parallel with $C_b$). The series capacitance $C_b$ is split into three parts representing the parallel-plate capacitance of the facing trace ends and two fringing capacitances. The frequency-dependent $\varepsilon_{eff}$ is used to calculate their values after [1]. The radiation resistance is given by

$$R_r = \frac{120\, \pi^2}{K}$$                     (5.5.19.10)

where

$$K = \frac{\sin 2\,a}{2.0\,a} + \cos 2.0\,a + 2.0\,a\ \text{Si}(2.0\,a) - 1.0$$                     (5.5.19.11)

$$a = \pi\, W_e(f) / \lambda_0$$                     (5.5.19.12)

$W_e(f)$ = the frequency-dependent effective width.  Calculate by equating the incremental capacitance of the frequency-dependent microstrip line to a parallel-plate capacitor having the same dielectric and plate spacing equal to the dielectric thickness.

$$(5.5.19.13)$$

$Si(x)$ = sine integral function

Garg and Bahl [4] recommend the following model

$$C_{even} / w = (\varepsilon_r / 9.6)^{0.9} (s / w)^{m_e} e^{K_e} \quad (pF / m) \tag{5.5.19.14}$$

$$C_{odd} / w = (\varepsilon_r / 9.6)^{0.8} (s / w)^{m_o} e^{K_o} \quad (pF / m) \tag{5.5.19.15}$$

where for $0.1 \leq s / w \leq 1.0$

$$m_o = \frac{w}{h}\left(0.619 \log \frac{w}{h} - 0.3853\right) \tag{5.5.19.16}$$

$$K_o = 4.26 - 1.453 \log \frac{w}{h} \tag{5.5.19.17}$$

and for $0.1 \leq s / w \leq 0.3$

$$m_e = 0.8675 \tag{5.5.19.18}$$

$$K_e = 2.043 (w / h)^{0.12} \tag{5.5.19.19}$$

For $0.3 \leq s / w \leq 1.0$

$$m_e = \frac{1.565}{(w / h)^{0.16}} - 1.0 \tag{5.5.19.20}$$

$$K_e = 1.97 - \frac{0.03}{w / h} \tag{5.5.19.21}$$

which is valid for

$$2.5 \leq \varepsilon_r \leq 15.0$$

The accuracy is within 7%.

REFERENCES

[1]     Benedek, Peter, and P. Silvester, "Equivalent Capacitances for Microstrip Gaps and Steps," *IEEE Transactions on Microwave Theory and Techniques*, Vol. MTT-20, No. 11, November 1972, pp. 729–733.

[2]     Biswas, Animesh, and Vijai K. Tripathi, "Modeling of Asymmetric and Offset Gaps in Shielded Microstrips and Slotlines," *IEEE Transactions on Microwave Theory and Techniques*, Vol. MTT-38, No. 6, June 1990, pp. 818–822.

[3]     Farrar, Andrew, and A.T. Adams, "Computation of Lumped Microstrip Capacities by Matrix Methods—Rectangular Sections and End Effect," *IEEE Transactions on Microwave Theory and Techniques*, Vol. MTT-19, No. 5, May 1971, pp. 495–496.

[4]     Garg, Ramesh, and I.J. Bahl, "Microstrip Discontinuities," *International Journal of Electronics*, Vol. 45, No. 1, 1978, pp. 81–87.

[5]     Hammerstad, E., "Computer-Aided Design of Microstrip Couplers with Accurate Discontinuity Models,"*1981 IEEE MTT-S International Microwave Symposium Digest*, pp. 54–56.

[6]     Hammerstad, E.O., and F. Bekkadal, *Microstrip Handbook*, The University of Trondheim, ELAB report STF44 A74169, February 1975.

[7]     Kumar, D., *et al.,* "A Study of Gaps in Microstrip Transmission Lines," *International Journal of Electronics*, Vol. 41, No. 6, 1976, pp. 617–620.

[8]     Maeda, Minora, "An Analysis of Gap in Microstrip Transmission Lines," *IEEE Transactions on Microwave Theory and Techniques*, Vol. MTT-20, No. 6, June 1972, pp. 390–396.

[8]     Özmehmet, K., "New Frequency Dependent Equivalent Circuit for Gap Discontinuities of Microstriplines," *IEE Proceedings*, Vol. 134, Pt. H., No. 3, June 1987, pp. 333–335.

[10]    Yang, Hung-Yu, *et al.,* "Microstrip Open-End and Gap Discontinuities in a Substrate-Superstrate Structure," *IEEE Transactions on Microwave Theory and Techniques*, Vol. MTT-37, No. 10, October 1989, pp. 1542–1546.

## 5.5.20 Asymmetric Microstrip Line Gap

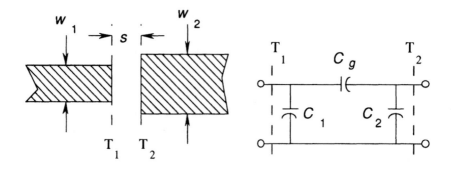

**Figure 5.5.20.1:    Asymmetric Microstrip Line Gap**

Kirschning *et al.* [2] gives these equations for the asymmetric microstrip line gap:

$$B_g = Q_1 \, \pi f h \, e^{(-1.86 \, g / h)} \left\{ 1.0 + 4.19 \left[ 1.0 - e^{-0.785 \, (h / w_1)^{0.5} \, w_2 / w_1} \right] \right\}$$

$$(\text{mS})$$

$$(5.5.20.1)$$

$$B_1 = \frac{2.0 \, \pi f \, C_1 \, (Q_2 + Q_3)}{1.0 + Q_2} \quad (\text{mS}) \tag{5.5.20.2}$$

$$B_2 = \frac{2.0 \, \pi f \, C_2 \, (Q_2 + Q_4)}{1.0 + Q_2} \quad (\text{mS}) \tag{5.5.20.3}$$

where $C_1$ and $C_2$ are the open end capacitances of microstrip lines having widths $w_1$ and $w_2$ in pF. The frequency $f$ is in GHz and $h$ is in mm. The $Q_n$'s are given by

$$Q_1 = 0.04598 \left\{ 0.03 + (w_1 / h)^{1.23 / \left[ 1.0 + 0.12 \, (w_2 / w_1 - 1.0)^{0.9} \right]} \right\}$$

$$\times (0.272 + 0.07 \, \varepsilon_r) \tag{5.5.20.4}$$

$$Q_2 = 0.10700 \, (w_1 / h + 9.0) \, (g / h)^{3.23} + 2.09 \, (g / h)^{1.05} \left( \frac{1.5 + 0.3 \, w_1 / h}{1.0 + 0.6 \, w_1 / h} \right)$$

$$(5.5.20.5)$$

$$Q_3 = e^{\left[-0.5978 \, (w_2/w_1)^{1.35}\right]} - 0.55 \qquad\qquad (5.5.20.6)$$

$$Q_4 = e^{\left[-0.5978 \, (w_1/w_2)^{1.35}\right]} - 0.55 \qquad\qquad (5.5.20.7)$$

and the equations are valid for

$0.1 \leq w_1 / h \leq 3.0$

$0.1 \leq w_2 / h \leq 3.0$

$1.0 \leq w_2 / w_1 \leq 3.0$

$0.2 \leq g / h \leq \infty$

$6.0 \leq \varepsilon_r \leq 13.0$

Error is less than 0.1 mS for frequencies up to 18 GHz and $f \cdot h$ products up to 12 GHz $\cdot$ mm.

## REFERENCES

[1]   Biswas, Animesh, and Vijai K. Tripathi, "Modeling of Asymmetric and Offset Gaps in Shielded Microstrips and Slotlines," *IEEE Transactions on Microwave Theory and Techniques*, Vol. MTT-38, No. 6, June 1990, pp. 818–822.

[2]   Kirschning, M., *et al.*, "Measurement and Computer-Aided Modeling of Microstrip Discontinuities by an Improved Resonator Method," *IEEE Microwave Theory and Techniques Symposium Digest 1983*, pp. 495–497.

[3]   Koster, Norbert H.L., and Rolf H. Jansen, "The Equivalent Circuit of the Asymmetrical Series Gap in Microstrip and Suspended Substrate Lines," *IEEE Transactions on Microwave Theory and Techniques*, Vol. MTT-30, No. 8, August 1982, pp. 1273–1279.

## 5.5.21 Microstrip Line Finger Break

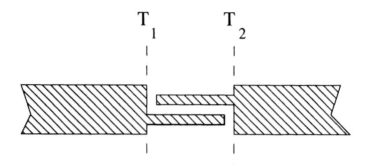

**Figure 5.5.21.1:**   Microstrip Line Finger Break

## 5.5.22 Microstrip Line Zig-Zag Slit

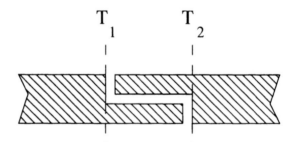

**Figure 5.5.22.1:**   Microstrip Line Zig-Zag Slit

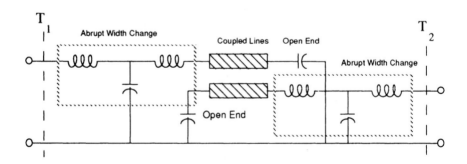

**Figure 5.5.22.2:**   Finger Break and Zig-Zag Slit Model

In [1] data are presented showing that the above two structures can be modeled by the combination of 2 coupled lines, abrupt change in widths, and the end effect capacitances. Lancombe and Cohen [2] reports on experiments with the number of fingers in an interdigital capacitor. It was found that the two-finger configuration resulted in > octave bandwidth (1.4:1 VSWR) whereas the multiple finger configurations did not (0.78 octave).

REFERENCES

[1]     Free, C.E., and C.S. Aitchison, "Excess Phase in DC Blocks," *Electronics Letters*, October 11, 1984, Vol. 20, No. 21, pp. 892–893.

[2]     Lacombe, David, and Jerome Cohen, "Octave-Band Microstrip DC Blocks," *IEEE Transactions on Microwave Theory and Techniques*, Vol. MTT-20, No. 8, August 1972, pp. 555–556.

## 5.5.23 Arbitrary Angle Microstrip Line Bend

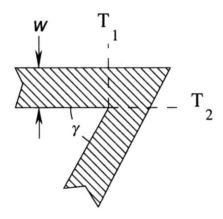

**Figure 5.5.23.1:**     Arbitrary Angle Microstrip Line Bend

The bend in microstrip line is useful whenever we need to modify the signal path to accommodate a packaging or layout scheme. The compensated bend has been described earlier. Mehran [1] gives graphs of s-parameters for the uncompensated structure, but no closed form equations are given. For small angles he found that coupling can be significant. Menzel [2] gives similar curves for the particular case of the 120° bend.

REFERENCES

[1]     Mehran, Reza, "Calculations of Microstrip Bends and Y-Junctions with Arbitrary Angle," *IEEE Transactions on Microwave Theory and Techniques*, Vol. MTT-26, No. 6, June 1978, pp. 400–405.

[2]   Menzel, Wolfgang, "Frequency Dependent Transmission Properties of Microstrip Y-Junctions and 120° Bends," *Microwaves, Optics and Acoustics*, Vol. 2, No. 2, March 1978, pp. 55–59.

### 5.5.24 Microstrip Line Y Junction

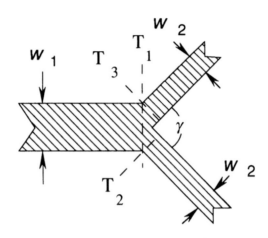

**Figure 5.5.24.1:**   **Microstrip Line Y Junction**

This structure is useful in design of combiner/dividers. Mehran [2] and Menzel [3] give s-parameter curves for the discontinuity introduced.

Closed-form equations for the radiation of this structure when $w_1 = 2.0\, w_2$ have been found by Lewin [1]. The radiated power is

$$P = 60.0\,(k\,h)^2\,F \tag{5.5.24.1}$$

where

$$k = \text{free-space wavenumber} \tag{5.5.24.2}$$

$$F = \frac{\varepsilon_r + 1.0 - 2.0\,c^2}{4.0\,c^2}\;\frac{s}{\sqrt{\varepsilon_r - c^2}}\;\ln\left(\frac{\sqrt{\varepsilon_r - c^2} + s}{\sqrt{\varepsilon_r - c^2} - s}\right)$$

$$-\,\frac{1.0 - c}{1.0 + c}\left[\frac{(\varepsilon_r + 1.0)\,(1.0 - c)\,(1.0 + 3.0\,c)}{4.0\,c^2}\right.$$

$$\frac{1.0}{\sqrt{\varepsilon_r}} \ln \frac{\sqrt{\varepsilon_r} + 1.0}{\sqrt{\varepsilon_r} - 1.0} + \frac{(2.0 \, \varepsilon_r - c)}{\Gamma} \ln \frac{\varepsilon_r - c + \Gamma}{\varepsilon_r - 1.0} \Bigg] \qquad (5.5.24.3)$$

and

$$c = \cos \beta \qquad (5.5.24.4)$$

$$s = \sin \beta \qquad (5.5.24.5)$$

$$\beta = \frac{\gamma}{2.0} \qquad (5.5.24.6)$$

$$\Gamma = \sqrt{2.0 \, \varepsilon_r \, (1.0 - c) - s^2} \qquad (5.5.24.7)$$

The equations were found by integration of the far fields of the structure. Coupling between the arms was neglected and can be expected to have an effect for small $\gamma$.

## REFERENCES

[1]    Lewin, Leonard, "Spurious Radiation from a Microstrip Y Junction," *IEEE Transactions on Microwave Theory and Techniques*, Vol. MTT-26, No. 11, November 1978, pp. 893–894.

[2]    Mehran, Reza, "Calculations of Microstrip Bends and Y-Junctions with Arbitrary Angle," *IEEE Transactions on Microwave Theory and Techniques*, Vol. MTT-26, No. 6, June 1978, pp. 400–405.

[3]    Menzel, Wolfgang, "Frequency Dependent Transmission Properties of Microstrip Y-Junctions and 120° Bends," *Microwaves, Optics and Acoustics,* Vol. 2, No. 2, March 1978, pp. 55–59.

[4]    Tsuji, Mikio, *et al.,* "Low-Loss Design Method for a Planar Dielectric-Waveguide Y Branch: Effect of a Taper of Serpentine Shape," *IEEE Transactions on Microwave Theory and Techniques*, Vol. MTT-39, No. 1, January 1991, pp. 6–13.

## 5.5.25 Slit in Microstrip Line

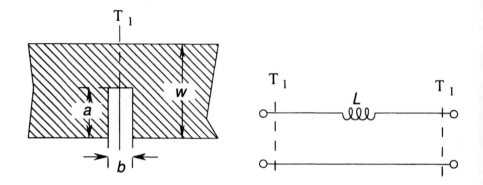

**Figure 5.5.25.1:     Microstrip Transverse Slit**

The microstrip line slit is used for tuning, as a circuit element in filters, and in directional couplers for mode velocity equalization. A simple inductive model (Figure 5.5.25.1) is presented in Hoefer [1]. His equations are

$$L = \frac{\mu_0 \, \pi}{2.0} \left(\frac{a'}{A}\right)^2 \tag{5.5.25.1}$$

where

$$\frac{a'}{A} = 1.0 - \frac{Z_{0,air}}{Z'_{0,air}} \tag{5.5.25.2}$$

and $Z_{0,air}$ is the impedance of a microstrip line of width $w$ and dielectric replaced by air, and $Z'_{0,air}$ is the impedance of a microstrip line of width $w' = w - a$ and dielectric also replaced by air. The capacitance of the gap which is in parallel with the inductor may be calculated as half the capacitance of a microstrip line gap having dimensions $s = b$ and $w = 2.0\, a$ (see Figure 5.5.19.1).

The above equations are valid for

$b < h$

$b << \lambda_g$

$0.0 \leq a / w \leq 0.9$

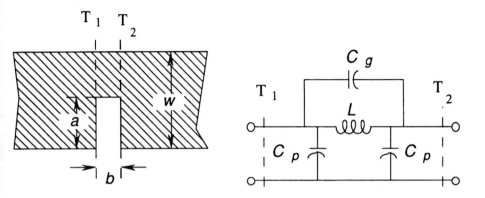

**Figure 5.5.25.2:** **Microstrip Transverse Slit**

Mosig [2] presents a more complete model (shown in Figure 5.5.25.2), which includes the capacitance across the slit and to ground. Design curves are given in this reference.

REFERENCES

[1] Hoefer, Wolfgang J., "Equivalent Series Inductivity of a Narrow Transverse Slit in Microstrip," *IEEE Transactions on Microwave Theory and Techniques*, Vol. MTT-25, No. 10, October 1977, pp. 822–824.

[2] Mosig, Juan R., and F.E. Gardiol, "Equivalent Inductance and Capacitance of a Microstrip Slot," *7th European Microwave Conference Proceedings*, 1977, Copenhagen, Denmark, pp. 455–459.

## 5.5.26 Microstrip Line Spur Line

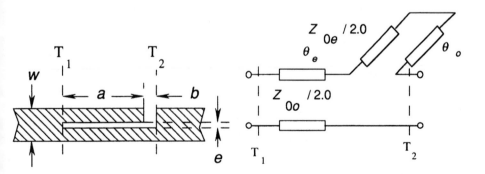

**Figure 5.5.26.1:** **Spur Line**

The spur line is a useful filter element first adapted to microstrip lines in Bates [1] for lines having equal mode velocities. This is rarely the case and the analysis of Nguyen [2] extended this to lines of unequal mode velocities. Nguyen finds a model for the spur line in terms of equivalent transmission lines as shown above. The parameters $Z_{0,e}$, $Z_{0,o}$, $\theta_e$, and $\theta_o$ are calculated with the equations for coupled microstrip lines.

REFERENCES

[1]     Bates, R.N., "Design of Microstrip Spur-Line Band-Stop Filters," *Microwaves, Optics and Acoustics*, Vol. 1, No. 6, November 1977, pp. 209–214.

[2]     Nguyen, C., *et al.*, "Millimeter Wave Printed Circuit Spurline Filters," *1983 IEEE MTT-S Symposium Digest*, pp. 98–100.

## 5.5.27 Air Bridges

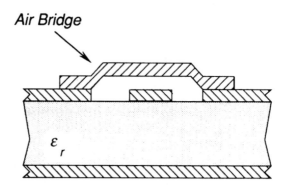

Figure 5.5.27.1:     Air Bridge

The air bridge is used in microwave integrated circuits to bridge a signal over other signal conductors. This occurs commonly in spiral inductors. The resulting structure couples energy between the signal paths. Wiemer [2] mentions that he has had success modeling this with the parallel-plate capacitance multiplied by 1.55 to 1.7 to include the effect of fringing or stray capacitances. This is in agreement with the results of Jansen *et al.* [1] which suggest the factor 1.7.

REFERENCES

[1]     Jansen, R.H., *et al.*, "Theoretical and Experimental Broadband Characterization of Multiturn Square Spiral Inductors in Sandwich Type GaAs MMICs," *15th European Microwave Symposium*, Paris, 1985, pp. 946–951.

[2]     Wiemer, L., and R.H. Jansen, "Determination of Coupling Capacitance of Underpasses, Air Bridges and Crossings in MICs and MMICs," *Electronics Letters*, Vol. 23, No. 7, March 26, 1987, pp. 344–346.

## 5.5.28 Coupled Microstrip Line Right-Angle Bends

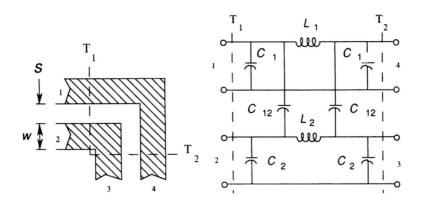

**Figure 5.5.28.1:** Coupled Microstrip Line Right-Angle Bends

This structure occurs in spiral inductors and is a significant error source in their simulation. Closed-form relations for CAE have been fit to the curves of Hill and Tripathi [2], which apply for $w = s$

$$\frac{C_{11,e}}{C_{11,\infty} h} = -0.18306061 + 5.0447137 \left(\frac{w}{h}\right) \tag{5.5.28.1}$$

$$\frac{C_{12,e}}{C_{12,\infty} h} = -0.039680602 + 2.4009002 \left(\frac{w}{h}\right) \tag{5.5.28.2}$$

$$\frac{C_{22,e}}{C_{22,\infty} h} = -0.12157902 + 1.0334401 \left(\frac{w}{h}\right) \tag{5.5.28.3}$$

$$\frac{L_{11,e}}{L_{11,\infty} h} = -0.50404018 + 4.4984334 \left(\frac{w}{h}\right) \tag{5.5.28.4}$$

$$\frac{L_{22,e}}{L_{22,\infty} h} = -0.53287219 + 0.48572951 \left(\frac{w}{h}\right) \tag{5.5.28.5}$$

$$\frac{L_{12,e}}{L_{12,\infty} h} = -0.74416701 + 0.44005033 \sqrt{\frac{w}{h}} \tag{5.5.28.6}$$

where

$$C_1 = \frac{C_{11,e} - C_{12,e}}{2.0} \tag{5.5.28.7}$$

$$C_2 = \frac{C_{22,e} - C_{12,e}}{2.0}$$

(5.5.28.8)

$$C_{12} = \frac{C_{12,e}}{2.0}$$

(5.5.28.9)

$$L_1 = L_{11,e} + L_{12,e}$$

(5.5.28.10)

$$L_2 = L_{22,e} + L_{12,e}$$

(5.5.28.11)

The variables subscripted with infinity ($\infty$) are the parameter values corresponding to the isolated and coupled line without bend. These curve fits are valid to within a few percent for

$$0.2 \leq w / h \leq 3.0$$

Curved bend coupling is described in [1].

REFERENCES

[1]     Abouzahra, Mohamed, and Leonard Lewin, "Theory and Appliction of Coupling Between Curved Transmission Lines," *IEEE Transactions on Microwave Theory and Techniques*, Vol. MTT-30, No. 11, November 1982, pp. 1988–1995 and "Corrections" Vol. MTT-33, No. 1, January 1985, pp. 74–75.

[2]     Hill, A., and V.K. Tripathi, "Analysis and Modeling of Coupled Right Angle Microstrip Bend Discontinuities," *1989 IEEE MTT-S Symposium Digest,* pp. 1143–1146.

## 5.6   STRIPLINE DISCONTINUITIES

### 5.6.1   Abrupt Stripline 90° Bend

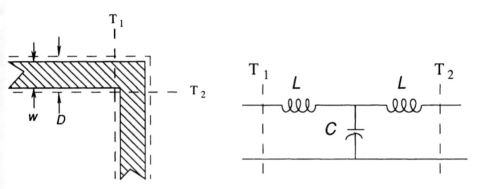

Figure  5.6.1.1:      Abrupt  Stripline  Bend

According to Howe [4], an abrupt bend in stripline can be approximated with 0.2 pF to ground.  In Oliner [5], a right-angle bend is modeled as the combination of series inductances and a shunt capacitance

$$L = \frac{2.0\, D\, Z_0 \left[ 0.878 + 2.0 \left( \frac{D}{\lambda_g} \right)^2 \right]}{\lambda_g\, \omega} \tag{5.6.1.1}$$

$$C = \frac{2.0\, \pi\, D}{\omega\, \lambda_g\, Z_0 \left[ 1.0 - 0.114 \left( \frac{2.0\, D}{\lambda_g} \right)^2 \right]} \tag{5.6.1.2}$$

where

$$D = \frac{b\, K(k)}{K(k')} \cong w + \frac{2.0\, b}{\pi} \ln 2.0 \tag{5.6.1.3}$$

$$k = \tanh \left( \frac{\pi}{2} \frac{w}{b} \right) \tag{5.6.1.4}$$

and $Z_0$ is calculated with the stripline equations from Chapter 2.

REFERENCES

[1]   Altschuler, H.M., and A.A. Oliner, "Discontinuities in the Center Conductor of Symmetric Strip Transmission Line," *IRE Transactions on Microwave Theory and Techniques*, Vol. MTT-8, May 1960, pp. 328–339, and "Addendum to 'Discontinuities in the Center Conductor of Symmetric Strip Transmission Line,'" Vol. MTT-10, No. 2, March 1962, p. 143.

[2]   Bahl, I.J., and Ramesh Garg, "A Designer's Guide to Stripline Circuits," *Microwaves*, January 1978, pp. 90–96.

[3]   Bhat, Bharathi, and Shiban K. Koul, *Stripline-Like Transmission Lines for Microwave Circuits*, John Wiley & Sons, New York, 1989.

[4]   Howe, Jr., Harlan, *Stripline Circuit Design*, Artech House, Norwood, MA, 1974.

[5]   Oliner, Arthur A., "Equivalent Circuits for Discontinuities in Balanced Strip Transmission Line," *IRE Transactions on Microwave Theory and Techniques*, Vol. MTT-3, No. 2, March 1955, pp. 134–143.

## 5.6.2   Arbitrary Angle Stripline Bend

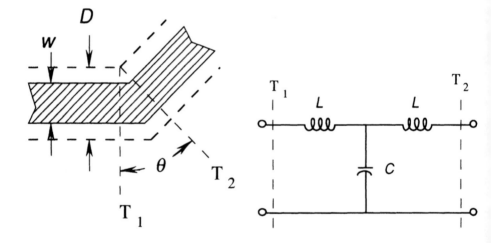

**Figure 5.6.2.1:**   **Arbitrary Angle Stripline Bend**

Oliner [4] gives

$$L = \frac{2.0 \, D \, Z_0}{\omega \, \lambda_g} \left\{ \Psi\left[ -0.5 \left( 1.0 - \frac{\theta}{\pi} \right) \right] - \Psi(-0.5) \right\} \tag{5.6.2.1}$$

$$= \frac{2.0\ D\ Z_0}{\omega\ \lambda_g} \left[ \Psi\left( \frac{\theta}{2.0\ \pi} - 0.5 \right) - \Psi(-0.5) \right] \tag{5.6.2.2}$$

$$C = \frac{2.0\ \pi\ D}{\lambda_g\ Z_0\ \omega\ \cot\left( \frac{\theta}{2.0} \right)} \tag{5.6.2.3}$$

where

$$D = \frac{b\ K(k)}{K(k')} \cong w + \frac{2.0\ b}{\pi} \ln 2.0 \tag{5.6.2.4}$$

$$k = \tanh\left( \frac{\pi}{2.0}\frac{w}{b} \right) \tag{5.6.2.5}$$

$\theta$ = bend angle in radians

The psi or digamma function, $\Psi(x)$, is tabulated in [1, Table 6.1]

$\Psi(x)$ = psi or digamma function = logarithmic derivative of gamma function

$$\equiv \frac{\partial\ [\ln \Gamma(x)]}{\partial x} \tag{5.6.2.6}$$

A closed-form equation was found that is accurate to within 0.001% for $0.5 \leq x \leq 1.0$

$$\Psi(x) \cong 0.144202 + 0.023591\ (x + 1.0) + 0.870648 \ln (x + 1.0)$$

$$- 0.744174\ /\ (x + 1.0) - 1.0\ /\ x \tag{5.6.2.7}$$

## REFERENCES

[1] Abramowitz, Milton, and Irene A. Stegun, ed., *Handbook of Mathematical Functions with Formulas, Graphs, and Mathematical Tables*, Dover Publications, 1972.

[2] Altschuler, H.M., and A.A. Oliner, "Discontinuities in the Center Conductor of Symmetric Strip Transmission Line," *IRE Transactions on Microwave Theory and Techniques*, Vol. MTT-8, May 1960, pp. 328–339 and "Addendum to 'Discontinuities in the Center Conductor of Symmetric Strip Transmission Line'," Vol. MTT-10, No. 2, March 1962, p. 143.

[3]  Bhat, Bharathi, and Shiban K. Koul, *Stripline-Like Transmission Lines for Microwave Circuits*, John Wiley & Sons, New York, 1989.

[4]  Oliner, Arthur A., "Equivalent Circuits for Discontinuities in Balanced Strip Transmission Line," *IRE Transactions on Microwave Theory and Techniques*, Vol. MTT-3, No. 2, March 1955, pp. 134–143.

### 5.6.3  Optimal Stripline Mitered 90° Bend

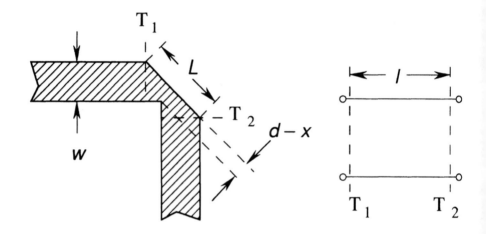

**Figure 5.6.3.1:**  **Optimal Stripline Mitered Bend**

Mathaei, Young, and Jones [1] give the optimal mitre dimensions for $b / \lambda = 0.0847$ and negligible $t$. Howe [2] states that in his experience this curve applies over frequencies through C-band. An equation fitting this graph has been derived for CAE use

$$\frac{a}{w} = 1.33068 - 0.38174 \frac{w}{b} + 0.152215 \left(\frac{w}{b}\right)^2 - 0.024477 \left(\frac{w}{b}\right)^3 \qquad (5.6.3.1)$$

which fits the curve to within 0.5% for $0.2 \leq w / b \leq 3.0$. The width of the narrow section of trace used for the bevel is

$$d - x = \frac{2.0 \, w - a}{\sqrt{2.0}} \qquad (5.6.3.2)$$

where

$$a = \frac{L}{\sqrt{2.0}} \qquad (5.6.3.3)$$

The effective length of the mitre is necessary for modeling and is useful when matching line electrical lengths. Some designers will use very large radius (> 4→5 $w$) bends when matching electrical lengths rather than use mitred corners. A CAE equation has been derived for this curve also:

$$l / w = \left[ 1.7854346 - 0.015055579 \frac{w}{b} - 0.11334006 \left( \frac{w}{b} \right)^2 \right.$$

$$\left. + 0.050432954 \left( \frac{w}{b} \right)^3 - 0.0097883487 \left( \frac{w}{b} \right)^4 \right]^{-1} \tag{5.6.3.4}$$

which fits the curve to within 0.3% for $0.2 \leq w / b \leq 3.0$.

### REFERENCES

[1]   G. Mathaei, L. Young and E.M.T. Jones, *Microwave Filters, Impedance-Matching Networks, and Coupling Structures*, Artech House, Norwood, MA, 1980.

[2]   Howe, Jr., Harlan, *Stripline Circuit Design*, Artech House, Inc., Norwood, MA, 1974.

[3]   Oliner, Arthur A., "Equivalent Circuits for Discontinuities in Balanced Strip Transmission Line," *IRE Transactions on Microwave Theory and Techniques*, Vol. MTT-3, No. 2, March 1955, pp. 134–143.

## 5.6.4   Optimal Stripline Rounded Bend

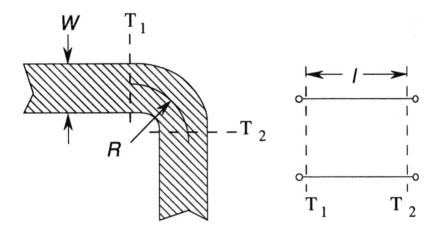

**Figure   5.6.4.1:       Optimal Stripline Rounded Bend**

The rounded bend is an alternative to the mitred corner when a controlled impedance bend is required and space is not a constraint. For best controlled impedance, Howe [1] recommends

$$R > 3 \times w$$

The length of the section of arc can be approximated by its centerline length

$$l \cong \frac{\pi R}{2.0} \tag{5.6.4.1}$$

REFERENCES

[1]     Howe, Jr., Harlan, *Stripline Circuit Design*, Artech House, Norwood, MA, 1974.

### 5.6.5   Stripline Abrupt Width Change

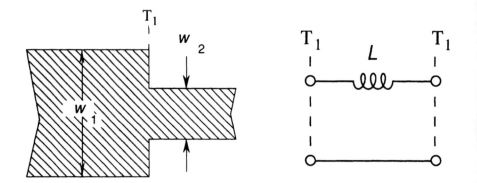

**Figure 5.6.5.1:       Stripline Abrupt Width Change**

References [1, 2, 6] give the most commonly seen relation for the inductance of this discontinuity.

$$L = \frac{2.0\,D}{\omega\,\lambda_g} \ln\left[ \csc\left( \frac{\pi\,Z_{0,2}}{2.0\,Z_{0,1}} \right) \right] \tag{5.6.5.1}$$

Which is valid for

$$b / \lambda_g \sim 0$$

$$w_1 \gg w_2$$

The impedances of isolated strips of width $w_1$ and $w_2$ are $Z_{0,1}$ and $Z_{0,2}$, respectively. In Nalbandian and Steenaart [5] an equation that has validity over a wider range of $w_1$ and $w_2$ is given:

$$L \cong \frac{Z_{0,1} \; 2.0 \; D \; K}{\lambda_g} \tag{5.6.5.2}$$

where

$$D' = \frac{b \; K(k)}{K(k')} \cong w_1 + \frac{2.0 \; b}{\pi} \ln 2.0 \tag{5.6.5.3}$$

$$D = \frac{b \; K(k)}{K(k')} \cong w_2 + \frac{2.0 \; b}{\pi} \ln 2.0 \tag{5.6.5.4}$$

$$k = \tanh\left(\frac{\pi \; w}{2.0 \; b}\right) \tag{5.6.5.5}$$

$$K = \ln\left[\left(\frac{1.0 - \alpha^2}{4.0 \; \alpha}\right)\left(\frac{1.0 + \alpha}{1.0 - \alpha}\right)^{\frac{\alpha + 1.0}{2.0 \; \alpha}}\right] + \frac{2.0}{A} \tag{5.6.5.6}$$

$$A = \left(\frac{1.0 + \alpha}{1.0 - \alpha}\right)^{2.0 \; \alpha}\left[\frac{1.0 + \sqrt{1.0 - D^2 / \lambda^2}}{1.0 - \sqrt{1.0 - D^2 / \lambda^2}}\right] - \frac{1.0 + 3.0 \; \alpha^2}{1.0 - \alpha^2} \tag{5.6.5.7}$$

$$\alpha = \frac{D'}{D} < 1.0 \tag{5.6.5.8}$$

## REFERENCES

[1] Altschuler, H.M., and A.A. Oliner, "Discontinuities in the Center Conductor of Symmetric Strip Transmission Line," *IRE Transactions on Microwave Theory and Techniques*, Vol. MTT-8, May 1960, pp. 328–339, and "Addendum to 'Discontinuities in the Center Conductor of Symmetric Strip Transmission Line'," Vol. MTT-10, No. 2, March 1962, p. 143.

[2] Bahl, I.J., and Ramesh Garg, "A Designer's Guide to Stripline Circuits," *Microwaves*, January 1978, pp. 90–96.

[3] Bhat, Bharathi, and Shiban K. Koul, *Stripline-Like Transmission Lines for Microwave Circuits*, John Wiley & Sons, New York, 1989.

[4]   Mathaei, G., L. Young, and E.M.T. Jones, *Microwave Filters, Impedance-Matching Networks, and Coupling Structures*, Artech House, Norwood, MA 1980.

[5]   Nalbandian, Vahakn, and Willem Steenaart, "Discontinuities in Symmetric Striplines Due to Impedance Steps and Their Compensations," *IEEE Transactions on Microwave Theory and Techniques*, Vol. MTT-20, No. 9, September 1972, pp. 573–578.

[6]   Oliner, Arthur A., "Equivalent Circuits for Discontinuities in Balanced Strip Transmission Line," *IRE Transactions on Microwave Theory and Techniques*, Vol. MTT-3, No. 2, March 1955, pp. 134–143.

### 5.6.6   Compensated Stripline Step Discontinuities

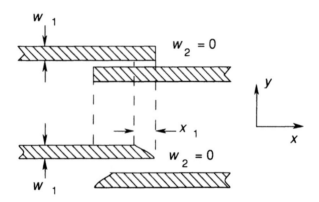

**Figure 5.6.6.1:**      Compensated Step Discontinuities

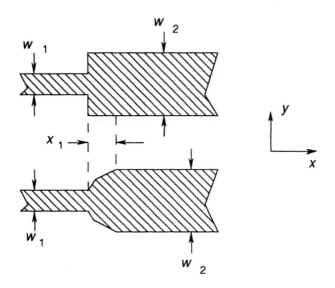

**Figure 5.6.6.2:** **Compensated Step Discontinuities**

The presence of a step discontinuity in any TEM line can be compensated for by the addition of a taper near the discontinuity as shown in the figures above (upper figures are uncompensated, lower are compensated). The cross-section is varied continuously to exchange the transmission line's parallel plate capacitance gradually for fringing capacitance, thus keeping the impedance constant until the step is complete.

For stripline, the line taper begins $x_1$ units back from the nominal position of the step:

$$x_1 = \frac{c\,b}{2} \cosh^{-1}\left(\frac{2w}{c\,b} + 1.0\right) \tag{5.6.6.1}$$

and the taper profile is given by:

$$y\,(x) = w - \frac{b\,c}{2}\left\{\cosh\left[\cosh^{-1}\left(\frac{2w}{b\,c} + 1.0\right) - \frac{2x}{b\,c}\right] - 1.0\right\} \tag{5.6.6.2}$$

where

$w = w_1 - w_2 =$ the difference between the two trace widths

$b =$ ground plane spacing

$$c = \frac{C_{fe}'}{\varepsilon} \tag{5.6.6.3}$$

The calculation can be completed with the equations in the stripline section of this book.

REFERENCES

[1]    Collin, R.E., "The Optimum Tapered Transmission Line Matching Section,"
       *Proceedings of the IRE*, Vol. 44, No. 4, April 1956, pp. 539–548, and
       "Correction," *Proceedings of the IRE*, Vol. 44, No. 12, December 1956, p. 1753.

[2]    Hutchings, J.L., *et al.*, "Contour Program Smoothes 'Strip' Discontinuities,"
       *Microwaves & RF*, November 1987, pp. 129–139.

[3]    Larsson, Mats A., "Compensation of Step Discontinuities in Stripline," *10th
       European Microwave Conference Proceedings*, 1980, pp. 367–371.

[4]    Malherbe, J.A.G., and Andre F. Steyn, "The Compensation of Step Discontinuities
       in TEM-Mode Transmission Lines," *IEEE Transactions on Microwave Theory and
       Techniques*, Vol. MTT-26, No. 11, November 1978, pp. 883–885.

[5]    Pramanick, P., and P. Bhartia, "Tapered Microstrip Transmission Lines," *1983
       IEEE MTT-S International Microwave Symposium Digest*, pp. 242–244.

[6]    STRIP, "Compensated Discontinuities Stripline and Microstripline," Design
       Program.

## 5.6.7    Stripline Open End

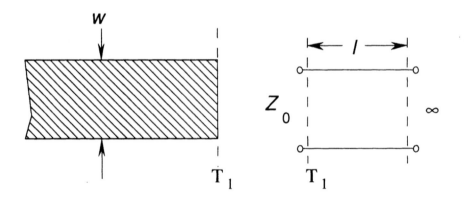

**Figure 5.6.7.1:**    Stripline Open End Effect

This structure occurs in designs, such as filters that have $\lambda / 4.0$ open-circuited lines.
The open-circuit fringing capacitance increases the electrical length of the line, which must
be corrected for proper design of the filter. This fringing effect can modeled by either a
length of transmission line or by a lumped capacitance.

For the infinitely wide strip the additional length is calculated with [1, 5]

$$\Delta l_\infty = \frac{b \ln 2.0}{\pi} \tag{5.6.7.1}$$

For the more realistic, finite-width stripline the additional length is [1]

$$\Delta l = \frac{1.0}{\kappa} \cot^{-1} \left[ \frac{4.0 \ \Delta l_\infty + 2.0 \ w}{\Delta l_\infty + 2.0 \ w} \cot \left( \kappa \ \Delta l_\infty \right) \right] \tag{5.6.7.2}$$

where

$$\kappa = \frac{2.0 \ \pi}{\lambda_g} \tag{5.6.7.3}$$

This equation was derived from (5.6.7.1) and measured data. For $\kappa \ \Delta l_\infty \ll 1.0$ (5.6.7.2) can be approximated with

$$\Delta l \cong \Delta l_\infty \left( \frac{\Delta l_\infty + 2.0 \ w}{4.0 \ \Delta l_\infty + 2.0 \ w} \right) \tag{5.6.7.4}$$

## REFERENCES

[1]   Altschuler, H.M., and A.A. Oliner, "Discontinuities in the Center Conductor of Symmetric Strip Transmission Line," *IRE Transactions on Microwave Theory and Techniques*, Vol. MTT-8, May 1960, pp. 328–339 and "Addendum to 'Discontinuities in the Center Conductor of Symmetric Strip Transmission Line,'" Vol. MTT-10, No. 2, March 1962, p. 143.

[2]   Bahl, I.J., and Ramesh Garg, "A Designer's Guide to Stripline Circuits," *Microwaves,* January 1978, pp. 90–96.

[3]   Boix, Rafael R., and Manuel Horno, "Lumped Capacitance and Open End Effects of Striplike Structures in Multilayered and Anisotropic Substrates," *IEEE Transactions on Microwave Theory and Techniques*, Vol. MTT-37, No. 10, October 1989, pp. 1523–1527.

[4]   Fusco, Vincent F., *Microwave Circuits: Analysis and Computer-aided Design,* Prentice-Hall, NJ, 1987.

[5]   Okoshi, T., *Planar Circuits for Microwaves and Lightwaves,* Springer-Verlag, New York, 1985. (This book has theory of how to solve planar circuits of various types.)

## 5.6.8 Stripline T Junction

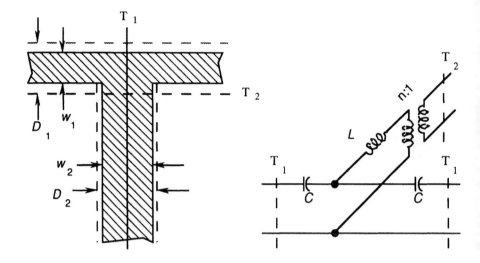

**Figure 5.6.8.1:** **Stripline T Junction**

The stripline T junction is described in [1] and [5]. These models and experimental data are compared in [4]. Models are obtained by equating the waveguide T model of [5] to an equivalent parallel plate guide model and then to stripline.

The equivalent circuit of the stripline T junction is calculated [6] with

$$n = n' \sqrt{\frac{D_2}{D}} \tag{5.6.8.1}$$

$$C = \frac{\lambda_g}{\omega D_2 Z_0 (0.785 \, \eta)^2} \tag{5.6.8.2}$$

for $D_2 / D < 0.5$

$$L = \frac{-Z_0 X_a}{2.0 \, \omega} + \frac{Z_0}{2.0 \, \eta \, \omega} \left\{ \frac{B_t}{2.0} + \frac{2.0 \, D}{\lambda_g} \left[ \ln 2.0 + \frac{\pi D}{6.0 \, D} + \frac{3.0}{2.0} \left( \frac{D}{\lambda_g} \right)^2 \right] \right\} \tag{5.6.8.3}$$

and for $D_2 / D > 0.5$

$$L = \frac{-Z_0 X_a}{2.0 \, \omega} + \frac{2.0 \, Z_0 D}{\lambda_g \, \eta^{12} \, \omega} \left[ \ln \left( \frac{1.43 \, D}{D_2} \right) + 2.0 \left( \frac{D}{\lambda_g} \right)^2 \right]$$

$$(5.6.8.4)$$

where

$$n' = \frac{\sin (\pi D_2 / \lambda_g)}{\pi D_2 / \lambda_g} \qquad (5.6.8.5)$$

$$B_t = \frac{2.0 \, D_1}{\lambda_g} \left\{ \ln \left[ \csc \left( \frac{\pi D_2}{2.0 \, D_1} \right) \right] + \frac{1.0}{2.0} \left( \frac{D_1}{\lambda_g} \right)^2 \cos^4 \left( \frac{\pi D_2}{2.0 \, D} \right) \right\} \qquad (5.6.8.6)$$

$$\frac{D_1}{D_2} = \frac{Z_2}{Z_1} \qquad (5.6.8.7)$$

$$D_1 = \frac{b \, K(k)}{K(k')} \cong w + \frac{2.0 \, b}{\pi} \ln 2.0 \qquad (5.6.8.8)$$

$$k = \tanh \left( \frac{\pi \, w}{2.0 \, b} \right) \qquad (5.6.8.9)$$

## REFERENCES

[1] Altschuler, H.M., and A.A. Oliner, "Discontinuities in the Center Conductor of Symmetric Strip Transmission Line," *IRE Transactions on Microwave Theory and Techniques*, Vol. MTT-8, May 1960, pp. 328–339 and "Addendum to 'Discontinuities in the Center Conductor of Symmetric Strip Transmission Line,'" Vol. MTT-10, No. 2, March 1962, p. 143.

[2] Bahl, I.J., and Ramesh Garg, "A Designer's Guide to Stripline Circuits," *Microwaves,* January 1978, pp. 90–96.

[3] Bhat, Bharathi, and Shiban K. Koul, *Stripline-Like Transmission Lines for Microwave Circuits*, John Wiley & Sons, New York, 1989.

[4] Franco, A.G., and A.A. Oliner, "Symmetric Strip Transmission Line Tee Junction," *IRE Transactions on Microwave Theory and Techniques*, Vol. MTT-10, March 1962, pp. 118–124. (Compares accuracy of several models with available experimental data.)

[5] Marcuvitz, N., *Waveguide Handbook*, McGraw-Hill, New York, 1955. Reprint with errata and preface by N. Marcuvitz. Peter Peregrinus Ltd., 1986.

[6] Oliner, Arthur A., "Equivalent Circuits for Discontinuities in Balanced Strip Transmission Line," *IRE Transactions on Microwave Theory and Techniques*, Vol. MTT-3, No. 2, March 1955, pp. 134–143.

## 5.6.9   Stripline Gap in Center Conductor

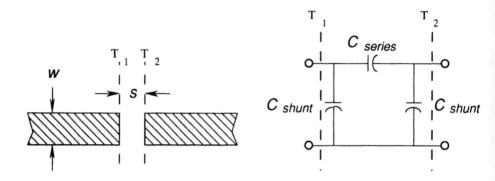

**Figure 5.6.9.1:**   Stripline Gap in Center Conductor

The stripline gap can be used for impedance matching and as a series capacitive element. Two models are possible, depending on the choice of reference plane. The capacitive $\pi$ model has reference planes at each end of the gap and the values are [4] calculated with

$$C_{series} = \frac{1.0}{2.0\,Z_0\,\omega} \left[ \frac{1.0 + (2.0\,Z_0\,B_b + Z_0\,B_a) \cot\left(\dfrac{\pi\,s}{\lambda_g}\right)}{\cot\left(\dfrac{\pi\,s}{\lambda_g}\right) - (2.0\,Z_0\,B_b + Z_0\,B_a)} \right] - \frac{B_A}{2.0\,\omega}$$

(5.6.9.1)

$$C_{shunt} = \frac{Y_0 + B_a\,\cot\left(\pi\,s\,/\,\lambda_g\right)}{\omega\,\cot\left(\pi\,s\,/\,\lambda_g\right) - \omega\,B_a\,/\,Y_0}$$

(5.6.9.2)

for $s \gg w$

$$C_{shunt} \cong \frac{\tan\left(\dfrac{2.0\,b\,\ln\,2.0}{\lambda_g}\right)}{Z_0\,\omega}$$

(5.6.9.3)

where $B_a$ and $B_b$ are the center-line model elements calculated with

$$B_a = \frac{-2.0\,b\,\ln\left[\cosh\left(\dfrac{\pi\,s}{2.0\,b}\right)\right]}{\lambda_g}$$

(5.6.9.4)

$$B_b = \frac{b \ln \left[ \coth \left( \frac{\pi s}{2.0\, b} \right) \right]}{\lambda_g} \tag{5.6.9.5}$$

Oliner, in fact, indicates that the approximation for $C_{shunt}$ should be used for $s > w$.

### REFERENCES

[1] Altschuler, H.M., and A.A. Oliner, "Discontinuities in the Center Conductor of Symmetric Strip Transmission Line," *IRE Transactions on Microwave Theory and Techniques*, Vol. MTT-8, May 1960, pp. 328–339 and "Addendum to 'Discontinuities in the Center Conductor of Symmetric Strip Transmission Line'," Vol. MTT-10, No. 2, March 1962, p. 143.

[2] Bahl, I.J., and Ramesh Garg, "A Designer's Guide to Stripline Circuits," *Microwaves,* January 1978, pp. 90–96.

[3] Bhat, Bharathi, and Shiban K. Koul, *Stripline-Like Transmission Lines for Microwave Circuits,* John Wiley & Sons, New York, 1989.

[4] Oliner, Arthur A., "Equivalent Circuits for Discontinuities in Balanced Strip Transmission Line," *IRE Transactions on Microwave Theory and Techniques,* Vol. MTT-3, No. 2, March 1955, pp. 134–143.

## 5.6.10 Stripline Round Hole in Center Conductor

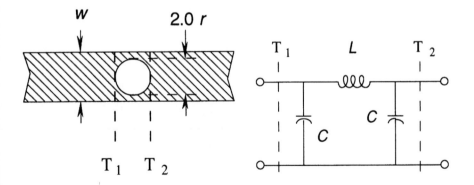

**Figure 5.6.10.1:** **Stripline Round Hole**

The model for a hole in stripline is given in Oliner [4]

$$C = \frac{1.0 + (B_a Z_0) \cot \left( \pi d / \lambda_g \right)}{\omega Z_0 \left[ \cot \left( \pi d / \lambda_g \right) - B_a Z_0 \right]} \tag{5.6.10.1}$$

$$L = \frac{-1.0}{\omega \, B_B \, Z_0} \tag{5.6.10.2}$$

where $B_a$ and $B_b$ are the model elements of the center-line model

$$B_a = \frac{1.0}{4.0 \, B_b \, Z_0^2} \tag{5.6.10.3}$$

$$B_b = - \frac{3.0}{4.0 \, \pi \, Z_0} \frac{\lambda_g \, b \, D}{d^3} \tag{5.6.10.4}$$

REFERENCES

[1]    Altschuler, H.M., and A.A. Oliner, "Discontinuities in the Center Conductor of Symmetric Strip Transmission Line," *IRE Transactions on Microwave Theory and Techniques*, Vol. MTT-8, May 1960, pp. 328–339 and "Addendum to 'Discontinuities in the Center Conductor of Symmetric Strip Transmission Line'," Vol. MTT-10, No. 2, March 1962, p. 143.

[2]    Bahl, I.J., and Ramesh Garg, "A Designer's Guide to Stripline Circuits," *Microwaves,* January 1978, pp. 90–96.

[3]    Bhat, Bharathi, and Shiban K. Koul, *Stripline-Like Transmission Lines for Microwave Circuits*, John Wiley & Sons, New York, 1989.

[4]    Oliner, Arthur A., "Equivalent Circuits for Discontinuities in Balanced Strip Transmission Line," *IRE Transactions on Microwave Theory and Techniques*, Vol. MTT-3, No. 2, March 1955, pp. 134–143.

## 5.6.11 Stripline Slot in Center Conductor

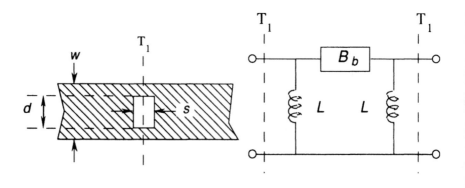

Figure  5.6.11.1:    Stripline  Slot

The model parameters for the stripline slot can be calculated with

$$L = \frac{1.0 + P}{\dfrac{-2.0 \, b \, \omega}{\lambda_g} \, \ln\left(\cosh \dfrac{\omega \, \tau}{b}\right)} \tag{5.6.11.1}$$

$$B_b' = \frac{b}{\lambda_g} \ln\left(\sinh \frac{w \, \tau}{b}\right) - \frac{B_a'}{2.0} - \frac{b \, \lambda_g}{w^2}\left(1.0 + \frac{2.0 \, b}{\pi \, w} \ln 2.0\right) Q \tag{5.6.11.2}$$

where

$$Q \cong \frac{1.0}{4.0\tau}\left[\frac{-1.0}{\ln \beta} - \left(\frac{1.0 - \beta^2}{\ln \beta}\right)^2 \sum_{n=1.0}^{N} f(2.0 \, n \, \tau) \frac{X_n^2}{n}\right] \tag{5.6.11.3}$$

$$P \cong \frac{1.0}{4.0}\left[\left(\frac{1.0 - \beta^2}{\ln \beta}\right)^2 \sum_{n=1.0}^{5.0} g(2.0 \, n \, \tau) \frac{X_n^2}{n} + \tau^2 Q\right] \tag{5.6.11.4}$$

$$N = \text{int}\left[(0.7 \, / \, \tau - 1.0) + 0.5\right] \tag{5.6.11.5}$$

$$\tau = \frac{\pi \, s}{2.0 \, w} \tag{5.6.11.6}$$

$$\beta = \cos \frac{\pi \, d}{2.0 \, w} \tag{5.6.11.7}$$

and

$$X_1 = 1.0 \tag{5.6.11.8}$$

$$X_2 = -1.0 + 3.0 \, \beta^2$$

$$X_3 = 1.0 - 8.0 \, \beta^2 + 10.0 \, \beta^4$$

$$X_4 = -1.0 + 15.0 \, \beta^2 - 45.0 \, \beta^4 + 35.0 \, \beta^6$$

$$X_5 = 1.0 - 24.0 \, \beta^2 + 126.0 \, \beta^4 - 224.0 \, \beta^6 + 126.0 \, \beta^8$$

Curve fits to the functions $f(x)$ and $g(x)$ of Figure 8 in [1] were derived for calculation purposes

$$f(x) \cong -0.924663 + 0.244265\ x + 0.0130211\ /\ x + 1.62211\ e^{-x} \qquad (5.6.11.9)$$

$$g(x) \cong \exp\left(0.70411622 - 0.26293172\ x + 0.01308667\ x^2\right) \qquad (5.6.11.10)$$

It is difficult to determine the fit accuracy due to the small size of the original figures. It is believed that $f(x)$ and $g(x)$ are accurate to better than 5% and 1%, respectively, over the required ranges. Model equations are valid for

$$\tau \geq 0.15$$

$$d\ /\ w \geq 0.25$$

REFERENCES

[1]    Altschuler, H.M., and A.A. Oliner, "Discontinuities in the Center Conductor of Symmetric Strip Transmission Line," *IRE Transactions on Microwave Theory and Techniques*, Vol. 8, May 1960, pp. 328–339 and "Addendum to "Discontinuities in the Center Conductor of Symmetric Strip Transmission Line," Vol. MTT-10, No. 2, March 1962, p. 143.

### 5.6.12 Stripline Multistrip Junction

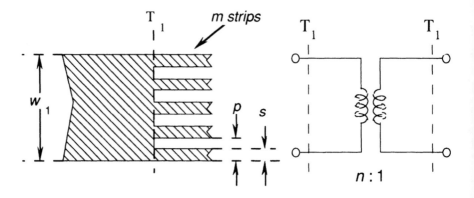

Figure 5.6.12.1:    Multi-strip Junction

Multiple slots are described in [1]. Design equations are

$$n = \sqrt{\dfrac{Z_{0,m}}{Z_0}} \qquad (5.6.12.1)$$

where

$$\frac{Z_{0,m}}{Z_0} = \frac{\dfrac{Z_{0,\infty}}{Z_0} + A}{1.0 + A} \qquad (5.6.12.2)$$

$$A = \frac{K(k')}{K(k)} \frac{p}{m} \frac{}{b} \qquad (5.6.12.3)$$

$$k = \tanh\left(\frac{\pi\, s}{2.0\, b}\right) \qquad (5.6.12.4)$$

$$k' = \sqrt{1.0 - k^2} \qquad (5.6.12.5)$$

$$\frac{Z_{0,\infty}}{Z_0} = 1.0 + \frac{2.0\, p}{\pi\, b}\ln\left[\csc\left(\frac{\pi\, s}{2.0\, p}\right)\right] \qquad (5.6.12.6)$$

These equations are valid for number of strips $m > 3$ with increasing accuracy with increasing $m$.

## REFERENCES

[1]    Altschuler, H.M., and A.A. Oliner, "Discontinuities in the Center Conductor of Symmetric Strip Transmission Line," *IRE Transactions on Microwave Theory and Techniques*, Vol. MTT-8, May 1960, pp. 328–339 and "Addendum to 'Discontinuities in the Center Conductor of Symmetric Strip Transmission Line,'" Vol. MTT-10, No. 2, March 1962, p. 143.

## 5.6.13 Stripline Slot in Ground Plane

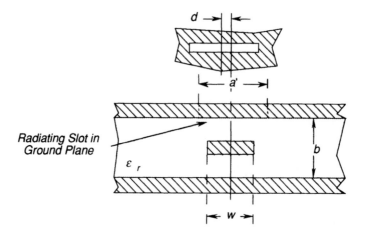

Figure 5.6.13.1:    Stripline Slot in Ground Plane

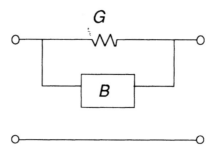

Figure 5.6.13.2:    Equivalent Circuit

For the slot in the ground plane of a stripline, Oliner [2] gives the equivalent circuit

$$\frac{G}{Y_0} = \frac{16.0}{3.0\ \pi} \frac{K(k')}{K(k)} \frac{a'}{\lambda} \left[ 1.0 - 0.374 \left(\frac{a'}{\lambda}\right)^2 + 0.130 \left(\frac{a'}{\lambda}\right)^4 \right] \tag{5.6.13.1}$$

Breithaupt [1] returned to Oliner's original equations and included offset (off-center) slots by adding a factor $(l_0/l)^2$

$$\frac{G}{Y_0} = \frac{16.0}{3.0\ \pi} \left(\frac{l_0}{l}\right)^2 \frac{K(k')}{K(k)} \frac{a'}{\lambda} \left[ 1.0 - 0.374 \left(\frac{a'}{\lambda}\right)^2 + 0.130 \left(\frac{a'}{\lambda}\right)^4 \right] \tag{5.6.13.2}$$

where

$$I = \int_{a'/2-d}^{-a'/2-d} \frac{\cos(\pi (x + d) / a') \ dx}{\sqrt{1.0 + k'^2 \sinh^2(\pi x / b)}} \tag{5.6.13.3}$$

$I_0 = I$ with $d = 0.0$ \hfill (5.6.13.4)

The distance between the centerline of the stripline and the aperture is $d$.

## REFERENCES

[1]   Breithaupt, Robert W., "Conductance Data for Offset Series Slots in Stripline," *IEEE Transactions on Microwave Theory and Techniques*, Vol. MTT-16, No. 11, November 1968, pp. 969–970. (Modifies Oliner to include off-center slots.)

[2]   Oliner, Arthur A., "Equivalent Circuits for Discontinuities in Balanced Strip Transmission Line," *IRE Transactions on Microwave Theory and Techniques*, Vol. MTT-3, No. 2, March 1955, pp. 134–143.

## 5.7   SUSPENDED MICROSTRIP LINE

### 5.7.1   Abrupt Width Change and Open End

Although the references [1, 2] do not give any closed-form relations, it is heartening to note that work has begun in the characterization of discontinuities of lines other than stripline and microstrip line.

REFERENCES

[1]   Achkar, J., *et al.*, "Frequency-Dependent Characteristics of an Asymmetric Suspended Stripline and its Open End Discontinuity for Millimeter Wave Applications," *19th European Microwave Conference Proceedings*, September 4–7, 1990, London, England, pp. 767–772.

[2]   Hong, Jia Sheng, *et al.*, "Exact Computation of Generalised Scattering Matrix of Suspended Microstrip Step Discontinuity," *Electronics Letters*, Vol. 25, No. 5, March 2, 1989, pp. 335–336.

### 5.7.2   Asymmetric Gap

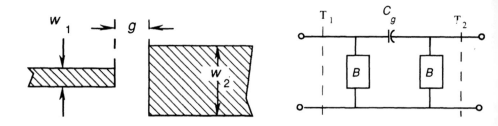

**Figure   5.7.2.1:**      Suspended Substrate Asymmetric Gap

Koster and Jansen [1] give design curves for the circuit parameters of this configuration. Values of the discontinuity are given for

$$0.15 \le g / h \le 3.0$$

$$0.1 \le w_2 / w_1 \le 10.0$$

## REFERENCES

[1] Koster, Norbert H.L., and Rolf H. Jansen, "The Equivalent Circuit of the Asymmetrical Series Gap in Microstrip and Suspended Substrate Lines," *IEEE Transactions on Microwave Theory and Techniques*, Vol. MTT-30, No. 8, August 1982, pp. 1273–1279.

## 5.8   FINLINE DISCONTINUITIES

Finline discontinuities are not covered in this book, however the references for fin-line in Chapter 3 and the following references may provide a good starting point.

| Discontinuity | References |
|---|---|
| Double steps | [2, 5] |
| Gaps in ground | [2, 4] |
| Open circuits | [1, 2] |
| Short circuits | [1, 2] |
| Steps | [2, 5, 8, 10] |
| Taper | [9, 10] |
| Transition between unilateral and bilateral fin-line | [3] |
| Transverse slits | [6–8] |

REFERENCES

[1]   Burton, Miles, and Wolfgang J.R. Hoefer, "An Improved Model for Short- and Open-Circuited Series Stubs in Fin Lines," *1984 IEEE MTT-S Symposium Digest,* pp. 3330–332.

[2]   El Hennaway, Hadia, and Klaus Schünemann, "New Structures for Impedance Transformation in Fin-Lines," *1982 IEEE MTT-S Symposium Digest,* pp. 198–200.

[3]   El Hennaway, Hadia, and Klaus Schünemann, "Computer-Aided Design of Semiconductor Mounts in Fin-Line Technology," *1981 IEEE MTT-S Symposium Digest,* pp. 307–309.

[5]   El Hennaway, Hadia, and Klaus Schünemann, "Hybrid Fin-Line Matching Structures," *IEEE Transactions on Microwave Theory and Techniques*, Vol. MTT-30, No. 12, December 1982, pp. 2132–2139.

[4]   El Hennaway, H., *et al.,* "Impedance Transformation in Fin-Line," *IEE Proceedings*, Part H, Vol. 129, No. 6, April 1983, pp. 342–350.

[6]   Meier, Paul J., "Integrated Fin-Line Millimeter Components," *1981 IEEE MTT-S Symposium Digest,* pp. 195–197.

[7]   Meier, Paul J., "Integrated Fin-Line Millimeter Components," *IEEE Transactions on Microwave Theory and Techniques*, Vol. MTT-22, No. 12, December 1974, pp. 1209–1216.

[8]   Pic, Etienne, and W.J.R. Hoefer, "Experimental Characterization of Fin Line Discontinuities Using Resonant Techniques," *1981 IEEE MTT-S Symposium Digest,* pp. 108–110.

[9] Pramanick, Protap, "Design Tapered Finlines Using a Calculator," *Microwaves & RF*, June 1987, pp. 111–114.

[10] Saad, A.M.K., "Analysis of Fin-Line Tapers and Transitions," *IEE Proceedings*, Part H, Vol. 130, No. 3, April 1983, pp. 230–235.

## 5.9  OTHER DISCONTINUITIES

### 5.9.1  Plated-Through Hole or Via

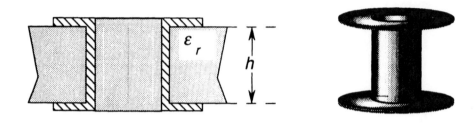

Figure  5.9.1.1:     Plated-Through  Hole

A via is used whenever we wish a signal to change layers in a two-sided or multilayer circuit. A *blind* or *buried* via occurs when the via is completely internal to the substrate and never reaches the surface. Vias may be created by drilling and plating, by an eyelet, or a wire.

There are three different cases commonly encountered for the plated-through hole or via. First, the via is used in series with the signal line when changing layers as shown in Figure 5.9.1.2. Second, the via can be used to connect surface mount components to a ground plane on another layer (Figure 5.9.1.3). Finally, vias are used to tie together ground planes on different layers of a PCB (Figure 5.9.1.4). When the via is in series with the signal path, we would like it to have the same characteristic impedance as the transmission lines it is interconnecting. When it is used for grounding, we want it to have as low an impedance as possible.

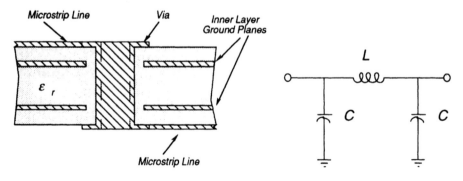

Figure  5.9.1.2:     Via  Connecting  Microstrip  Lines

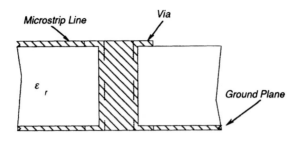

**Figure 5.9.1.3:** **Via to Ground Plane**

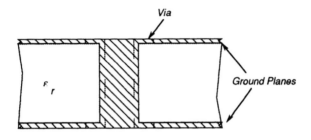

**Figure 5.9.1.4:** **Via Connecting Ground Planes**

The simplest model for vias is simply a round wire or cylindrical wire inductor. Equations of Chapter 6 can be used to calculate these inductances [8, 11, and others]. As frequency increases, the model may need to include resistance and inductance caused by current restriction at the transition from the planar line to the via, radiation losses, and stray capacitances to nearby ground planes and across the via.

The effect of the restriction of current at the transition from planar line to via is analyzed in Abbas and Weeks [1]. Design data is presented in a group of tables. Brewitt-Taylor [2] calculates the radiation resistance and a mutual impedance between two vias in the same substrate. As might be suspected, the via is found to radiate more than an equivalent length microstrip trace. The radiation resistance can be approximated with

$$R = \frac{k_0^3 \, h^3 \, Z_0 \left(1.0 - 1.0 \, / \, \varepsilon_r\right)}{2.0 \, \varepsilon_r} \tag{5.9.1.1}$$

where

$$k_0 = \frac{\omega}{c_0} \quad (\text{m}^{-1}) = \text{wave number}[1] \text{ in free space} \tag{5.9.1.2}$$

---

[1]The *wave number* is the number of wavelengths per unit length.

Kasten, *et al.* [6] found model parameters for the via configuration of Figure 5.9.1.2 with a 0.0625 in long via and inner ground planes spaced 0.0056 in from the microstrip lines. The measured values of $C$ and $L$ were approximately 0.5 pF and 3.4 nH. This lumped model was valid to approximately 2 GHz.

In Finch and Alexopoulos a planar waveguide model is used to analyze metallic and dielectric posts using a new rapidly convergent formulation. Quine, *et al.* model semiconductor vias. The coupling between vias is also examined.

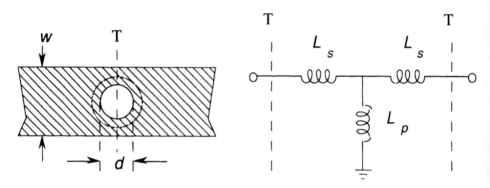

**Figure 5.9.1.5:** **Alternative Via Model**

A model for a via to ground in a microstrip line is shown in Figure 5.9.1.5 above. For $d \ll w$, Hoffman [5] states that the via is modeled well by $L_p$ the inductance of a cylindrical wire (see Chapter 6). As $d$ increases, the effect of $L_s$ becomes important. Hoffman presents graphs of the equivalent circuit element values for this structure which are included below. The figures give the value of the model elements for a range of $w / h$ and $d / w$ and $\varepsilon_r = 9.8$.

A new relation for the via was presented recently in the literature. Numerical simulation and measurements show good agreement with a relatively simple closed-form equation. The interested read should refer to Goldfarb and Pucel [4a]:

$$L = \frac{\mu_0}{2.0\ \pi}\left[ h\ \ln\left(\frac{h + \sqrt{r^2 + h^2}}{r}\right) + 1.5\ \left(r - \sqrt{r^2 + h^2}\right)\right] \tag{5.9.1.3}$$

$$R = R_{dc}\ \sqrt{1.0 + \frac{f}{f_\delta}} \tag{5.9.1.4}$$

$$f_\delta = \frac{1.0}{\pi\ \mu_0\ \sigma\ t^2} \tag{5.9.1.5}$$

$$r = d\ /\ 2.0 \tag{5.9.1.6}$$

$t$ = hole metallization thickness (5.9.1.7)

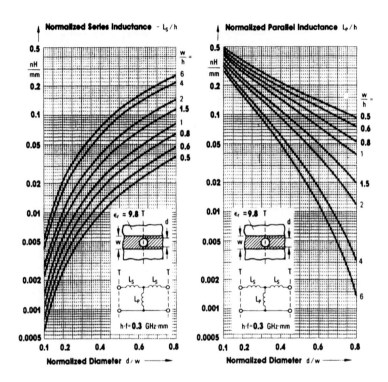

Figure  5.9.1.6:  [5,  Figure  10.20[2])

---

[2]Figures from Hoffman, Reinmut K., *Handbook of Microwave Integrated Circuits*, Artech House, Norwood, MA, 1987. Reprinted by permission of the publisher.

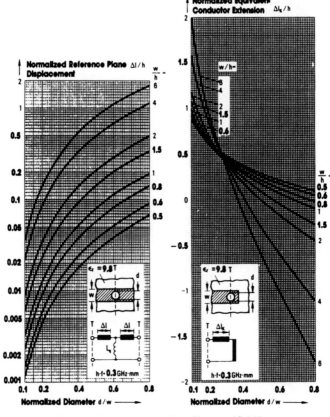

Figure 5.9.1.7:         [5, Figure 10.21]

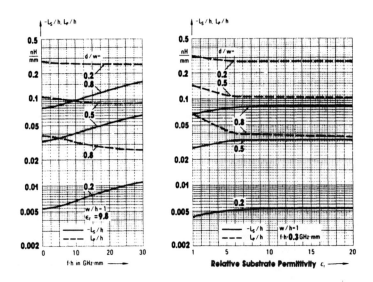

Figure 5.9.1.8:　　[5, Figure 10.22)

## REFERENCES

[1]　Abbas, S.A., and William T. Weeks, "Analytical Calculation of Metallized-Hole Resistance," *IEEE Transactions on Parts, Materials, and Packaging*, Vol. PMP-6, No. 1, March 1970, pp. 26–35.

[2]　Brewitt-Taylor, C.R., "Radiation from Via-Holes in MMICs," *17th European Microwave Conference Proceedings*, 1987, Rome, Italy, pp. 433–436.

[3]　Djordjevic, A.R., and T.K. Sarkar, "Computation of Inductance of Simple Vias Between Two Striplines Above a Ground Plane," *IEEE Transactions on Microwave Theory and Techniques*, Vol. MTT-33, No. 3, March 1985, pp. 265–269. (Models via as a flat wire connecting the two striplines. No closed-form equations.)

[4]    Finch, Kevin L., and Nicolaos G. Alexopoulos, "Shunt Posts in Microstrip
       Transmission Lines," *IEEE Transactions on Microwave Theory and Techniques*,
       Vol. 38, No. 11, November 1990, pp. 1585–1594.

[4a]   Goldfarb, Marc E., and Robert A. Pucel, "Modeling Via Hole Grounds in
       Microstrip," *IEEE Microwave and Guided Wave Letters*, Vol. 1, No. 6, June
       1991, pp. 135–137.

[5]    Hoffman, Reinmut, *Handbook of Microwave Integrated Circuits*, Artech House,
       Norwood, MA, 1987.

[6]    Kasten, Jeffery S., *et al.*, "Enhanced Through-Reflect-Line Characterization of
       Two-Port Measuring Systems Using Free-Space Capacitance Calculation," *IEEE
       Transactions on Microwave Theory and Techniques*, Vol. MTT-38, No. 2,
       February 1990, pp. 215–217.

[7]    Quine, John P., *et al.*, "Characterization of Via Connections in Silicon Circuit
       Boards," *IEEE Transactions on Microwave Theory and Techniques*, Vol. MTT-36,
       No. 1, January 1988, pp. 21–27.

[8]    Ramo, Simon, and John R. Whinnery, *Fields and Waves in Modern Radio*, Wiley,
       London, 1944.

[9]    U.S. Department of Commerce, National Bureau of Standards, *Radio Instruments
       and Measurements*, Circular C74, "Calculation of Inductance," January 1, 1937.

[10]   Wang, Taoyun, *et al.*, "Quasi-static Analysis of a Microstrip Via Through a Hole in
       a Ground Plane," *IEEE Transactions on Microwave Theory and Techniques*, Vol.
       MTT-36, No. 6, June 1988, pp. 1007–1013.

[11]   Wang, Taoyun, *et al.*, "The Excess Capacitance of a Microstrip Via in a Dielectric
       Substrate," *IEEE Transactions on Computer-Aided Design*, Vol. 9, No. 1, January
       1990, pp. 48–56.

## 5.9.2   Optimal Plated-Through Hole or Via

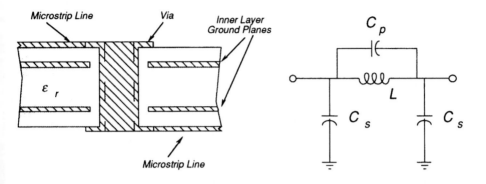

**Figure 5.9.2.1:**      **Optimal Via**

As described in Wang, *et al.* [1], it is possible to design a via for a desired impedance. The discontinuity is made up of a series inductive tube shunted by a capacitance. Some of the capacitance is due to the open ends on the outer layers. Inner layer ground planes create a capacitance to ground along the inductor. It may be possible to minimize any discontinuity (over some bandwidth) by equating these to the distributed inductance and capacitance of a transmission line with the desired characteristic impedance.

Wang gives design charts which allow one to synthesize arbitrary impedance (matched) vias passing through a ground plane from wires on either side.

### REFERENCES

[1]   Wang, Taoyun, *et al.*, "Quasi-static Analysis of a Microstrip Via Through a Hole in a Ground Plane," *IEEE Transactions on Microwave Theory and Techniques*, Vol. MTT-36, No. 6, June 1988, pp. 1007–1013.

[2]   Wang, Taoyun, *et al.*, "The Excess Capacitance of a Microstrip Via in a Dielectric Substrate," *IEEE Transactions on Computer-Aided Design*, Vol. 9, No. 1, January 1990, pp. 48–56.

### 5.9.3 Wraparound Ground

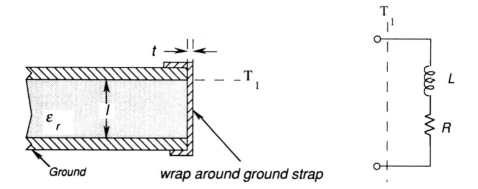

**Figure 5.9.3.1:** **Wraparound Ground**

On some circuits the ground plane is extended around the edges of the board. This can be done with a strap made of a thin piece of metal or by plating a slot in the dielectric. Components may also be strapped to the ground plane with this technique through a rectangular opening in the dielectric. This discontinuity may be modeled with the equations for the flat inductor found in Chapter 6.

# Chapter 6

# Inductors

In this chapter closed-form equations for inductor design are presented. We have already seen a number of techniques for creating inductive reactances. The incremental inductance of any of the structures in Chapter 3 may be calculated from $Z_0$ using the equations of Chapter 2. The techniques described in Chapter 2 can make inductive elements from open- and short-circuited lines. Many of the discontinuities in Chapter 5 are used as inductors. In this chapter we describe the type of structure that is optimized for its inductance. These equations enable the calculation of the inductance from physical dimensions.

Although the accuracy of these equations is quite good in calculating the inductance, the impedance of the inductor is made up of the resistance and stray capacitance as well. Since these strays are not included the equations can only provide a starting point for design. Typically several iterations of measurement and adjustment are required before the desired $Q$ and resonance-free inductance are achieved over the frequency range. Once the dimensions are determined, the inductance is found to be repeatable.

This chapter includes many useful configurations; consult the references, particularly Grover [2] for others.

## 6.1   ABCD MATRICES

Series inductor:

$$\begin{bmatrix} 1.0 & j\,\omega\,L \\ 0.0 & 1.0 \end{bmatrix}$$

Shunt inductor:

$$\begin{bmatrix} 1.0 & 0.0 \\ \dfrac{-j}{\omega\,L} & 1.0 \end{bmatrix}$$

Ideal transformer:

$$\begin{bmatrix} \dfrac{n_1}{n_2} & 0.0 \\[2ex] 0.0 & \dfrac{n_2}{n_1} \end{bmatrix}$$

where $n_1$ and $n_2$ are number of primary and secondary turns, respectively.

## 6.2   WIRE INDUCTORS

### 6.2.1   Round Wire Inductor

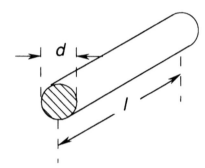

**Figure  6.2.1.1:**          **Straight  Round  Wire**

The self-inductance of a straight wire at high frequency due to Grover [2] is:

$$L = 0.002 \, l \left[ \ln \left( \frac{4.0 \, l}{d} \right) - 1.00 + \frac{d}{2.0 \, l} + \frac{\mu_r \, T(x)}{4.0} \right] \quad (\mu H) \qquad (6.2.1.1)$$

where

   $r = d / 2.0$  (cm)

   $l$ = wire length  (cm)

   $f$ = frequency (Hz)

The function $T$ corrects the inductance for ac effects.  Values of $T$ are tabulated in Grover [2, Table 52, p. 266] and in [5, Table 8, p. 282] where the variable $\delta$ is $T / 4$.  Grover's tables are more comprehensive and precise.  For convenience, we have developed an approximation for $T$ by curve fitting the data. It is accurate to a few percent over the range $0.0 \le x \le 100.0$. For large $x$, $T(x)$ approaches $0.0$.

$$T(x) \cong \sqrt{\frac{0.873011 + 0.00186128\,x}{1.0 + -0.278381\,x + 0.127964\,x^2}} \qquad (6.2.1.2)$$

where

$$x = 2.0\,\pi\,r \sqrt{\frac{2.0\,\mu\,f}{\sigma}} \qquad (6.2.1.3)$$

The impedance of curved bond wires is found in Chapter 3.

## REFERENCES

[1]    Greenhouse, H.M., "Design of Planar Rectangular Microelectronic Inductors," *IEEE Transactions on Parts, Hybrids, and Packaging*, Vol. PHP-10, No. 2, June 1974, pp. 101–109.

[2]    Grover, F.W., *Inductance Calculations*, Van Nostrand, Princeton, NJ, 1946; reprinted by Dover Publications, NY, 1962. (A concise book based on work at NBS.)

[3]    Terman, F.E., *Radio Engineer's Handbook*, McGraw-Hill, New York, 1945. (A classic.)

[4]    U.S. Department of Commerce and Labor, *Miscellaneous Pamphlets on Inductance*, U.S. Bureau of Standards, 1920. (A tremendously comprehensive collection of inductance equations unavailable anywhere else.)

[5]    U.S. Department of Commerce, Circular of the National Bureau of Standards C74, *Radio Instruments and Measurements*, "Calculation of Inductance," January 1, 1937.

## 6.2.2    Flat Wire Inductor

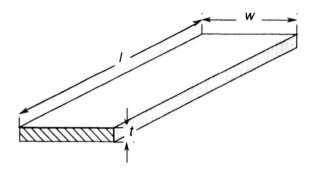

**Figure  6.2.2.1:        Flat  Wire  Inductor**

Flat or ribbon wires are used as bond wires. Flat wire provides a low-inductance, low-resistance technique for interconnections. A similar option is mesh. A mesh provides the added advantage of being perforated at regular distances. This can be used to provide a more repeatable bond length. The inductance of circuit traces can also modeled with this structure.

The equation below is often quoted in the literature as coming from [7]:

$$L = 2.0 \times 10^{-3} \times l \left[ \ln\left(\frac{l}{w+t}\right) + 1.193 + 0.2235 \left(\frac{w+t}{l}\right) \right] \quad \text{(nH)}$$

$$(6.2.2.1)$$

Terman, in fact, gives:

$$L = 2.0 \times 10^{-3} \times l \left[ \ln\left(\frac{2.0\ l}{w+t}\right) + 0.5 + 0.2235 \left(\frac{w+t}{l}\right) \right] \quad \text{(nH)}$$

$$(6.2.2.2)$$

where $l$, $w$, $t$ are in cm. This is the same equation with the factor of 2.0 removed from the numerator of the ln() term and added to the constant 0.5 giving 1.193 (0.5 + ln 2.0). This is mentioned because the above equations are found in the literature in many different but related forms.

The factor 0.2235 that appears above in fact varies between 0.22313 for $t = 0$ and 0.22352 for $t = w$. A frequency-corrected equation for inductance can be found in Greenhouse [3]

$$L = 0.002\ l \left[ \ln \left(\frac{2.0\ l}{w+t}\right) + 0.25049 + \left(\frac{w+t}{3.0\ l}\right) + \frac{\mu\ T(x)}{4} \right] \quad (6.2.2.3)$$

Again, $T(x)$ can be found with the closed-form relation (6.2.1.2). The resistance is found with:

$$R = \frac{K_c R_s l}{2.0 (w + t)} \quad (\Omega) \tag{6.2.2.4}$$

The surface resistivity $R_s$ is given in Chapter 2. The factor $K_c$ is a current crowding factor found in Caulton [1, Figure 2] (original reference is Terman [4]). The curve has been fit with a closed-form equation:

$$K_c = 1.048312 + 0.21768001 \ln\left(\frac{w}{t}\right) + 0.77176189 \, e^{-w/t} \tag{6.2.2.5}$$

The fit is valid to within ±0.5% over the range

$$1.0 \leq w/t \leq 100.0$$

REFERENCES

[1]     Caulton, M., et al., "Hybrid Integrated Lumped-Element Microwave Amplifiers," *IEEE Transactions Electron Devices*, Vol. ED-15, July 1968, pp. 459–466.

[2]     Djordjević, Antonije R., et al., "Inductance of Perfectly Conducting Foils Including Spiral Inductors," *IEEE Transactions on Microwave Theory and Techniques*, Vol. MTT-38, No. 10, October 1990, pp. 1407–1414.

[3]     Greenhouse, H.M., "Design of Planar Rectangular Microelectronic Inductors," *IEEE Transactions on Parts, Hybrids, and Packaging*, Vol. PHP-10, No. 2, June 1974, pp. 101–109.

[4]     Grover, F.W., *Inductance Calculations*, Van Nostrand, Princeton, NJ, 1946; reprinted by Dover Publications, NY, 1962. (A concise book based on his work at NBS.)

[5]     Pettenpaul, Ewald, et al., "CAD Models of Lumped Elements on GaAs up to 18 GHz," *IEEE Transactions on Microwave Theory and Techniques*, Vol. MTT-36, No. 2, February 1988, pp. 294–304.

[6]     Sobol, Harold, "Applications of Integrated Circuit Technology to Microwave Frequencies," *Proceedings of the IEEE*, Vol. 59, No. 8, August 1971, pp. 1200–1211. (Losses due to surface roughness from unpublished Sperry data, discussion of lumped elements.)

[7]     Terman, F.E., *Radio Engineer's Handbook*, McGraw-Hill, New York, 1945. (A classic.)

[8]     U.S. Department of Commerce, Circular of the National Bureau of Standards C74, *Radio Instruments and Measurements*, "Calculation of Inductance," January 1, 1937.

[9]     U.S. Department of Commerce and Labor, *Miscellaneous Pamphlets on Inductance*, U.S. Bureau of Standards, 1920. (A tremendously comprehensive collection of inductance equations unavailable anywhere else.)

### 6.2.3   Cylinder or Round Tubular Inductor

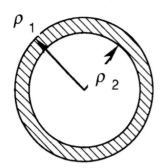 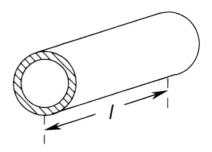

**Figure  6.2.3.1:        Tubular Inductor**

The tubular inductor closely approximates the structure of a plated-through hole or via. The ratio of the ac resistance to dc resistance of a tubular conductor is more constant than for an equivalent solid conductor. This is useful where precise frequency response is to be maintained. Also see Chapter 5 for more information on the via as a discontinuity.

The equation of Grover [2] is

$$L = 0.002\, l \left[ \ln \left( \frac{2.0\, l}{\rho_1} \right) + \ln\, \zeta - 1.0 \right] \quad (\mu H) \tag{6.2.3.1}$$

where

$l, \rho_1, \rho_2$ are in cm

The value of $\ln\,\zeta$ is in Grover [2, Table 4]. A polynomial has been fit for convenience:

$$\ln\,\zeta = 0.25009128 - 0.0017049618 \left( \frac{\rho_2}{\rho_1} \right) - 0.51598981 \left( \frac{\rho_2}{\rho_1} \right)^2$$

$$+ 0.37420782 \left( \frac{\rho_2}{\rho_1} \right)^3 - 0.10669571 \left( \frac{\rho_2}{\rho_1} \right)^4 \tag{6.2.3.2}$$

The polynomial fit is within ±0.3% of the tabulated data for

$$0.0 \le \frac{\rho_2}{\rho_1} \le 1.0$$

## REFERENCES

[1]   Djordjević, Antonije R., *et al.*, "Inductance of Perfectly Conducting Foils Including Spiral Inductors," *IEEE Transactions on Microwave Theory and Techniques*, Vol. MTT-38, No. 10, October 1990, pp. 1407–1414.

[2]   Grover, F.W., *Inductance Calculations*, Van Nostrand, Princeton, NJ, 1946; reprinted by Dover Publications, NY, 1962. (A concise book based on his work at NBS.)

[3]   U.S. Department of Commerce and Labor, *Miscellaneous Pamphlets on Inductance*, U.S. Bureau of Standards, 1920. (A tremendously comprehensive collection of inductance equations unavailable anywhere else.)

## 6.3 SOLENOIDAL OR HELICAL INDUCTORS

### 6.3.1 Round or Square Helical Coil Inductor

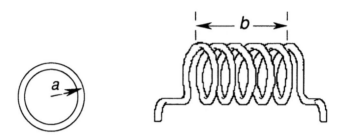

**Figure 6.3.1.1:** Round Helical Coil

Helical coils are conveniently wound on drill bits or other metal rod of specified diameter. If two wires are wound together on the form and one removed, a fixed distance between turns is easily achieved. When insulated wire is used, the insulation often determines the wire spacing. Spacing can also be checked in manufacturing using a spark plug style thickness gauge. Small PTFE spacers or solder mask on the PCB prevents shorting to any traces or ground plane near the coil and ensures repeatable positioning.

$$L = \frac{\mu_0\, n^2\, \pi\, a^2}{b}\, K_N \qquad (6.3.1.1)$$

where

$$K_N = \frac{(d\, b\, /\, a^2)\,[F(k) - E(k)] + (4.0\, d\, /\, b)\, E(k) - (8.0\, a\, /\, b)}{3.0\, \pi} \qquad (6.3.1.2)$$

$$k = \frac{2.0\, a}{d} \qquad (6.3.1.3)$$

$$d = \sqrt{4.0\, a^2 + b^2} \qquad (6.3.1.3a)$$

$F(k), E(k)$ = complete elliptic integral of the first and second kind, respectively

The radius, $a$, is measured from the helix center to the center of the wire. Nagaoka's constant $K_N$ is an adjustment factor that corrects for the effects of the coil's length. Wheeler gives an approximation formula to avoid the use of complete elliptic integral functions in the above; however as Miller [5] points out these are quite simple to calculate to any desired accuracy. See Chapter 12 for more information.

If a polynomial expression is desired, Lundin [2] presents a formula accurate to 1 ppm. For $2.0\,a \leq b$ (long coil):

$$L = \frac{\mu_0\, n^2\, \pi\, a^2}{b}\left[ f_1\!\left(\frac{4.0\,a^2}{b^2}\right) - \frac{4.0}{3.0\,\pi}\frac{2.0\,a}{b}\right] \tag{6.3.1.4}$$

and for $2.0\,a > b$ (short coil):

$$L = \mu_0\, n^2\, a\left\{ \left[\ln\left(\frac{8.0\,a}{b}\right) - 0.5\right] f_1\!\left(\frac{b^2}{4.0\,a^2}\right) + f_2\!\left(\frac{b^2}{4.0\,a^2}\right)\right\} \tag{6.3.1.5}$$

where for $0 \leq x \leq 1.0$

$$f_1(x) = \frac{1.0 + 0.383901\,x + 0.017108\,x^2}{1.0 + 0.258952\,x} \tag{6.3.1.6}$$

$$f_2(x) = 0.093842\,x + 0.002029\,x^2 - 0.000801\,x^3 \tag{6.3.1.7}$$

Grover [1] notes that the maximum inductance for a given length of wire occurs at

$$\frac{b}{2.0\,a} = 0.408$$

The maximum is broad so the exact ratio is not critical.

For the square helical inductor the equations of Wheeler [8] are

$$L = \frac{4.0\,\mu_0\, n^2\, a}{\pi}\left\{\ln\left[1.0 + \frac{\pi\,a}{b} + \frac{1.0}{3.6443 + 2.0\,(b/a) + 0.5098\,(b/a)^2}\right]\right\} \tag{6.3.1.8}$$

This relation is accurate to within $0.001\,L$.

For coils wound on polygonal forms with an even number of sides, $N$, the NBS [7] recommends replacing $a$ with $a_0$ where

$$a_0 = r\cos^2\frac{\pi}{2.0\,N} \tag{6.3.1.9}$$

The variable $r$ is the radius of a circle that circumscribes the polygon.

### 6.3.1.1    Inductor Stray Capacitance

An empirical formula for the stray capacitance of this inductor is available in Medhurst [4]. This equation is

$$C = 2.0\,a\,H \tag{6.3.1.1.1}$$

where *H* is tabulated in [4]. A CAE equation has been fit to this data:

$$H = 0.299267 + 0.1012 \left(\frac{b}{2.0\,a}\right) + \frac{0.0705822}{\left(\dfrac{b}{2.0\,a}\right)} + \frac{3.568348\text{e-}6}{\left(\dfrac{b}{2.0\,a}\right)^2} \qquad (6.3.1.1.2)$$

## REFERENCES

[1]    Grover, F.W., *Inductance Calculations*, Van Nostrand, Princeton, NJ, 1946; reprinted by Dover Publications, NY, 1962. (A concise book based on his work at NBS.)

[2]    Lundin, Richard, "A Handbook Formula for the Inductance of a Single-Layer Circular Coil," *Proceedings of the IEEE*, Vol. 73, No. 9, September 1985, pp. 1428–1429.

[3]    Medhurst, R.G., "H.F. Resistance and Self-Capacitance of Single-Layer Solenoids," *Wireless Engineer*, February 1947, pp. 35–43.

[4]    Medhurst, R.G., "H.F. Resistance and Self-Capacitance of Single-Layer Solenoids," *Wireless Engineer*, March 1947, pp. 80–92.

[5]    Miller, H. Craig, "Inductance Formula for a Single-Layer Circular Coil," *Proceedings of the IEEE*, Vol. 75, No. 2, February 1987, pp. 256–257.

[6]    Siocos, Christos A., "New Tables for the Mutual Inductance of Coaxial Cylindrical Coils," *IEEE Transactions on Parts, Materials and Packaging*, Vol. PMP-6, No. 1, March 1970, pp. 12–19, and "Correction to 'New Tables for the Mutual Inductance of Coaxial Cylindrical Coils,'" p. 70.

[7]    U.S. Department of Commerce, Circular of the National Bureau of Standards C74, *Radio Instruments and Measurements*, "Calculation of Inductance," January 1, 1937.

[8]    Wheeler, Harold A., "Inductance Formulas for Circular and Square Coils," *Proceedings of the IEEE*, Vol. 70, No. 12, December 1982, pp. 1449–1450.

## 6.4  PLANAR OR PANCAKE INDUCTORS

### 6.4.1  Planar Circular Spiral Inductors

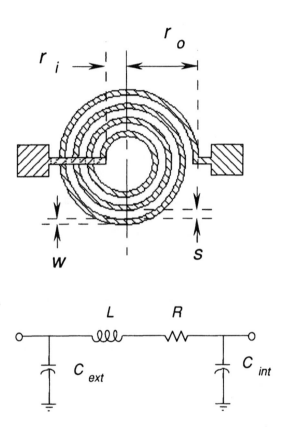

**Figure  6.4.1.1:**      **Planar  Circular  Spiral  and  Model**

The spiral inductor is a technique of forming a planar inductor in a small space. The shape is described by an increasing radius with angle. The most common types of spirals have the radius functions given below[1] (the spiral of Figure 6.4.1.1 is an Archimedes spiral). These are the equations used to algorithmically generate the spiral inductor patterns.

$$r = r_i + k\,\theta \qquad\qquad \text{Archimedes spiral} \qquad\qquad (6.4.1.1)$$

---

[1] Olivei [8] determined that the hyperbolic spiral has the lowest skin effect loss and the highest inductance per turn.

$$r = r_i e^{h\theta} \qquad\qquad \text{Logarithmic spiral} \qquad\qquad (6.4.1.2)$$

$$r = \frac{r_i}{1.0 - k\, r_i\, \theta} \qquad\qquad \text{Hyperbolic spiral} \qquad\qquad (6.4.1.3)$$

The equivalent circuit of the spiral inductor is also shown in Figure 6.4.1.1. The resistance is calculated with the skin effect resistance equations of the equivalent rectangular conductor.

$$R = \frac{K' \pi n a R_s}{w} \quad (\Omega) \qquad\qquad (6.4.1.4)$$

where

$R_s$ = sheet resistance of the trace   ($\Omega$ / square)

$K'$ = 1.5, empirically

Finite resistance of the coil will limit the $Q$. Circular spirals have higher $Q$'s (up to 100) than rectangular spirals; however, they also have lower inductance for an equivalent area. It has been found that $r_i / r_o = 1/5$ optimizes the $Q$.

The stray capacitance is distributed between turns and to any ground plane beneath the inductor. Burkett indicates that for spiral inductors with a lower ground plane the capacitance is 2 to 4 times the calculated parallel-plate capacitance due to fringing effects. This is probably because the fringing capacitance from the large outermost turn, $C_{ext}$, dominates. Wolff and Kapusta [16] successfully modeled the capacitance of a spiral inductor with a ground plane using coupled microstripline equations. This neglects the effects of the curved lines and coupling of nonadjacent turns. See Sections 5.5.4 and 6.6.2 for information on these effects.

The inductance is made up of three parts:

$$L = L_{external} + L_{internal} + L_{mutual} - L_{image} \qquad\qquad (6.4.1.5)$$

The internal inductance is the self-inductance of the spiral created by magnetic flux internal to the conductor and is negligible. External inductance is the self-inductance created by flux external to the conductor. The self-inductance is increased by mutual inductance to adjacent turns (see Section 6.6.2). If a ground plane is present beneath the coil, the total inductance is reduced by an image of the coil that appears in the ground plane.

A number of relations for the inductance are found in the literature. Remke and Burdick [12] compare these formulas and their ranges of applicability. It is found that many of these relations are not valid for the range of parameters in which they are commonly used. A new formulation is presented

$$L_T = 2.0 \sum_{k=1.0}^{n-1.0} \sum_{j=k+1.0}^{n} \mu \sqrt{a\,b} \left[ \left( \frac{2.0}{k_1} - 1.0 \right) K(k_1) - \frac{2.0}{k_1} E(k_1) \right]$$

$$+ \sum_{k=1.0}^{n} \mu\,(2.0\,c - w_1) \left[ (1.0 - \frac{k_2^2}{2.0}) K(k_2) - E(k_2) \right] \qquad (6.4.1.6)$$

where

$$k_1 = \sqrt{\frac{4.0\,a\,b}{(a+b)^2}} \qquad (6.4.1.7)$$

$$k_2 = \sqrt{\frac{4.0\,c\,(c - w_1)}{(2.0\,c - w_1)^2}} \qquad (6.4.1.8)$$

$$a = \text{mean radius} = \frac{r_o + r_i}{2.0} \qquad (6.4.1.9)$$

$$b = \text{thickness} = t$$

$$c = r_o - r_i \qquad (6.4.1.10)$$

$$n = \text{number of turns}$$

$$w_1 = w / 2.0 \qquad (6.4.1.11)$$

The functions $K(k)$ and $E(k)$ are the complete elliptic integrals of the first and second kind. See Chapter 12 for more information. This inductance equation is valid for $f \ll f_{resonance}$. As the resonance is approached the effect of the capacitances in the model need to be considered.

For the case of a spiral inductor having a lower ground plane the equation is

$$L_T = 2.0 \sum_{k=1.0}^{n-1.0} \sum_{j=k+1.0}^{n} \mu \sqrt{a\,b} \left[ \left( \frac{2.0}{k_1} - 1.0 \right) K(k_1) - \frac{2.0}{k_1} E(k_1) \right]$$

$$+ \sum_{k\,=\,1.0}^{n} \mu \ (2.0 \ c - w_1) \left[ (1.0 - \frac{k_2^2}{2.0}) \ K(k_2) - E(k_2) \right]$$

$$- \sum_{i\,=\,1.0}^{n} \sum_{j\,=\,1.0}^{n} \mu \ \sqrt{a_R \ b_R} \left[ \left( \frac{2.0}{k_3} - 1.0 \right) K(k_3) - \frac{2.0}{k_3} E(k_3) \right]$$

$$(6.4.1.12)$$

where the variables are as defined above and

$$a_R = r_i + (i - 0.5) \ (w + s) \qquad\qquad (6.4.1.13)$$

$$b_R = r_i + (j - 0.5) \ (w + s) \qquad\qquad (6.4.1.14)$$

$$k_3 = \sqrt{\frac{4.0 \ a_R \ b_R}{(2.0 \ h)^2 + (a_R + b_R)^2}} \qquad\qquad (6.4.1.15)$$

$h$ = substrate thickness

### Other Inductance Relations

Other equations found in the literature are listed here for reference as they are still commonly encountered and it may be desirable to adapt them for proprietary models. Before working with these, consult [12, 16].

Terman [14] gives the following relation, *Stefan's formula*, for the inductance of a flat wire or pancake spiral inductor:

$$L = 0.03193 \ a \ n^2 \left[ \left( 1.0 + \frac{b^2}{32.0 \ a^2} + \frac{c^2}{96.0 \ a^2} \right) \ln \left( \frac{8.0 \ a}{d} \right) - y_1 + \frac{y_3 \ c^2}{16.0 \ a^2} \right]$$

$$+ \ 0.03193 \ a \ n \left( \ln \frac{D}{d} + 0.155 \right) \ (\mu H) \qquad\qquad (6.4.1.16)$$

where the dimensions (in inches) are

$$D = w + s \qquad\qquad (6.4.1.17)$$

$$d = t + w \qquad\qquad (6.4.1.18)$$

and $y_1$ and $y_3$ are in Table 16 of [14]. For a thin trace relative to $c$ (the usual case for planar circuits), $y_1 = 0.5$ and $y_3 = 0.597$. For other cases the following equations were fit for CAE use:

$$y_1 = 3.74219 - 2.21817 \left(\frac{b}{c}\right) + 0.51724 \left(\frac{b}{c}\right)^2 - 3.24182 \, e^{-b/c} \qquad (6.4.1.19)$$

$$y_3 = 0.59631105 + 0.038756584 \left(\frac{b}{c}\right) + 0.31720149 \left(\frac{b}{c}\right)^2 - 0.17592173 \left(\frac{b}{c}\right)^3 3$$

$$+ 0.039713327 \left(\frac{b}{c}\right)^4 \qquad (6.4.1.20)$$

The accuracy of the curve fits ($y_1$, $y_3$) is to within $\pm 0.08\%$ and $\pm 1.5\%$, respectively. The accuracy of the inductance equation at dc is stated to be within 1%; however, this is not borne out in practice with planar structures.

A commonly quoted, simpler relation [4] due to Wheeler is:

$$L = \frac{n^2 a^2}{8.0 \, a + 11.0 \, c} \quad (\text{nH}) \qquad (6.4.1.21)$$

where all of the dimensions are in mils. An improved equation that is more appropriate for use with planar spiral inductors was derived by Burkett [1]

$$L = \frac{0.8 \, n^2 a^2}{6.0 \, a + 10.0 \, c} \quad (\text{nH}) \qquad (6.4.1.22)$$

Burkett's inductors were fabricated with a ground plane.

The connection from the center of the inductor to the outside can be made with a via and a trace on the opposite side of the substrate, an air bridge, or a bond wire. These may need to be included in the model as well.

Spiral inductors that have been ion beam milled having 0.5 mil lines and spaces can be purchased as SMD components. Olivei [8] overlaid the inductor patterns with a ferrite film to get large value inductances.

Easier to lay out versions of the circular spiral inductor are shown below [5]. This layout allows circular sections to be used rather than true spirals.

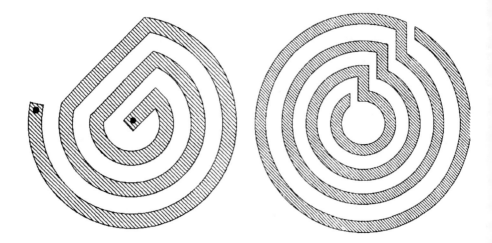

**Figure 6.4.1.2:** Alternate Circular Spiral Inductors

## REFERENCES

[1] Burkett, Frank S., Jr., "Improved Designs for Thin Film Inductors," *21st Electronic Components Conference Proceedings,* May 10–12, 1971, Washington, DC, pp. 184–194.

[2] Caulton, Martin, *et al.,* "Hybrid Integrated Lumped-Element Microwave Amplifiers," *IEEE Transactions on Electron Devices,* Vol. ED-15, No. 7, July 1968, pp. 459–466.

[3] Daly, Daniel A., *et al.,* "Lumped Elements in Microwave Integrated Circuits," *IEEE Transactions on Microwave Theory and Techniques,* Vol. MTT-15, No. 12, December 1967, pp. 713–721.

[4] Gupta, K. C., Ramesh Garg, and Rakesh Chadha, *Computer-Aided Design of Microwave Circuits,* Artech House, Norwood, MA, 1981.

[5] Hollomon, James K., Jr., *Surface-Mount Technology for PC Board Design,* SAMS, Indianapolis, IN, 1989.

[6] Hurt, John C., "A Computer-Aided Design System for Hybrid Circuits," *IEEE Transactions on Components, Hybrids, and Manufacturing Technology,* Vol. CHMT-3, No. 4, December 1980, pp. 525–535.

[7] Ion Beam Milling, Inc., "Spiral Chip Inductors," Advertisement.

[8] Olivei, Alfredo, "Optimized Miniature Thin-Film Planar Inductors, Compatible With Integrated Circuits," *IEEE Transactions on Parts, Materials and Packaging,* Vol. PMP-5, No. 2, June 1969, pp. 71–88.

[9]    Parisot, M., *et al.*, "Highly Accurate Design of Spiral Inductors for MMICs with Small Size and High Cut-off Frequency Characteristics," *1984 IEEE MTT-S International Microwave Symposium Digest*, May 29–June 1, 1984, San Francisco, CA, pp. 106–110.

[10]   Pengelly, R.S., and D.C. Rickard, "Design, Measurement and Application of Lumped Elements up to J-Band," *7th European Microwave Conference Proceedings*, 1977, Copenhagen, Denmark, pp.460–464.

[11]   Pettenpaul, Ewald, *et al.*, "CAD Models of Lumped Elements on GaAs up to 18 GHz," *IEEE Transactions on Microwave Theory and Techniques*, Vol. MTT-36, No. 2, February 1988, pp. 294–304.

[12]   Remke, Ronald L., and Glenn A. Burdick, "Spiral Inductors for Hybrid and Microwave Applications," *24th Electronic Components Conference Proceedings*, May 13–15, 1974, Washington, DC, pp. 152–161.

[13]   Rodriguez, Robert, *et al.*, "Modeling of Two-Dimensional Spiral Inductors," *IEEE Transactions on Components, Hybrids, and Manufacturing Technology*, Vol. CHMT-3, No. 4, December 1980, pp. 535–541.

[14]   Terman, F.E., *Radio Engineer's Handbook*, McGraw-Hill, New York, 1945.

[15]   U.S. Department of Commerce, Circular of the National Bureau of Standards C74, *Radio Instruments and Measurements*, "Calculation of Inductance," January 1, 1937.

[16]   Wolff, Ingo, and Hartmut Kapusta, "Modeling of Circular Spiral Inductors for MMICs," *1987 IEEE MTT-S International Microwave Symposium Digest*, June 9–11, 1987, Las Vegas, NV, pp. 123–126.

## 6.4.2 Planar Square Spiral Inductor

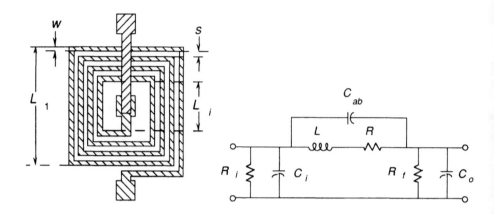

**Figure 6.4.2.1:** **Planar Square Spiral**

The square spiral inductor is an easier to lay out version of the circular spiral inductor. Again, connection from the inner turn to the outside is made with a via and a trace on the other side, an air bridge, or a bond wire. Modeling is somewhat complicated over the circular spiral by the addition of the bend discontinuities and coupling of the corners.

Shepherd [12] models this component with a cascade of coupled microstrip lines and abrupt bend discontinuities. The air bridge was modeled as parallel-plate capacitors coupling into the cascade of lines. Wolff and Kibuuka [14] show that the effect of the coupled corners is important, reducing the total inductance and modifying the frequency dependency. Design curves of the coupled corner discontinuity are available in Hill and Tripathi [7]. See Chapter 5 for CAE equations. The effect of a back-side ground plane (shunt capacitance) has been reduced by making the entire inductor a series of air bridges. The air under the bridges reduces the shunt capacitance.

It might be suggested that a concerted effort has not yet been made to simulate this structure rigorously and derive CAE equations that agree well with experimental data. Rather, most authors have taken an approach of adapting existing equations for wire inductors. Experimental data are sparse and agreement under one set of conditions has not guaranteed applicability for others. In fairness, this structure is complex—it has significant features in all three dimensions, can be used with several different techniques of connecting to the center, a ground plane may or may not be present and a multiple-turn inductor is not an extension of a single-turn inductor. The following categories can be identified as important to a model's broadband accuracy:

- Technique of connecting to center: via and backside trace, bond wire, air or dielectric bridge
- Dimensions $w$, $s$, substrate $\varepsilon_r$, overlay $\varepsilon_r$
- Number of turns

- Backside ground plane (and $h$) or no ground plane
- Presence of magnetic overlay or substrate
- Length of sections (from $L_1$) and frequency of operation
- Metallization losses and dielectric losses (tan $\delta$)

If simulation below resonance is all that is desired, available static inductance equations may suffice; however, if the inductor is to be used as a resonator or the performance needs to be predicted beyond resonance some experimental work should be expected. A number of the available EM solving programs have shown good results with modeling this structure. With these caveats we will proceed to present the best available information for the designer.

In the model of Figure 6.4.2.1, $C_{ab}$ represents the air bridge capacitance. Jansen *et al.* [9] found that this was calculable from the parallel-plate capacitance multiplied by a factor of 1.7 independent of turns. The resistor $R$ is calculable from the skin effect resistance of the rectangular conductor; $R_i$ and $R_o$ represent dielectric and radiation losses.

In Greenhouse [5], an accurate technique is described after Grover for calculating the inductance of the square spiral inductor. This method calculates the inductance of the individual sections and includes both the positive and negative mutual inductances between parallel sections. Accuracies ranging from 6% to 20% of the measured value were obtained. Krafcsik and Dawson [9] added frequency dependency of the mutual inductances to the analysis of Greenhouse by taking into account the phase difference between the coupling inductors. As is the case with the circular spiral, the total inductance is

$$L = L_{external} + L_{internal} + L_{mutual} - L_{image} \qquad (6.4.2.1)$$

The internal inductance is the self-inductance of the spiral created by magnetic flux internal to the conductor and is negligible. External inductance is the self-inductance created by flux external to the conductor. The self-inductance is increased by mutual inductance to adjacent turns. If a ground plane is present beneath the coil, the total inductance is reduced by an image of the coil that appears in the ground plane.

Greenhouse breaks the inductor into straight segments and sums the terms above for each of the segments. The internal inductance, $L_{internal}$, is calculated with (6.2.2.2). Mutual inductance is calculated for all possible pairs of parallel conductors with

$$L_{mutual} = 2.0\, l_n\, \ln\left( \frac{l_n}{GMD_n} + \sqrt{1.0 + \frac{l_n^2}{GMD_n^2}} - \sqrt{1.0 + \frac{GMD_n^2}{l_n^2}} + \frac{GMD_n}{l_n} \right)$$
$$\text{(nH)}$$

$$(6.4.2.2)$$

where the length of the $n$th set of parallel conductors is $l_n$ and the geometric mean distance[2] between them in terms of centerline distance $d$ is the variable $GMD_n$:

---

[2]Grover [6], p. 20.

$$GMD_n = \exp\left\{ \ln d - \left[ \frac{1.0}{12.0 \left(\frac{d}{w}\right)^2} + \frac{1.0}{60.0 \left(\frac{d}{w}\right)^4} \right.\right.$$

$$\left.\left. + \frac{1.0}{168.0 \left(\frac{d}{w}\right)^6} + \frac{1.0}{360.0 \left(\frac{d}{w}\right)^8} + \frac{1.0}{660.0 \left(\frac{d}{w}\right)^{10}} + \dots \right]\right\} \qquad (6.4.2.3)$$

Mutual inductances add when the current flow is in the same direction and subtract when they are opposite. The mutual inductance to the image inductor appearing in any ground plane is

$$L_{image} = 2.0\, l \left[ \ln\left(\frac{L}{D} + \sqrt{1.0 + \frac{L^2}{D^2}}\,\right) - \sqrt{1.0 + \frac{D^2}{L^2}} + \frac{D}{L} \right] \quad (\text{nH})$$

$$(6.4.2.4)$$

The mutual inductances can be modified to account for the difference in phase of the current in the segments following the suggestion of Krafcsik and Dawson

$$L_{mutual,KD} = L_{mutual} \cos \theta \qquad\qquad (6.4.2.5)$$

where

$\theta$ = phase between currents in the two segments

This can be calculated using linear distance (as if the coil were straightened out) between the two segments and the value of $\varepsilon_{eff}$ calculated for microstrip (ground plane present) or coplanar strips (no ground plane). These equations are given in Chapter 3.

A computer program is necessary to evaluate these relations for any but the simplest of spiral inductors. Once this is accomplished, evaluation and design is quick and straightforward.

### Other Square Spiral Inductor Equations

Liao [11] gives a simpler equation for the inductance

$$L = 8.5 \sqrt{A}\, n^{5/3} \quad (\text{nH}) \qquad\qquad (6.4.2.6)$$

where

$A$ = area occupied by inductor in $cm^2$

$n$ = number of turns

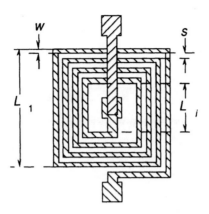

**Figure 6.4.2.2:** **Rectangular Planar Spiral Inductors**

NBS Circular C74 [13] gives a relation for the flat rectangular coil shown in Figure 6.4.2.2:

$$L_0 = 0.009210 \, n^2 \, (a + a_1) \left[ \log \left( \frac{2.0 \, a \, a_1}{w + n \, s} \right) - \frac{a}{a + a_1} \log (a + g) \right.$$

$$\left. - \frac{a_1}{a + a_1} \log (a_1 + g) \right] + 0.004 \, (a + a_1) \, n^2 \left[ 2.0 \, \frac{g}{a + a_1} - \frac{1}{2} \right.$$

$$\left. + 0.447 \, \frac{w + n \, s}{a + a_1} \right] - 0.004 \, n \, (a + a_1) \, (A_1 + B_1) \qquad (6.4.2.7)$$

The variables $a_0$ and $a_0'$ are the dimensions of the outside of the $n$ turn coil measured between center lines. The other variables are

$$g = \sqrt{a^2 + a_1^2} \qquad (6.4.2.8)$$

$$a = a_0 - (n - 1.0) \, s \qquad (6.4.2.9)$$

$$a_1 = a_0' - (n - 1.0) \, s \qquad (6.4.2.10)$$

$$A_1 = \ln \frac{v + 1.0}{v + \tau} \qquad (6.4.2.11)$$

$$B_1 = -2.0 \left[ \frac{\delta_{12} \, (n - 1.0)}{n} + \frac{\delta_{13} \, (n - 2.0)}{n} + \frac{\delta_{14} \, (n - 3.0)}{n} + \dots + \frac{\delta_{1n}}{n} \right]$$

$$(6.4.2.12)$$

$\delta_{1n}$ = functions of $v = \dfrac{w}{s}$ and $\tau = \dfrac{t}{s}$ tabulated in Table 15 of [14]

$$(6.4.2.13)$$

## REFERENCES

[1]    Cahana, David, "A New Transmission Line Approach for Designing Spiral Microstrip Inductors for Microwave Integrated Circuits," *1983 IEEE MTT-S International Microwave Symposium Digest*, pp. 245–247.

[2]    Caulton, Martin, *et al.*, "Hybrid Integrated Lumped-Element Microwave Amplifiers," *IEEE Transactions on Electron Devices*, Vol. ED-15, No. 7, July 1968, pp. 459–466.

[3]    Chow, Y.L., and X.Y. She, "Simple Formulas for Microstrip Rectangular Spiral Inductor," *1987 IEEE MTT-S International Microwave Symposium Digest*, pp. 583–588.

[4]    Djordjević, Antonije R., *et al.*, "Inductance of Perfectly Conducting Foils Including Spiral Inductors," *IEEE Transactions on Microwave Theory and Techniques*, Vol. 38, No. 10, October 1990, pp. 1407–1414.

[5]    Greenhouse, H.M., "Design of Planar Rectangular Microelectronic Inductors," *IEEE Transactions on Parts, Hybrids, and Packaging*, Vol. PHP-10, No. 2, June 1974, pp. 101–109.

[6]    Grover, F.W., *Inductance Calculations*, Van Nostrand, Princeton, NJ, 1946; reprinted by Dover Publications, NY, 1962. (A concise book based on his work at NBS.)

[7]    Hill, A., and V.K. Tripathi, "Analysis and Modeling of Coupled Right Angle Microstrip Bend Discontinuities," *1989 IEEE MTT-S Symposium Digest*, pp. 1143–1146.

[8]    Jansen, R.H., *et al.*, "Theoretical and Experimental Broadband Characterization of Multiturn Square Spiral Inductors in Sandwich Type GaAs MMICs," *15thEuropean Microwave Symposium*, Paris, 1985, pp. 946–951.

[9]    Krafcsik, David M., and Dale E. Dawson, "A Closed-Form Expression for Representing the Distributed Nature of the Spiral Inductor," *1986 Microwave and Millimeter-Wave Monolithic Circuits Symposium*, June 4–5, 1986, Baltimore, MD, pp. 87–92.

[10]   Lang, David, "Broadband Model Predicts S-parameters of Spiral Inductors," *Microwaves & RF*, January 1988, pp. 65–66.

[11]   Liao, Samuel Y., *Microwave Circuit Analysis and Amplifier Design*, Prentice-Hall, Englewood Cliffs, NJ, 1987.

[12]   Shepherd, Peter R., "Analysis of Square-Spiral Inductors for Use in MMIC's," *IEEE Transactions on Microwave Theory and Techniques*, Vol. MTT-34, No. 4, April 1986, pp. 467–472.

[13]   U.S. Department of Commerce, Circular of the National Bureau of Standards C74, *Radio Instruments and Measurements*, "Calculation of Inductance," January 1, 1937.

[14]   Wolff, Ingo, and Godfrey Kibuuka, "Computer Models for MMIC Capacitors and Inductors," *14th European Microwave Conference Proceedings*, Liège, 1984, Belgium, pp. 853–858.

### 6.4.3   Single-Turn Circular Planar Inductor

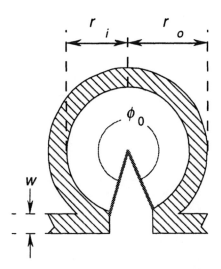

**Figure 6.4.3.1:**     **Single-Turn Planar Inductor**

Pettenpaul [2] calculates the inductance of this structure with

$$L = \frac{\mu_0}{2.0 \, \pi \, (w + t)^2} \int_0^{\phi_0 / \sqrt{2}} \left( \phi_0 \sqrt{2.0} - 2.0 \, \phi \right) \cos \left( \sqrt{2.0} \, \phi \right) F(\phi) \, d\phi$$

$$(6.4.3.1)$$

where

$$F(\phi) = G(r_o, r_o) - G(r_i, r_o) - G(r_o, r_i) + G(r_i, r_i) \qquad (6.4.3.2)$$

$$G(r_1, r_2) = \frac{R^3}{3.0} + \frac{2.0 \; r_1 \; r_2}{3.0} \, R \, \cos\left(\sqrt{2.0} \; \phi\right)$$

$$+ \frac{2.0 \; r_1{}^3 \cos\left(\sqrt{2.0} \; \phi\right)}{3.0} \, \text{asinh} \frac{r_2 - r_1 \cos\left(\sqrt{2.0} \; \phi\right)}{r_1 \; |\sin\left(\sqrt{2.0} \; \phi\right)|} \quad (6.4.3.3)$$

$$R = \sqrt{r_1 + r_2 - 2.0 \; r_1 \; r_2 \cos\left(\sqrt{2.0} \; \phi\right)} \quad (6.4.3.4)$$

### REFERENCES

[1]    Aitchison, Colin S., *et al.*, "Lumped-Circuit Elements at Microwave Frequencies," *IEEE Transactions on Microwave Theory and Techniques*, Vol. MTT-19, No. 12, December 1971, pp. 928–937.

[2]    Pettenpaul, Ewald, *et al.*, "CAD Models of Lumped Elements on GaAs up to 18 GHz," *IEEE Transactions on Microwave Theory and Techniques*, Vol. MTT-36, No. 2, February 1988, pp. 294–304.

## 6.4.4   Single-Turn Rectangular Planar Inductor

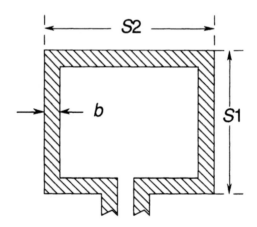

**Figure  6.4.4.1:**      **Single-Turn  Rectangular  Planar  Inductor**

$$L = 0.02339\left[(s_1 + s_2) \log\left(\frac{2.0 \; s_1 \; s_2}{b + c}\right) - s_1 \log\left(s_1 + g\right) - s_2 \log\left(s_2 + g\right)\right]$$

$$+ 0.01016\left[2.0 \; g - \frac{s_1 + s_2}{2.0} + 0.447 \; (b + c)\right] \quad (\mu\text{H}) \quad (6.4.4.1)$$

where all dimensions are in inches, and

$b = w$ = the trace width

$c = t$ = the trace thickness

$g = \sqrt{s_1{}^2 + s_2{}^2}$

For a square, $s_1 = s_2 = s$ ($g = \sqrt{2}\, s$), the above simplifies to:

$$L = 0.04678\, s \left\{ \log \left[ \frac{2.0\ s^2}{(s + \sqrt{2.0}\ s)(w + t)} \right] \right\}$$

$$+ 0.01016 \left[ 2\sqrt{2}\ s - s + 0.447(w + t) \right] \quad (\mu H) \qquad (6.4.4.2)$$

where all dimensions are in inches.

### REFERENCES

[1]    Djordjević, Antonije R., *et al.*, "Inductance of Perfectly Conducting Foils Including Spiral Inductors," *IEEE Transactions on Microwave Theory and Techniques*, Vol. MTT-38, No. 10, October 1990, pp. 1407–1414.

[2]    Terman, F.E., *Radio Engineer's Handbook*, McGraw-Hill, New York, 1945.

## 6.4.5   Octagonal Spiral Inductor

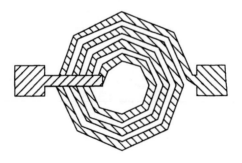

**Figure  6.4.5.1:**        **Octagonal  Spiral  Inductor**

The octagonal spiral inductor is an alternative structure to the circular spiral and the square spiral inductor shapes. Although the spiral is an easy shape to describe mathematically, the CAD layout tool may not allow straightforward mathematical programming of shapes. Most create metallization based on rectangular and circular primitives. This shape can be built up from rectangular shapes fairly easy.

Mondal [1], fit results of measurements to the inductance equations of Wheeler (1928) modifying the coefficients to the result

$$L = \frac{a^{1.54} n^{1.93}}{17.23 \, c^{0.55}} \quad (\text{nH}) \tag{6.4..5.1}$$

where $a$ and $c$ are in mils. The stated accuracy is 3.1% rms. Mondal also gives an interpolation equation for inductors with half turns.

$$L_{n+1/2} = \frac{(n+1/2)^2}{2} \left[ \frac{L_n}{n^2} + \frac{L_{n+1}}{(n+1)^2} \right] \tag{6.4.5.2}$$

where

$L_n$ = the inductance calculated with $n$ turns

$L_{n+1}$ = the inductance calculated with $n+1$ turns

$L_{n+1/2}$ = the inductance of an $n + 1/2$ turn inductor

$n \in I$ (integers)

The techniques of Greenhouse described in the above section on the square spiral inductor may also be adapted to this structure and other planar polygon inductors.

### REFERENCES

[1]    Mondal, Jyoti P., "Octagonal Spiral Inductor Measurement and Modelling for MMIC Applications," *International Journal of Electronics*, Vol. 68, No. 1, 1990, pp. 113–125.

[2]    Stubbs, Malcolm, "Simulation Tool Accurately Models MMIC Passive Elements," *Microwaves & RF*, January 1988, pp. 75–79.

## 6.5   TOROIDAL INDUCTORS

Inductance of toroids is generally calculated from manufacturer's data with the relation:

$$L = \left(\frac{N}{100}\right)^2 \times (\text{inductance} / 100 \text{ turns}) \tag{6.5.1}$$

Where the (inductance / 100 turns) term is a parameter of the chosen coil form given by the manufacturer for a given material and toroid size and $N$ is the number of turns. Curves of the $Q$ vs. frequency are also given for many particular cases.

Toroidal and other inductors that use high-permeability materials to increase their inductance will introduce distortion. The distortion is generated by core material nonlinearities. If this is a problem the effect can be minimized by:

- keeping signal levels low
- removing dc before it reaches the inductors
- using air cores where possible

For a complete listing of inductor core shapes, sizes, materials, and their parameters see manufacturers' catalogs (Micrometals, Nytronics, and others). Some common materials' parameters are summarized in Table 6.5.1.

The wire gauge used in winding the inductor should be chosen as large as possible for lowest coil resistance and still fit on the form. The wire should also be able to sustain the required current flow without the wire or its insulation melting. Tables of wire fusing current for various gauges and insulation melting temperatures are available in *Reference Data for Radio Engineers*.

| Mix | Material | Relative Permeability | Temperature (ppm/°C) | Frequency Range (MHz) | Color Code |
|-----|----------|----------------------|---------------------|----------------------|-----------|
| 0 | Phenolic | 1.0 | 0 | 50-250 | Tan |
| 1 | Carbonyl C | 20.0 | 280 | 0.15-2 | Blue |
| 2 | Carbonyl E | 10.0 | 95 | 0.25-10 | Red |
| 3 | Carbonyl HP | 35 | 370 | 0.02-1.0 | Gray |
| 3F | HP/Ferrite | 80.0 | 700 | 0.01-1.0 | Gray/Orange |
| 6 | Carbonyl SF | 8.5 | 35 | 2.0-30 | Yellow |
| 7 | Carbonyl TH | 9.0 | 30 | 1.0-20 | White |
| 8 | Carbonyl GQ4 | 35.0 | 255 | 0.02-1.0 | Orange |
| 10 | Carbonyl W | 6.0 | 150 | 10-100 | Black |
| 12 | Synthetic Oxide | 4.0 | 170 | 20-200 | Green/White |
| 15 | Carbonyl GS6 | 25.0 | 190 | 0.10-3.0 | Red/White |
| 17 | Carbonyl | 4.0 | 50 | 20-200 | Lavender |
| 22 | Synthetic Oxide | 4.0 | 410 | 20-200 | Green/Orange |
| 26 | | 75 | 822 | | Yellow/White |
| 28 | | 22 | 415 | | Gray/Green |
| 33 | | 33 | 635 | | Gray/Yellow |
| 40 | | 60 | 950 | | Green/Yellow |
| 50 | Ferrite | 125 | 1000 | 0.01-1.0 | Orange |

**Table 6.5.1: Some Common Core Materials [3]**

## 6.5.1 Toroidal Inductor on Form with Square Cross-Section

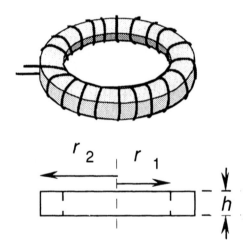

**Figure 6.5.1.1:    Toroidal Inductor with Square Cross-section**

For a coil form whose $\mu_r$ is 1.0 [2]:

$$L = 0.004606\, n^2\, h\, \log_{10}\left(\frac{r_2}{r_1}\right) \quad (\mu H) \qquad (6.5.1.1)$$

where

$n$ = number of turns

## 6.5.2 Toroidal Inductor on Form with Circular Cross-Section

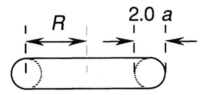

Figure 6.5.2.1: Toroidal Inductor with Circular Cross-Section

The toroid of circular cross-section can be calculated with [3]

$$L = 0.01257 \, n^2 \left( R - \sqrt{R^2 - a^2} \right) \tag{6.5.2.1}$$

### REFERENCES

[1] DeMaw, M.F., "Magnetic Cores in RF Circuits," *RF Design*, April 1980, pp. 12–16.

[2] Micrometals, Inc., *Inductor Catalog*.

[3] U.S. Department of Commerce, National Bureau of Standards, Circular C74, *Radio Instruments and Measurements*, "Calculation of Inductance," January 1, 1937.

## 6.6 MUTUAL INDUCTANCES

In previous sections we have calculated the inductance of various conductors in isolation. When multiple conductors are involved, the inductance of each is affected by the presence of the others. In the case of two conductors carrying the same current, the inductance is either increased or decreased by an amount known as the **mutual inductance**. The direction of inductance change is determined by the relative directions of the current flows.

The total inductance of a two wire circuit carrying the same current is therefore

$$L = L_1 + L_2 \pm 2 \, M_{12} \tag{6.6.1}$$

## 6.6.1 Parallel Wires

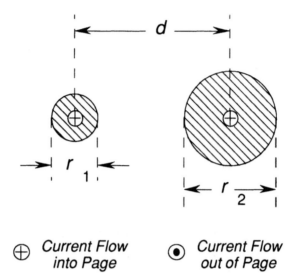

| $\oplus$ | Current Flow into Page | $\odot$ | Current Flow out of Page |

**Figure 6.6.1.1:** **Mutual Inductance of Parallel Wires**

For equal size ($r_1 = r_2$), parallel, wires with current flowing in the same direction the mutual inductance is

$$M_{12} = 0.002\, l \left[\ \ln\left(\frac{l}{d} + \sqrt{1.0 + \frac{l^2}{d^2}}\ \right) - \sqrt{1.0 + \frac{d^2}{l^2}} + \frac{d}{l}\ \right] \quad (\mu H)$$

(6.6.1.1)

where $d$, $l$ are in cm. The total inductance is

$$L_{total} = L_1 + L_2 + 2\,M_{12}$$

(6.6.1.2)

where $l$ is the length and $L_1$ and $L_2$ are calculated with (6.2.1.1).

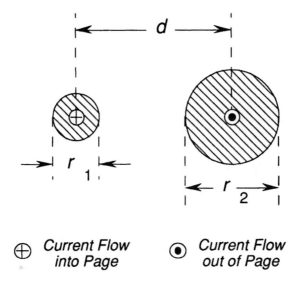

Figure 6.6.1.2:     Return Circuit of Parallel Wires

For a return circuit of parallel wires the total inductance is

$$L_{total} = L_1 + L_2 - 2\,M_{12} \qquad\qquad (6.6.1.3)$$

where $l$ is the length, $L_1$ and $L_2$ are calculated with (6.2.1.1), and $M_{12}$ is calculated with (6.6.1.1). This can be simplified to

$$L_{total} = 0.002\,l\left(\ln\frac{d^2}{r_1\,r_2} + 0.5\right) \qquad\qquad (6.6.1.4)$$

REFERENCE

[1]     Grover, F.W., *Inductance Calculations*, Van Nostrand, Princeton, NJ, 1946; reprinted by Dover Publications, NY, 1962. (A concise book based on work at NBS.)

## 6.6.2 Parallel Rectangular Wires

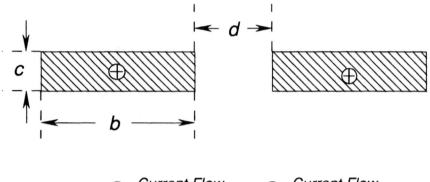

**Figure 6.6.2.1:** **Parallel Rectangular Conductors**

For parallel rectangles with current flowing in the same direction the mutual inductance is

$$M_{12} = 0.002 \, l \left( \ln \frac{2.0 \, l}{d} - \ln k - 1.0 + \frac{d}{l} - \frac{d^2}{4.0 \, l^2} \right) \qquad (6.6.2.1)$$

For values of $k$, see Tables 1 and 2 of Grover [1]. Curve fits to this data have been derived for several values of $b \, / \, c$. For $b \, / \, c = 0.0$:

$$\ln k = 0.019361421 - 0.080643435 \, (c \, / \, d)^3 - 0.018564917 \, e^{\, c \, / \, d} \qquad (6.6.2.2)$$

For $b \, / \, c = 0.1$:

$$\ln k = 0.020327157 - 0.077011572 \, (c \, / \, d)^3 - 0.019343819 \, e^{c \, / \, d} \qquad (6.6.2.3)$$

For $b \, / \, c = 1.0$:

$$\ln k = 0.0010533337 - 0.0080261331 \, (c \, / \, d)^3 - 0.00094568688 \, e^{c \, / \, d} \qquad (6.6.2.4)$$

These are accurate to approximately 4% over the range. The total inductance is

$$L_{total} = L_1 + L_2 + 2 \, M_{12} \qquad (6.6.2.5)$$

where $l$ is the length and $L_1$ and $L_2$ are calculated with (6.2.2.1).

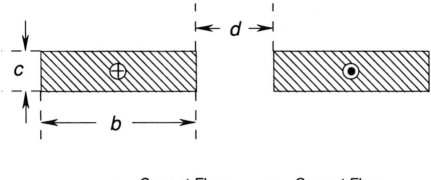

⊕ Current Flow into Page     ⊙ Current Flow out of Page

**Figure 6.6.2.2:**     **Return Circuit of Parallel Rectangular Inductors**

The total inductance of the return circuit is

$$L_{total} = L_1 + L_2 - 2\,M_{12} \tag{6.6.2.6}$$

where $M_{12}$ is calculated with (6.6.2.1) and $L_1$ and $L_2$ are calculated with (6.2.2.1). The total inductance may also be written as

$$L_{total} = 0.004\,l\left[\ln\frac{d}{b+c} + 1.5 + \ln k - \ln e\right] \tag{6.6.2.7}$$

where $\ln e$, and $\ln k$ are given in Grover [1] Tables 1, 2, and 3. Equations (6.6.2.2–4) are curve fits to $\ln k$ data. A curve fit to $\ln e$ is:

$$\ln e \cong \frac{7.3458933 \times 10^{-5} + 0.03476855\,x - 0.01272244\,x^2}{1.0 + 4.1576571\,x + 22.249414\,x^2 - 14.940776\,x^3} \tag{6.6.2.8}$$

where

$$x = b/c \text{ or } c/b \tag{6.6.2.9}$$

This equation is an accurate fit to the data of Grover to within approximately 1%.

The meander line inductor can be analyzed with the equations of this section and Section 6.2.2. Meander lines are used as resistive elements, inductors, and delay lines. Note that adjacent conductors have equal and opposite current flow which reduces the total inductance over the equivalent length of straight conductor.

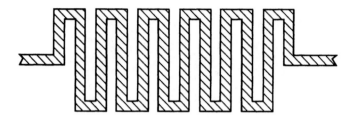

**Figure 6.6.2.3:** Meander Line

REFERENCE

[1] Grover, F.W., *Inductance Calculations*, Van Nostrand, Princeton, NJ, 1946; reprinted by Dover Publications, NY, 1962. (A concise book based on work at NBS.)

[2] Harokopus, William P., Jr., "Electromagnetic Coupling and Radiation Loss Considerations in Microstrip (M)MIC Design," *IEEE Transactions on Microwave Theory and Techniques*, Vol. MTT-39, No. 3, March 1991, pp. 413–421.

[3] Sato, Risaburo, "A Design Method for Meander-Line Networks Using Equivalent Circuit Transformations," *IEEE Transactions on Microwave Theory and Techniques*, Vol. MTT-19, No. 5, May 1971, pp. 431–442.

# Chapter 7

# Capacitors

The capacitances of the structures in Chapters 3, 4, and 5 can be calculated using the equations of those chapters and (2.3.4). Chapter 2 describes capacitors made from short- and open-circuited transmission lines. The structures of this chapter are optimized for maximum capacitive reactance.

## 7.1   ABCD MATRICES

Series capacitor:

$$\begin{bmatrix} 1.0 & \dfrac{-j}{\omega C} \\ 0.0 & 1.0 \end{bmatrix}$$

Shunt capacitor:

$$\begin{bmatrix} 1.0 & 0.0 \\ j\omega C & 1.0 \end{bmatrix}$$

## 7.2 PARALLEL-PLATE CAPACITOR

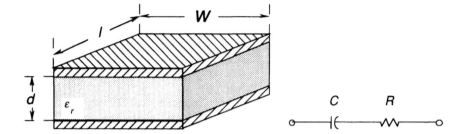

**Figure 7.2.1: Parallel-Plate Capacitor**

The equation for the parallel-plate capacitor is:

$$C = \frac{\varepsilon_0 \, \varepsilon_r \, l \, w}{d} \quad \text{(F)}$$

(7.2.1)

The equation assumes that there are no significant fringing fields or

$d \ll l$

$d \ll w$

$\varepsilon_r \gg \varepsilon_0$

The finite $Q$ of the capacitor is expressed in terms of its Dissipation factor:

$$DF = \tan \delta + \omega C R_e$$

(7.2.2)

where

$R_e$ = is the lead or electrode resistance

$\tan \delta$ = the loss tangent of the dielectric

For finite $l$ and $w$ and for unequal top and bottom plate areas, see [3] and [4].

### REFERENCES

[1]     Cohn, Seymour, "Thickness Corrections for Capacitive Obstacles and Strip Conductors," *IRE Transactions on Microwave Theory and Techniques*, Vol. MTT-8, No. 11, November 1960, pp. 638–644.

[2]     Herbert, J.M., *Ceramic Dielectrics and Capacitors*, Gordon and Breach Science
        Publishers, New York, 1983.

[3]     Langton, N.H., "The Parallel-Plate Capacitor with Symmetrical Placed Unequal
        Plates," *Journal of Electrostatics*, Vol. 9, 1981, pp. 289–305.

[4]     Lin, Weigan, "Computation of the Parallel-Plate Capacitor with Symmetrically
        Placed Unequal Plates," *IEEE Transactions on Microwave Theory and
        Techniques*, Vol. MTT-33, No. 9, September 1985, pp. 800-807.

[5]     Mondal, Jyoti P., "An Experimental Verification of a Simple Distributed Model
        of MIM Capacitors for MMIC Applications," *IEEE Transactions on Microwave
        Theory and Techniques*, Vol. MTT-35, No. 4, April 1987, pp. 403–408.

[6]     Natarajan, Sundaram, "Measurement of Capacitances and Their Loss
        Factors," *IEEE Transactions on Instrumentation and Measurement*, Vol.
        MTT-38, No. 6, December 1989, pp. 1083–1087.

## 7.3     MULTILAYER DIELECTRIC PARALLEL-PLATE CAPACITOR

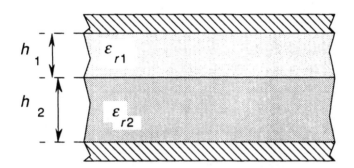

**Figure 7.3.1: Parallel-Plate Capacitor with Layered Dielectric**

The equations for a capacitor with layered dielectric are useful when calculating the
effects of air or water intrusion into the dielectric. Use (7.2.1) and replace the dielectric
constant, $\varepsilon_r$, with

$$\varepsilon_{r,layered} = \left( \sum_{n=1}^{\text{all layers}} \frac{h_n}{h_t \, \varepsilon_{r,n}} \right)^{-1} \qquad (7.3.1)$$

where

$$h_t = \text{total thickness} = \sum_{j=1}^{\text{all layers}} h_j$$

$\varepsilon_{r,n}$ = relative dielectric constant of the $n$th layer

## 7.4   THIN-FILM CAPACITORS

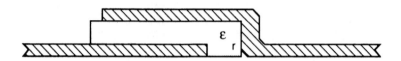

**Figure 7.4.1: Thin-Film Capacitor**

This structure may be calculated with the usual equations for a parallel-plate capacitor. Hurt [1] uses an equation that includes the effects of fringing capacitance:

$$C = \frac{\varepsilon_r \, \varepsilon_0 \, (l + \Delta_f) \, (w + \Delta_f)}{h} \tag{7.4.1}$$

where $l$ and $w$ are the dimensions of the top plate, which are effectively enlarged by an amount $\Delta_f$ due to fringing fields:

$$\Delta_f = \frac{4.0 \, h \, \ln \, (2.0)}{\pi} \tag{7.4.2}$$

REFERENCES

[1]   Hurt, John C., "A Computer-Aided Design System for Hybrid Circuits," *IEEE Transactions on Components, Hybrids, and Manufacturing Technology*, Vol. CHMT-3, No. 4, December 1980, pp. 525–535.

[2]   McLean, David A., "Dielectric Materials and Capacitor Miniaturization," *IEEE Transactions on Parts, Materials and Packaging*, Vol. PMP-3, No. 4, December 1967, pp. 163–169.

[3]   Pettenpaul, Ewald, *et al.,* "CAD Models of Lumped Elements on GaAs up to 18 GHz," *IEEE Transactions on Microwave Theory and Techniques*, Vol. MTT-36, No. 2, February 1988, pp. 294–304.

## 7.5   MULTILAYER CAPACITOR

**Figure  7.5.1:  PCB  Capacitor**

The design equations for multilayer capacitors are an extension of the parallel-plate calculations above.

$$C = \frac{0.2249 \; \varepsilon_r \; A \; (n - 1.0)}{d} \quad \text{(pF)} \tag{7.5.1}$$

where

$A$ = area of plates in square inches

$n$ = number of conductor layers

$d$ = plate spacing

The equation is valid for $d$ small.

REFERENCE

[1]   U.S. Department of Commerce, National Bureau of Standards, Circular C74, *Radio Instruments and Measurements*, "Calculation of Inductance," January 1, 1937.

## 7.6 INTERDIGITAL CAPACITOR

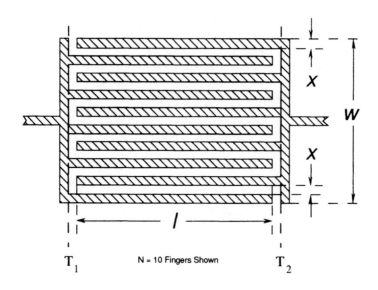

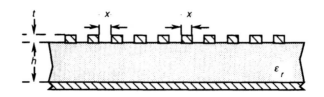

**Figure 7.6.1: Interdigital Capacitor**

The most often quoted relation for this structure is attributed to Alley [2]

$$C_2 = \frac{\varepsilon_r + 1.0}{w} l \ [(N - 3.0)A_1 + A_2] \quad (\text{pF}/\text{in}) \tag{7.6.1}$$

$$R_{series} = \frac{4.0 \, l \, R_s}{3.0 \, w} \tag{7.6.2}$$

in pF / in of width, $w$, and $N$ is the number of fingers. Equations are valid for

$$h > w/N$$

CAE equations have been fit to the curves of $A_1$ and $A_2$ in Alley:

$$A_1 = \left[ 0.3349057 - 0.15287116 \left( \frac{t}{x} \right) \right]^2 \quad (\text{pF} / \text{in}) \qquad (7.6.3)$$

$$A_2 = \left[ 0.50133101 - 0.22820444 \left( \frac{t}{x} \right) \right]^2 \quad (\text{pF} / \text{in}) \qquad (7.6.4)$$

These equations are valid to better than ±0.35% for

$3.0 \le t / x \le \infty$

The model element $C_1$ is calculated from the sum of the equivalent capacitance of the input (output) terminal and the shunt capacitance of the array of lines. The shunt capacitance can be calculated with the software presented in Smith [11]. The presence of fringing capacitance to the ground plane lowers inductance (due to the mutual inductance of the image); however, the $Q$ is also decreased.

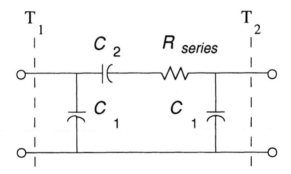

**Figure 7.6.2: Interdigital Capacitor Model**

Other models have been described [4, 12, 13] that are based on the parameters of arrays of lines and the discontinuities or numerical solutions. In particular, finger inductance, T junctions, open ends, and right-angle bends create significant effects.

Other interdigital capacitors are described in Crampagne *et al.* [3] (stripline interdigital capacitor) and Fouladian [5] (multiple generalized edge coupled striplines).

### REFERENCES

[1]     Aitchison, Colin S., *et al.*, "Lumped-Circuit Elements at Microwave Frequencies," *IEEE Transactions on Microwave Theory and Techniques*, Vol. MTT-18, No. 12, December 1971, pp. 928–937.

[2]     Alley, Gary D., "Interdigital Capacitors and Their Application to Lumped-Element Microwave Integrated Circuits," *IEEE Transactions on Microwave*

*Theory and Techniques*, Vol. MTT-18, No. 12, December 1970, pp. 1028–1033.

[3] Crampagne, Raymond, *et al.*, "Explicit Formulae for the Dispersion Diagram of a Triplate Interdigital Line in a Homogeneous Dielectric Medium," *International Journal of Electronics*, Vol. 41, No. 4, 1976, pp. 375–380.

[4] Esfandiari, Reza, "Design of Interdigitated Capacitors and Their Application to Gallium Arsenide Monolithic Filters," *IEEE Transactions on Microwave Theory and Techniques*, Vol. MTT-31, No. 1, January 1983, pp. 57-64.

[5] Fouladian, Jamshid, and Protap Pramanick, "Generalized Analytical Equations for Strip Transmission Lines," *Microwave Journal*, November 1989, pp. 165–168.

[6] Hikita, Mitsutaka, *et al.*, "Miniature SAW Antenna Duplexer for 800-MHz Portable Telephone Used in Cellular Radio Systems," *IEEE Transactions on Microwave Theory and Techniques*, Vol. MTT-36, No. 6, June 1988, pp. 1047–1056.

[7] Hobdell, John L., "Optimization of Interdigital Capacitors," *IEEE Transactions on Microwave Theory and Techniques*, Vol. MTT-27, No. 9, September 1979, pp. 788–791.

[8] Hurt, John C., "A Computer-Aided Design System for Hybrid Circuits," *IEEE Transactions on Components, Hybrids, and Manufacturing Technology*, Vol. CHMT-3, No. 4, December 1980, pp. 525–535.

[9] Pettenpaul, Ewald, *et al.*, "CAD Models of Lumped Elements on GaAs up to 18 GHz," *IEEE Transactions on Microwave Theory and Techniques*, Vol. MTT-36, No. 2, February 1988, pp. 294–304.

[10] She, X.Y., and Y.L. Chow, "Interdigital Microstrip Capacitor as a Four-Port Network," *IEE Proceedings*, Vol. 133, Pt. H, No. 3, June 1986, pp. 191–197.

[11] Smith, J. I., "The Even- and Odd-Mode Capacitance Parameters for Coupled Lines in Suspended Substrate," *IEEE Transactions on Microwave Theory and Techniques*, Vol. MTT-19, No. 5, May 1971, pp. 424–430.

[12] Wolff, Ingo, and Godfrey Kibuuka, "Computer Models for MMIC Capacitors and Inductors," *14th European Microwave Conference Proceedings*, 1984, Liège, Belgium, pp. 853–858.

[13] Yanyang, Shi, and Jiang Zongquiang, "A Hybrid Approach to Analyse a Complex Microwave Planar Network," 1990, pp. 1509–1512.

## 7.7 CONCENTRIC CYLINDERS

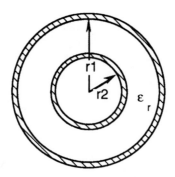

**Figure 7.7.1: Capacitance of Concentric Cylinders**

This equation can be used to calculate the capacitance of two concentric cylinders and to estimate the stray capacitance of spiral inductors.

$$C = 1.112 \, \varepsilon_r \frac{r_1 \, r_2}{r_1 - r_2} \quad (\text{pF} / \text{cm}) \tag{7.7.1}$$

Results are in pF/cm of length. The stated accuracy is 0.1%.

REFERENCE

[1]   U.S. Department of Commerce, National Bureau of Standards, Circular C74, *Radio Instruments and Measurements*, "Calculation of Inductance," January 1, 1937.

## 7.8 CAPACITOR DIELECTRIC CONSTANTS AND LOSS TANGENTS

Some of the dielectrics found in capacitors are tabulated in Table 7.8.1.

| Dielectric | $\varepsilon_r$ |
|---|---|
| Air | 1.0006 |
| Aluminum | 8.0 |
| Aluminum Oxide | 7.0 |
| Ceramic | 35.0–6000 |
| Glass | 6.11 |
| Kapton | 3.2 |
| Mica | 6.8 |
| Mylar | 3.0 |
| Polycarbonate | 2.7 |
| Polyethylene | 3.3 |
| Polypropylene | 2.5 |
| Polystyrene | 2.5 |
| Polysulfone | 2.7 |
| Quartz | 3.8 |
| Sapphire | 9.4 $(x, y)$ and 11.6 $(z)$ |
| Tantalum | 27.6 |
| Tantalum Oxide | 11.0 |
| Teflon | 2.1 |
| Vacuum | 1.00 |

Table 7.8.1:  Capacitor Dielectric Constants

# *Chapter 8*

# *Resistors*

## 8.1 ABCD MATRICES

The ABCD matrix of a shunt resistor is:

$$\begin{bmatrix} 1.0 & 0.0 \\ \dfrac{1.0}{R} & 1.0 \end{bmatrix}$$

The ABCD matrix of a series resistor is:

$$\begin{bmatrix} 1.0 & R \\ 0.0 & 1.0 \end{bmatrix}$$

## 8.2 RECTANGULAR PLANAR RESISTOR

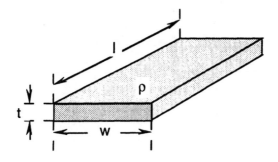

**Figure 8.2.1: Rectangular Planar Resistors**

This is the relation for resistance on which most equations are based.  Chip resistors are formed by a rectangular film on a ceramic substrate with metal end caps. Trimming will introduce strays [1].  At high frequencies the skin effect effectively reduces the amount of the conductor used to carry the current.  See Chapter 2 for more information.

$$R = \frac{\rho \, l}{w \, t} \quad (\Omega)$$

(8.2.1)

where the resistivity, $\rho$, is in $\Omega$-units of $l$, $w$, and $t$ .  The inverse of resistivity is the conductivity $\sigma$.  See Chapter 9 for a tabulation of common conductor resistivities.

For other cross-sectional geometries the resistance is calculated by replacing $w\,t$ in the above equation with the area.  Calculation of the high frequency resistance can be calculated by multiplying the length of the surface by the skin depth and using this for the area.  This approximation is not valid for surfaces with tight bends relative to the skin effect.

The above formula must be modified in many situations due to the effects of termination of the resistive film to the metallization at the ends and other process variables.  Hybrid and resistor manufacturers will have empirical modifications of the (8.2.1).  See [2] for an example of what has been done at Tektronix, Inc.

### REFERENCES

[1]    Demurie, Stefaan N., and Gilber de Mey, "Parasistic Capacitance Effects of Planar Resistors," *IEEE Transactions on Components, Hybrids, and Manufacturing Technology*, Vol. CHMT-12, No. 3, September 1989, pp. 348–351.

[2]    Garg, Raj, "Empirical Approach to Resistor Design," *International Journal for Hybrid Microelectronics,"* Vol. 6, No. 1, October 1983, pp. 122–126.

[3]    Tischler, Oscar, "Resistor Technology Assessed at Microwave Frequencies," *Microwave System News & Communications Times*, February 1985, pp. 106–112.

## 8.3    LEADED RESISTORS

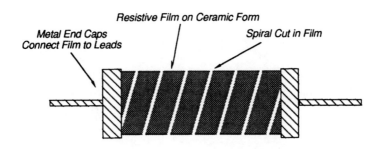

**Figure  8.3.1: Metal  Film  Resistor**

The metal film resistor is formed by coating a ceramic form with a resistive film.
The film is then physically cut to form a helix of film.  The ends are then capped to
make connections to wire leads.  The length, thickness, and resistivity of the film
determine the total resistance.

This type of resistor is useful into the low hundreds of MHz range and possible
further in less stringent applications.  The capacitance of the end caps and the spiral gap
are represented by a capacitor shunting the resistor.  The leads and the spiral cut add an
inductance.  A family of resistors can be characterized by a resistance value below
which the dominant strays are inductive and above which the dominant strays are
capacitive.

For high frequency designs this type of component has two main problems.  First,
the lead inductance is a function of the PCB layout and the way the component is
loaded in the board.  The second problem is that there are different ways to build the
resistor that give the same resistance but different strays.  A different resisitivity
material can be substituted and the helix lengthened or shortened to correct for the
change.  This can wreak havoc with designs that depend on the strays being repeatable.

## 8.4   COAXIAL RESISTORS

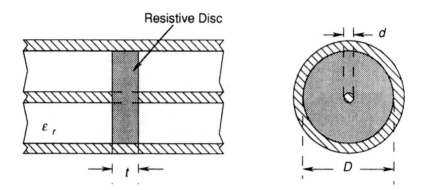

**Figure  8.4.1:  Shunt  Coaxial  Resistor**

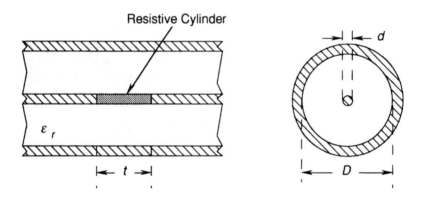

**Figure  8.4.2:  Coaxial  Series  Resistor**

As described in Section 8.1 the resistance is calculated with (where $d$ is the diameter of the resistive disk):

$$R = \frac{4.0 \, \rho \, t}{\pi \, d^2} \quad (\Omega)$$

(8.4.1)

If the diameter of the bend is small relative to the skin depth the high frequency resistance can be calculated with

$$R_{ac} = \frac{4.0 \, \rho \, t}{\pi \, \delta \, (2.0 \, d - 1.0)} \quad (\Omega)$$

(8.4.2)

The shunt and series coaxial resistors are distributed resistances used for high frequency attenuators and terminations. The shunt resistor is plated on a disk of dielectric. Resistive films are plated thin relative to the skin depth over the operating frequency range. This ensures a constant resistance vs. frequency. The lengths are kept electrically short to prevent discontinuity reactances.

By the appropriate combination of elements, a T or π pad attenuator can be built. A termination can be either a series or shunt element ending in a short.

## 8.5    MEANDER LINE RESISTORS

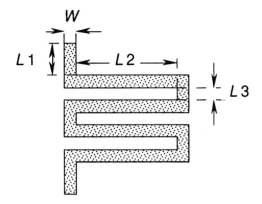

**Figure 8.5.1: Meander Line Resistor**

Meander line resistors are used to increase the amount of resistance that can be achieved in a given space. A similar structure can result from resistance trimming material or removing jumpers across resisitive loops. This structure is not suitable for high frequency design because of the capacitive and inductive reactances of the bends and capacitances of the parallel coupled lines.

The resistance of meander lines can be calculated with a modification of the standard planar resistor equation. The resistance of the sections ($L_1$, $L_2$, $L_3$) is calculated as above and then the corner resistances are calculated [5]:

$$R_{corner} = 0.441 \frac{\rho}{w} \frac{l}{t} \quad (\Omega) \tag{8.5.1}$$

The resistance of the corners is thus lower than an equivalent straight section. This effect can be eliminated by a slightly different construction which uses metal at the corners leaving only straight resistive sections (Figure 8.5.2 below).

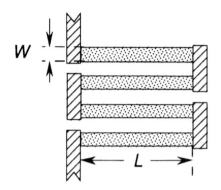

**Figure 8.5.2: Shorted Meander Line Resistor**

In this configuration metal foil shorts the corners, minimizing their effect on the resistance equations.

## REFERENCES

[1]    Barlage, F. Michael, "An Analysis of Current Crowding at 180° Bends in Film Resistors," *IEEE Transactions on Parts, Hybrids, and Packaging*, Vol. PHP-9, No. 4, December 1973, pp. 262–263.

[2]    Barlage, F.M., "Optimizing Serpentine Resistor Geometry," *Solid State Technology*, May 1973, pp. 43–59.

[3]    Demurie, Stefaan N., and Gilber de Mey, "Parasistic Capacitance Effects of Planar Resistors," *IEEE Transactions on Components, Hybrids, and Manufacturing Technology*, Vol. MTT-12, No. 3, September 1989, pp. 348–351.

[4]    Ellis, Ed, "Process Specific Thick Film Materials," *Hybrid Circuit Technology*, November 1990, pp. 21–25.

[5]    Rogers Corporation, "Design/Process Guide:  Ohmega-Ply® Clad RT/duroid® Laminates," 1986.

# Chapter 9

# Printed Circuit Fabrication

## 9.1 PCB CONSTRUCTION

Inherent in any PCB design are a number of electrical performance trade-offs. For boards having no controlled impedance lines, the choice of the materials and layering is usually left up to the PCB fabrication house. For controlled impedance design it is important to understand the trade-offs implicit in the layup. In this chapter the components used to build a printed-circuit board are described together with the parameters relevant to transmission line design. In particular, the effect of dimensional tolerances and material properties on electrical performance can be calculated with the information of this chapter using the relations of the previous chapters. In order to design a successful printed-circuit board transmission line it is also necessary to have a good understanding of the process. With this information, and an experienced printed-circuit board fabrication house, a design can be both electrically correct and cost effective.

The basic building blocks of a PCB are illustrated in Figure 9.1.1.

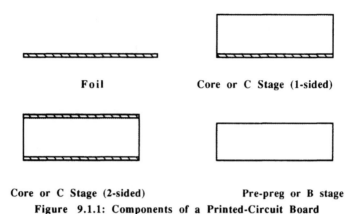

Foil                            Core  or  C  Stage  (1-sided)

Core or  C  Stage  (2-sided)              Pre-preg  or  B  stage
Figure  9.1.1:  Components  of  a  Printed-Circuit  Board

Metal foil is available as separate sheets or preapplied to the dielectric material. Metal foil is applied under heat and pressure to the dielectric. The dielectric flows around **dendrites** that were formed during creation of the metal foil and create a strong physical bond. To create traces and component pads, a chemically resistant mask is applied to the metal. The undesired metal is then selectively removed by etching. Through-holes are plated by additive techniques which build up metal on the dielectric inside the holes, creating plated-through holes (also known as vias).

The dielectric layers can be solid pieces of material or a combination of dielectric and a binder such as epoxy.[1] The dielectric is often a woven fabric or, less commonly, chips or loose fibers. A **pre-preg** layer is a dielectric layer whose epoxy has not been fully cured. This allows it to reflow easily when adhering layers together. This flow of epoxy is important to bond the layers and to prevent shorts during plating of any plated-through holes.

When the patterns for each metal layer have been created, a sandwich is made up of the desired components. This is then pressed together under heat and vacuum to bond the layers together and create the final circuit board. This stack of metal and dielectric layers is known as a **layup**.

---

[1]For example, G-10 and FR-4 are dielectrics constructed from a fiberglass fabric and epoxy.

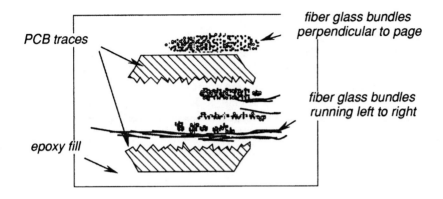

PCB traces

fiber glass bundles
perpendicular to page

fiber glass bundles
running left to right

epoxy fill

**Figure 9.1.2: Cross-Section of PCB Inner Layer**

In Figure 9.1.2, the cross-section of part of a laminated sandwich of materials is diagrammed. Part of a cross-section of an inner layer of an FR-4 printed-circuit board is shown (additional layers would continue vertically on the page). Notice the roughened surfaces of the copper traces that grip the fiber glass and epoxy. There are only a few layers of fiber glass cloth between the metallization layers, indicating that the layer is thin (on the order of ten mils). In this case, we see that a great deal of the space is filled with epoxy, rather than a homogenous mix of glass fibers and epoxy. Because it is the epoxy which actually flows to fill in surface irregularities in any layup, inner layer pc traces will generally be surrounded by epoxy.

### 9.1.1 Designing a Layup

Designing a layup begins with a number of physical constraints.

The thickness of the finished board is determined by system requirements (connectors, card guides, etc.).

The layers of the layup must be made symmetric about the center line. Different thicknesses of dielectric or metallization on either side of the pc board center line can cause board warpage. Board warpage is also minimized by equalizing the density of traces and balancing ground planes.

Inner layer dielectric thickness must be at least 1.5 times as large as the sum of the opposing metallization thicknesses. These inner layers must also have sufficient epoxy to fill the bumps created by the PC traces when the layers are compressed.

Finally, only a finite number of material thicknesses and dielectric constants are available off-the-shelf, so we are limited to selecting from these. The vendor may not have all of these in stock, so it is good practice to call the vendor who will actually build the board for a list of the materials on hand. The layup is then designed and line widths are calculated with these materials from this list. There is no point in carefully designing a layup with catalog materials only to discover that the vendor does not have what you wanted on hand.

For electrical purposes the starting point of any controlled impedance design should be C stage or core material. C stage has less variation because it is pre-cured and has copper already applied. The alternatives—built up B stage or pre-preg—introduce additional error sources and variations. These create a greater variation in impedance and propagation velocity.

In FR-4 up to 44 layers are possible at some vendors, 14 is a more typical maximum for production runs. Microwave laminations of 50 layers (Rogers RO2800) and 14 layers (RTI Duroid) have been reported in Markstein [1].

It is important to work with the PCB fabrication shop to develop a layup that meets electrical requirements and can be built.

## 9.2   SUBSTRATE DIELECTRICS

### 9.2.1   Properties of Commercial Dielectrics

The two parameters of a dielectric material which are critical in controlled impedance designs are $\varepsilon_r$ and tan $\delta$ (also called loss tan, or tan delta). These determine the line dimensions for a particular $Z_0$ and the dielectric losses, respectively. Table 9.2.1.1 summarizes these parameters for many of the popular commercial dielectrics.

| Company, Dielectric | $\varepsilon_r$ | tan $\delta$ | Description | Appearance |
|---|---|---|---|---|
| **Generic** | | | | |
| G-10 | 4.3±.05 | 0.008 | Epoxy/ Glass | woven |
| **Norplex/Oak** | | | | |
| FR-4 ED 130 | 4.3 | 0.020 | Epoxy/ Glass | woven |
| FR-4 EM 145 | 4.3 | 0.020 | Epoxy/ Glass | woven |
| FR-4 G 50 | 4.3 | 0.020 | Epoxy/ Glass | woven |
| Cyanate Ester CE 245 | 3.5 | 0.005 | | |
| Polyimide G 30 | 4.2 | 0.015 | | |
| Paper/Phenolic NP 492 | 4.8 | 0.045 | | |
| Paper/Phenolic NP 930 | 4.7 | 0.025 | | |
| **Mica** | | | | |
| FR-4 EG 150 | 4.5 | 0.020 | | |
| Polyimide PG 418 | 4.2 | 0.005 | | |
| **Mitsubishi Plastics** | | | | |
| E002 | 5.10 | 0.022 (1 MHz) 0.020 (1 GHz) | | |
| K002 | 3.6 | 0.0035 (1 MHz) 0.0035 (1 GHz) | | |
| K012 | 3.5 | 0.0021 (1 MHz) | | |
| **Polyclad** | | | | |
| FR-4 PCL-FR-204 | 4.6 | 0.20 | | |
| Polyimide PCL-GI-702 | 4.6 | 0.010 | | |
| **Hi-Tek** | | | | |
| Cyanate Ester AroCy M-40S | 3.63 | 0.002 | | |
| **Arlon** | | | | |
| DiClad 810 | 10.2 | 0.0027 (10 GHz) | | |
| Cu Clad 217 | 2.17±0.04 | 0.0009 (1 MHz) | PTFE/Glass | woven, bendable |
| Cu Clad 233 | 2.33±0.02 | 0.0014 | | |
| Cu Clad 250 | 2.45, 2.50, 2.55±0.4 | 0.0008 (1 MHz) 0.0018 (10 GHz) | | |
| **Rogers Corp.** | | | | |

| | | | | |
|---|---|---|---|---|
| Duroid 5500 | 2.5±0.04 | | Ceramic/<br>PTFE | smooth,<br>uniform |
| Duroid 5870 | 2.33±0.02 | 0.0005 (1 MHz)<br>0.0012 (10 GHz) | PTFE/Random<br>Glass | |
| Duroid 5880 | 2.20±0.02 | 0.0004 (1 MHz)<br>0.0009 (10 GHz) | PTFE/Random<br>Glass | |
| Duroid 6002 | 2.94±0.04 | | | |
| Duroid 6006 | 6.00±0.2 | 0.0025 (10 GHz) | | |
| Duroid 6010.2 | 10.2±0.25 | | Ceramic/PTFE | |
| Duroid 6010.5 | 10.5±0.25 | 0.0028, max<br>(10 GHz) | | |
| RO2800 | 2.88±0.06 | | | |
| Polyimide Fleximid | 3.4 | 0.003 | | |
| TMM-3 | 3.24 | 0.0018 (10 GHz) | | |
| TMM-4 | 4.5 | 0.0018 (10 GHz) | | |
| TMM-6 | 6.5 | 0.0018 (10 GHz) | | |
| TMM-10 | 9.8 | 0.0017 (10 GHz) | | |
| TMM-13 | 12.85 | 0.0019 (10 GHz) | | |
| **Keene Corp.** | | | | |
| 522-50 | 2.5±0.05 | 0.0010 (1 MHz) | | |
| 522-45 | 2.45±0.05 | 0.0010 (1 MHz) | | |
| 522-48 | 2.48±0.05 | 0.0010 (1 MHz) | | |
| 522-55 | 2.55±0.05 | 0.0010 (1 MHz) | | |
| 527-45 | 2.45±0.04 | 0.0019 (10 GHz) | | |
| 527-50 | 2.50±0.05 | 0.0019 (10 GHz) | | |
| 527-55 | 2.55±0.04 | 0.0019 (10 GHz) | | |
| 527-68 | 2.68±0.04 | 0.0019 (10 GHz) | | |
| 870-33 | 2.33±0.04 | 0.0012 (10 GHz) | | |
| 880-20 | 2.20±0.04 | 0.00085<br>(10 GHz) | | |
| 810-20 | 10.2±0.25 | 0.002 (10 GHz) | | |
| 810-50 | 10.5±0.25 | 0.002 (10 GHz) | | |
| Epsilam 6 | 6.00 | 0.0018 (10 GHz) | | |
| Epsilam 10 | 10.2±.25 | 0.002 (10 GHz) | | |
| **Crane Polyflon** | | | | |
| CuFlon | 2.1 | 0.0001<br>0.00045 (1 GHz)<br>0.00045<br>(18 GHz) | | |
| **Others** | | | | |

| Polysulfone | 3.5-3.9 (1 MHz) | 0.0056-0.009 |
| | 3.0 (10 GHz) | (1MHz) |
| | | 0.005 (10 GHz) |

Table 9.2.1.1: Properties of Dielectrics

REFERENCES

[1]    Angeniéniex, G., "Broadband Dielectric Characterization of Aluminum Nitride," *Microwave Journal*, October 1990, pp. 91–98.

[2]    Bahl, Inder, and Kevin Ely, "Modern Microwave Substrate Materials," *Microwave Journal*, 1990 State of the Art Reference, pp. 131–146.

[3]    Faulkner, J., *et al.*, "Soft Microstrip with Integral Ground Plane Aids in Supercomponenet Integration," *Microwave Journal*, November 1983, pp. 105-115.

[4]    Frish, David, and Francis J. Nuzzi, "Improving PCB Performance with Thermoplastic and Metal-Core Substrates," *PC Fab*, February 1985, pp. 90–98.

[5]    Grace, Frank E., "Tolerance Study of Printed-Circuit Process Steps for a Microwave Application," *IEEE Transactions on Parts, Materials and Packaging*, Vol. PMP-1, No. 1, June 1965, pp. S-16–S-27.

[6]    Gupta, Chandra, John Levantis, and Nancy Jane Bailey, "Design of Microstrip Receiver Supercomponents at mm-Wave Frequencies," *Microwave Journal*, September 1986, pp. 161-175.

[7]    Hales, Neal, "Multilayer Printed Circuit Boards and Their Utilization in State-of-the-Art Digital Products," Wescon/82, Session 9, pp. 1–8.

[8]    Liao, Samuel Y., *Microwave Circuit Analysis and Amplifier Design*, Prentice-Hall, Englewood Cliffs, NJ, 1987, p. 199.

[9]    MIL-P-13949F, March 10, 1981, "General Specification for Plastic Sheet, Laminated, Metal Clad (for Printed Wiring Boards)."

[10]   Ogden, Cameron, "The Impact of Process Capability on the Manufacturability of Microwave Circuit Boards," *Microwave Journal*, April 1990, pp. 267–274.

[11]   Pirocanac, Diane, "Fine Line Printed Circuit Boards," *Electronic Packaging and Production*, three-part series, November 1989, pp. 118–125; December 1989, pp. 87–92; February 1990, pp. 124–131.

[12]   Polyflon Company, "CuFlon Pure PTFE Ultra Low Loss Microwave Substrate," product brochure.

[13]    Rogers Corporation, "Make the Most of Your Microwave Designs," Advertisement.

[14]    Rogers Corporation, Product Bulletins.

[15]    Spitz, S. Leonard, "Specialty Laminates Fill Special Needs," *Electronic Packaging and Production*, February 1989, pp. 108-111.

[16]    Spitz, S. Leonard, "Toward a Better Beryllia," *Electronic Packaging and Production*, August 1989, p. 15.

[17]    3M, Electronic Products Division, "ε-10 High Dielectric Constant Microwave Substrate," Product Brochure.

[18]    3M, Electronic Products Division, Product Bulletin.

[19]    Union Carbide Corporation, Engineered Plastics & Carbon Fibers Division, "Advantages and Processing of Polysulfone Printed Wiring Boards," November 1982.

[20]    Vidano, Ron, *et al.*, "Packaging a Microwave Antenna with GaAs Chips as Radiating Elements," *Electronic Packaging & Production*, December 1989, pp. 52–55.

### 9.2.2    Effect of Resin/Glass Mix on $\varepsilon_r$ of FR-4

In FR-4 and G-10 boards, a fabric of woven glass fibers is impregnated with epoxy and cured to form the substrate.  Depending on the amount of glass used in a given PCB layer, the dielectric constant will be different.  The effect of the amount of epoxy in the dielectric on the dielectric constant is shown in Figure 9.2.2.1 [1-3] .  Table 9.2.2.1 lists the parameters for some commonly available materials.  The overall trend is the introduction of greater amounts of glass for thicker layers, which increases the dielectric constant.

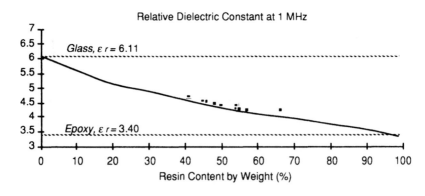

Figure 9.2.2.1:    $\varepsilon_r$ **vs. Resin Content**

The amount of resin and glass in a particular layer is dependent on a number of variables under the control of the vendor. A sheet of material will have 2 to 4% more epoxy if it is a B stage or pre-preg than if it is a C stage. Some of this extra epoxy will be pressed out during lamination. A variation of approximately ±2% in epoxy content can be expected, which produces a variation in $\varepsilon_r$ of about ±0.06. This mix will also vary from vendor to vendor..

| Material | Resin Content (%) | Nominal Thickness (mils) | $\varepsilon_r$ |
|---|---|---|---|
| 1080 | 57 | 2.50 | 4.22 |
| 2113 | 48 | 3.75 | 4.46 |
| 106/1080 | 60 | 4.42 | 4.20 |
| 2 ply 1080 | 57 | 4.90 | 4.22 |
| 2116 | 50 | 4.91 | 4.41 |
| 106/2113 | 55 | 5.42 | 4.27 |
| 1080/2113 | 55 | 6.50 | 4.27 |
| 106/2116 | 55 | 6.55 | 4.27 |
| 1080/2116 | 52 | 6.80 | 4.35 |
| 7628 | 41 | 7.13 | 4.68 |
| 2 plies 2113 | 48 | 7.30 | 4.46 |
| 106/7628 | 46 | 7.80 | 4.52 |
| 2113/2116 | 46 | 8.25 | 4.52 |
| 2 plies 2116 | 50 | 9.35 | 4.40 |
| 1080/2116/1080 | 54 | 9.90 | 4.30 |
| 2116/1080/2116 | 46 | 10.00 | 4.50 |
| 1080/7628 | 46 | 10.00 | 4.52 |
| 1080/2116/1080 | 54 | 10.10 | 4.30 |
| 106/7628/106 | 50 | 11.30 | 4.40 |
| 2116/7628 | 45 | 11.90 | 4.55 |
| 1080/7628/1080 | 48 | 12.20 | 4.46 |
| 2 plies 7628 | 41 | 14.30 | 4.68 |
| 3 plies 2116 | 50 | 14.30 | 4.40 |
| 2113/7628/2113 | 41 | 14.50 | 4.55 |
| 7628/106/7628 | 43 | 16.10 | 4.62 |
| 7628/1080/7628 | 44 | 16.20 | 4.58 |
| 2 plies 2113/2 plies 2116 | 46 | 16.20 | 4.58 |
| 2116/7628/2116 | 47 | 17.00 | 4.49 |
| 7628/2113/7628 | 43 | 17.50 | 4.62 |
| 2 plies 1080/2 plies 7628 | 46 | 19.50 | 4.52 |
| 2 plies 1080/2 plies 7628 | 46 | 19.50 | 4.52 |
| 2 plies 1080/2628 | 44 | 20.00 | 4.70 |
| 3 plies 7628 | 41 | 20.50 | 4.68 |
| 2 plies 2113/2 plies 7628 | 44 | 20.00 | 4.70 |
| 2 plies 2116/2 plies 7628 | 45 | 23.20 | 4.55 |
| 2 plies 1080/3 plies 7628 | 43 | 25.00 | 4.75 |
| 4 plies 7628 | 40 | 28.00 | 4.90 |

**Table 9.2.2.1:** FR-4 Characteristics

REFERENCES

[1]    Coombs, C.F., Jr., ed., *Printed Circuits Handbook*, McGraw-Hill, New York, 1979.

[2]   Nelco, Inc., "Typical Laminate Characteristics," Product Literature.

[3]   Polyclad Laminates, Inc., Product Literature.

[4]   Waite, John M., "Impedance Control Discussion," Unpublished, December 6, 1989.

## 9.2.3   Ceramic Dielectrics

| Ceramics | $\varepsilon_r$ | tan $\delta$ |
|---|---|---|
| Alumina ($Al_2O_3$ 85%) | 8.3 | 0.0058 |
| Alumina ($Al_2O_3$ 96%) | 10 | 0.0002 |
| Alumina ($Al_2O_3$ 99.5%) | 9.70 | 0.0002 (1 MHz) |
| | | 0.0003 (10 GHz) |
| Aluminum Nitride (AlN) | 8.8 | 0.0005–0.001 |
| Beramic Z  (BeO 97.9%) | 6.65 (1 MHz) | 0.0006 |
| Beryllia (BeO 97%) | 6.90 | 0.0002 (1 MHz) |
| | | 0.0003 (10 GHz) |
| Boron Nitride (BN) | 4.0 | 0.0004 |
| Fused Silica | 3.78 | 0.0001 (1 MHz) |
| | | 0.00025 (25 GHz) |
| Magnesium Titanate | 16 | 0.0002 |
| Quartz, Fused Silica | 3.78 | 0.0004 (10 GHz) |
| Rutile ($TiO_2$ 99.5%) | 85 | 0.0002 |
| Silicon Carbide (SiC) | 40.0 | 0.5 |

**Table   9.2.3.1:**       **Ceramic   Dielectric   Properties**

REFERENCES

[1]   Manz, Barry, "Aluminum Nitride Challenges BeO for RF Packaging," *Microwaves & RF*, September 1989, pp. 48–55.

[2]   Spitz, S. Leonard, "Ceramics Keep Circuits Cool," *Electronic Packaging and Production*, July 1989, pp. 35–41.

[3]   Spitz, S. Leonard, "Toward a Better Beryllia," *Electronic Packaging and Production*, August 1989, p. 15.

## 9.2.4   Flexible PCB Dielectrics

Printed-circuit boards can be made from thin sheets of dielectric and metallization which are flexible. This allows the board to be folded to conform to odd packaging demands and to replace cabling. The table below lists the important dielectric properties of these materials. Flexible dielectrics are commonly available in thicknesses of 0.5, 1, 2, 3, and 5 mils.

| Dielectric | $\varepsilon_r$ | tan $\delta$ |
|---|---|---|
| Dacron / Epoxy | 3.2 | 0.015 |
| Kapton® (Polyimide) | 3.5 (1 kHz) | 0.0025 (1 kHz) |
| Mylar® (Polyester) | 3.45 (1 kHz) | 0.0028 (1 kHz) |
| | 3.35 (1 MHz) | 0.008 (100 kHz) |
| | 3.32 (10 MHz) | 0.0115 (10 MHz) |
| Nomex® (Aramid fibers) | 1.6–3.7 (higher for thicker) | 0.004–0.007 (higher for thicker) |
| | (60 Hz) | (60 Hz) |

**Table 9.2.4.1:** **Flexible PCB Dielectric Properties**

REFERENCE

[1]    Gurley, Steve, *Flexible Circuits: Design and Applications*, Marcel Dekker, NY, 1984.

## 9.2.5   Semiconductor Dielectrics

| Dielectric | $\varepsilon_r$ | tan $\delta$ | $\sigma$ (S / m) |
|---|---|---|---|
| Gallium Arsenide (GaAs) | 13.10 | 0.0016 (10 GHz) | $8 \times 10^{-3}$ |
| Germanium (Ge) | 16.0 | | |
| InSb | 15.9 | | |
| Sapphire ($Al_2O_3$) | 10 | 0.0001 (10 GHz) | $5.5 \times 10^{-4}$ |
| Silicon (Si) | 11.7 | 0.005 | $4.39 \times 10^{-4}$ |
| SiO | 6 | 0.007 | |
| $SiO_2$ | 3.9 | 0.007 | |
| $Ta_2O_5$ | 25 | 0.003 | |

**Figure 9.2.5.1:** **Semiconductor Dielectric Properties**

## 9.2.6   Properties of Other Dielectric Materials

Occasionally it is desired to build a transmission line structure into the mechanical structure of the system it is in. For example, a controlled impedance line might be molded into a case.

The plastics vendor literature provides a good start to finding the properties of the material. Dielectric constants and power factors are usually given at 60 Hz and some higher frequency (1 kHz or 1 MHz). Plastics with common trade names come in a variety of forms with slightly different electrical properties—try to determine the exact type you are using. The most variation will be found in materials that are loaded or filled with a second material to varying degrees in order to achieve different physical parameters such as rigidity.

It should not be assumed that the low frequency parameters will hold at microwave or even RF frequencies. High frequency data or sample testing will be necessary.

REFERENCE

[1]    GE Company Plastics Group, *Plastics Properties Guide.*

### 9.2.7   Dielectric Dimensions and Tolerances

To determine the effects of mechanical tolerances on the design we need to know the vendor specifications for these materials. Although these will vary depending on the material quality and the vendor, the numbers in this section can be used for estimation purposes. Together with the equations of the previous chapters the effect of variations on the electrical parameters $Z_0$, $\varepsilon_r$, $v_p$, and electrical length can be calculated.

### *9.2.7.1    Finished PCB Dielectric Thicknesses*

As discussed in Section 9.1, the printed-circuit board is built out of a number of layers sandwiched together. A large number thicknesses can be achieved from the available component materials; however, system physical requirements will usually dictate the exact value of the overall thickness. The following overall thicknesses are the most commonly used.

| | |
|---|---|
| Backplanes | 0.125 in ±10% |
| Daughter Cards | 0.03125 in, 0.0625 in, |
| | 0.093 in |

**Table  9.2.7.1:**        **Finished  PCB  Dielectric  Thicknesses**

Standard microwave dielectric thicknesses (overall) vary, depending on material. Some typical values are:

| |
|---|
| $0.010 \pm 0.0007$ in |
| $0.031 \pm 0.001$ in |
| $0.062 \pm 0.002$ in |
| $0.093 \pm 0.003$ in |
| $0.125 \pm 0.004$ in |

**Table  9.2.7.1.2:**        **Microwave  Dielectric  Thicknesses**

### 9.2.7.2 Finished B-Stage Dielectric Tolerances

| Dielectric Thickness (in) | Tolerance (±in) |
|---|---|
| 0.005 to 0.008 | 0.0015 |
| 0.009 to 0.016 | 0.002 |
| 0.017 to 0.030 | 0.003 |
| 0.031 to 0.060 | 0.005 |

**Table 9.2.7.2.1:** Dielectric Thickness Tolerances

### 9.2.7.3 C-Stage Dielectric Tolerances

Tolerances are a function of the dielectric thickness and are given in Table 9.2.7.3.1. Class 4 applies to microwave dielectrics; classes 2 and 3 are two grades of glass reinforced dielectrics (FR-4, G-10).

| dielectric thickness (in) | Class 2 (± in) | Class 3 (± in) | Class 4 (± in) |
|---|---|---|---|
| 0.0010 to 0.0045 | 0.00075 | 0.00075 | |
| 0.0045 to 0.0065 | 0.0010 | 0.00075 | |
| 0.0066 to 0.0120 | 0.0015 | 0.0010 | |
| 0.0121 to 0.0200 | 0.0020 | 0.0015 | |
| 0.0201 to 0.0299 | 0.0025 | 0.0020 | |
| 0.0300 to 0.040 | 0.0040 | 0.0030 | 0.0020 |
| 0.041 to 0.065 | 0.0050 | 0.0030 | 0.0020 |
| 0.066 to 0.100 | 0.0070 | 0.0040 | 0.0030 |
| 0.101 to 0.140 | 0.0090 | 0.0050 | 0.0035 |
| 0.141 to 0.250 | 0.0120 | 0.0060 | 0.0040 |

**Table 9.2.7.3.1:** Dielectric Thickness Tolerances

### REFERENCES

[1] Hornig, Carl F., "Multilayer Backplanes are Designed for Efficiency," *Electronic Packaging and Production*, June 1985, pp. 162-164.

[2] Markstein, Howard W., "Multilayer Board Fabrication Sheds Previous Limitations," *Electronics Packaging and Production*, July 1989, pp. 24–28.

## 9.3   METALLIZATION

### 9.3.1   Standard Conductive Layers and Tolerances

C stage materials usually begin the lamination process with copper material already attached by the vendor. On some layers, sheets of copper are added. These conductive layers of the PCB are made from thin sheets of electrodeposited or rolled copper. The sheets are purchased in rolls or in flat sheets prepunched with alignment holes.

Metal backings can also be provided which are quite thick (*e.g.*, 0.125"). This extra thickness is useful for improved grounding, mechanical rigidity, and thermal management.

### 9.3.2   Metal Thickness

Inner metallization layers are bare copper. Outer layers are usually solder-plated copper although they may be solder mask over bare copper. Metallization thickness is specified in ounces of copper per square foot.

| Copper  Weight | Inner Layer | Outer Layer (after  plating) |
|---|---|---|
| 1/2 oz | $0.0007 \pm 50 \ \mu in$ | |
| 1 oz | $0.0014 \pm 50 \ \mu in$ | $0.0028 \pm 190 \ \mu in$ |
| 2 oz | $0.0028 \pm 50 \ \mu in$ | $0.0042 \pm 190 \ \mu in$ |
| 3 oz | $0.0042 \pm 50 \ \mu in$ | $0.0056 \pm 190 \ \mu in$ |

Table   9.3.1.1.1:      Metallization   Thicknesses

It is preferable to use a single weight of copper on inner layers (core layers) due to possible etching restrictions (some vertical etchers cannot etch each side at a different rate). Outer copper layers can be applied separately and do not have this restriction.

### 9.3.3   Trace Width Tolerances

Table 9.3.1.2 tabulates line width tolerances as a function of copperweight. Greater metal thickness requires a longer etch time and creates a corresponding increase in uncertainty of line widths and spacings. You should specify the thinnest metal that meets your other requirements (losses, power handling).

Another large influence on the tolerances of trace widths and spacings is the density of metal over the board's surface. It is best to keep the density constant over the board's surface. This can be done by filling empty areas of the board with dummy traces or cross hatching to keep ensure an even etching rate. These areas are known as *thieving*.

| Copper Weight | I | II, III |
|---|---|---|
| 0.5 oz | ±1 mil | ±0.75 mil |
| 1 oz | 2 | 1.5 |
| 2 oz | 3 | 2 |
| 3 oz | 4 | 3 |

**Table 9.3.1.2:** **Trace Width and Spacing Tolerances**

## 9.3.4   Minimum Trace Widths and Line Spacings

There are limitations to the thinnest line widths that can be achieved repeatably. Some vendors report that 2 to 4 mil lines and spaces are possible [1].

| Copper Weight | Nominal Thickness (in) | Min. Spacing (in) | Min. Width (in) |
|---|---|---|---|
| 1/4–1/2 oz. | 0.00035 | 0.006 | 0.006 |
| 1/2–1 oz. | 0.0007 | 0.009 | 0.010 |
| 1 oz. | 0.0014 | 0.012 | 0.013 |
| 2 oz. | 0.0028 | 0.015 | 0.010 |

**Table 9.3.4.1:** **Minimum Trace Width and Spacings**

REFERENCE

[1]    Markstein, Howard W., "Multilayer Board Fabrication Sheds Previous Limitations," *Electronics Packaging and Production*, July 1989, pp. 24–28.

## 9.3.5   Plating

Metal is usually plated over the etched circuit traces and into holes as summarized in Table 9.3.5.1. It is becoming more common to have solder mask over bare copper on outer layers.

This construction works particularly on SMD boards and boards with large areas of ground plane. In plated designs, the ground plane will collect solder during wave soldering or if solder masked the plating will reflow and crinkle the solder mask.

Plating is applied by electrodeposition or by an electroless (chemical) process.

Gold plating (≥99.7% Au) is typically applied as a greater than 30 μin layer over 100 μin of electroplated nickel.

| Plating Type | Usage | Thickness |
|---|---|---|
| Copper | Good conductivity | 1–2 mils |
| Gold | Wire bondable, corrosion resistant, low-loss | 10–50 μin |
| Tin-Lead | Corrosion resistant, solderable | 0.1–0.5 mils |
| Nickel | Barrier between Cu and Au | 50–300 μin |
| Solder | Corrosion resistant, solderable | 0.1–0.3 mil |

Table 9.3.5.1:    Plating Properties

## 9.3.6 Electrical Properties of Metals

The electrical properties of commonly used metals are given in Table 9.3.2. The relative resistivity numbers are provided for reference when using certain CAE programs. The resistivity of thin films of metal will be substantially higher than the values in the table.

| Material | $\mu_r$ | $\varepsilon_r$ | $\sigma$(S / m) | $\rho / \rho_{Cu}$ (unitless) | $\rho / \rho_{Au}$ (unitless) |
|---|---|---|---|---|---|
| Silver | 0.99998 | 1 | $6.17 \times 10^7$ | 0.9335 | 0.6645 |
| Copper | 0.999991 | 1 | $5.76 \times 10^7$ | 1.0000 | 0.7118 |
| Aluminum | 1.00002 | 1 | $3.72 \times 10^7$ | 1.5484 | 1.1022 |
| Sodium | 1 | 1 | $2.1 \times 10^7$ | 2.7429 | 1.9524 |
| Brass | 1 | 1 | $2.56 \times 10^7$ | 2.2500 | 1.6016 |
| Tin | 1 | 1 | $0.87 \times 10^7$ | 6.6207 | 4.7126 |
| Graphite | 1 | 1 | $0.01 \times 10^7$ | 576.0 | 410.0 |
| Gold | 1 | 1 | $4.10 \times 10^7$ | 1.4049 | 1.000 |
| Steel[1] | 1,000 | | $0.14$–$0.77 \times 10^7$ | 7.481-41.143 | 5.324-29.286 |
| Stainless Steel | 1,000 | | $0.11 \times 10^7$ | 52.364 | 37.273 |
| Mumetal | 100,000 | | | | |
| Solder | 1 | | $0.7 \times 10^7$ | 8.2286 | 5.8571 |
| Nickel | 600 | | $1.45 \times 10^7$ | 3.9724 | 2.8276 |

Table 9.3.2:  Electrical Properties of Common Metals

REFERENCE

[1]    *Reference Data for Radio Engineers*, Howard W. Sams & Co., Inc., 1956.

[1]Steel comes in a variety of alloys with varying permeabilities. This is a representative value.

## 9.3.7 Metal Roughness

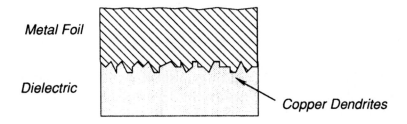

**Figure 9.3.3.1:** Metal Roughness

As discussed earlier, the surface roughness of substrate metallization entwines with the dielectric to form a mechanical bond. This same roughness introduces additional losses at high frequencies by increasing the path length when the skin depth approaches the surface roughness.

There are two different techniques used to create the thin layer of copper used on printed-circuit boards. *Electrodeposited copper* is made by electrically attracting copper from a bath onto a rotating drum. This gives a rough surface, which is desirable for good adhesion to the dielectric. The conductor side nearest the drum is smoother than the electrochemically applied surface. *Rolled copper* is cast, milled smooth, and then cold rolled to the desired thickness. This gives a very smooth surface that has lower electrical losses.

| Material | Surface Roughness | Resistivity |
|---|---|---|
| *Electrodeposited Copper* | | $<1.5940 \ \Omega\text{-g/m}^2$ |
| 1/2 oz | 75 μin, rms | |
| 1 oz | 95 μin, rms | |
| 2 oz | 115 μin, rms | |
| *Rolled Copper* | | $1.52361 \ \Omega\text{-g/m}^2$ |
| 1/2 oz | 55 μin, rms | |
| 1 oz | 55 μin, rms | |
| 2 oz | 55 μin, rms | |

**Table 9.3.3.1:** Metal Foil Roughness

Although rolled copper has less surface roughness, the conductivity is not anisotropic whereas with electrodeposited copper it is. The conductivity of rolled copper is about 20% lower against the direction of the roll [3].

Thick metal layers laminated to the dielectric have surface roughnesses of approximately 70 μin., rms.

REFERENCES

[1]     Bonfield, Robin, "Choose Metals for Optimum PTFE Performance,"
        *Microwaves & RF*, March 1990, pp. 121–124.

[2]     Gurley, Steve, *Flexible Circuits: Design and Applications*, Marcel Dekker,
        New York, 1984.

[3]     Hiroyuki Tanaka, and Fumiaki Okada, "Precise Measurements of Dissipation
        Factor in Microwave Printed Circuit Boards," *IEEE Transactions on
        Instrumention and Measurement*, Vol. IM-38, No. 2, April 1989, pp. 509–514.

[4]     Swansen, William K., "Good Planning Ensures Clean Microwave Boards,"
        *Microwaves & RF*, December 1985, pp. 79–86.

## 9.4  PLATED-THROUGH HOLES OR VIAS

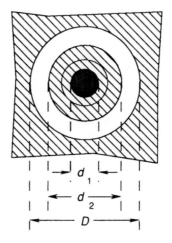

**Figure 9.4.1: Plated-Through Hole Pad Dimensions**

Plated-through holes can have diameters of approximately 0.032 to 0.189". A ring of metallization and a ring of clearance must be provided on each layer the hole passes through. The required dimensions are given in the table below for preferred plated-through hole sizes. Remember that the outer conductor can introduce a capacitive discontinuity to the pad if it is too close. Vias can pass completely through the board; be buried internally (*buried via*); or start on the outside and stop inside the board (*blind via*).

Plated-through holes should have a 5:1 or lesser ratio of board thickness to hole diameter (9:1 is possible) and should have diameters of at least 20 mil to avoid drilling and plating difficulties. Plated-through holes can be provided that connect to a thick metal backing layer. The plating thickness in the hole is approximately 1 mil.

Pins, rivets, and eyelets are possible alternatives to plated-through holes; however, they introduce an extra step in the assembly process.

| Plated-Through Hole Diameter, $d_1$ (in) | Minimum Annular Ring Width, $d_2$ (in) | Metallization Clearance, $D$ (in) |
|---|---|---|
| 0.036 | 0.080 | 0.150 |
| 0.058 | 0.100 | 0.170 |
| 0.082 | 0.130 | 0.200 |
| 0.036 | 0.060 | 0.100 |
| 0.058 | 0.080 | 0.130 |
| 0.082 | 0.110 | 0.160 |

Table 9.4.1: Plated-Through Hole Dimensions

*Thermal relief vias* (Figure 9.4.2) isolate a pad thermally from a solid ground or power plane. This allows the pad to be heated more easily for soldering and un-soldering component leads. This pattern introduces an inductive discontinuity at high frequencies.

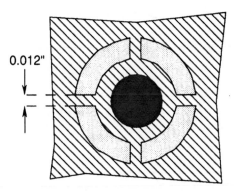

0.012"

Figure 9.4.2: Thermal Relief Via

## 9.5 SOLDER MASK

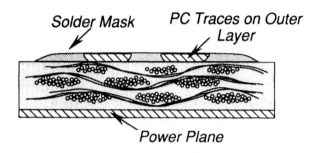

**Figure 9.5.1: Solder Mask**

Solder mask is applied over a finished printed-circuit board's traces to restrict the flow of solder to the desired areas during wave soldering. Solder mask also prevents the flow of small leakage currents between traces caused by impurities introduced during handling. Hard solder mask also prevents fine traces from being damaged during handling. Solder mask can be applied dry in sheets or wet by screening or spraying.

Electrically, our concern is that the addition of the solder mask reduces outer layer trace impedance. The exact amount of the reduction depends on the dielectric constants of the solder mask and the substrate, and the $w/h$ ratio. See Chapter 3 for information on calculating this effect.

| Type | Description | $\varepsilon_r$ | tan $\delta$ | thickness |
|---|---|---|---|---|
| Standard (PC401) | epoxy | 3.3–4.0 | 0.03–0.05 | 1–2 mil |
| Dry Film | | 3.80 | | 3–3-1/2 mil |
| Photoimaged | Screened | 3.8 | | 1 mil |
| Conformask™ | | 4.0–4.2 | | 2 mil |

**Table 9.5.1: Solder Mask Properties**

The thickness of the solder mask is controlled by the screen mesh used in the screening process. Excessively thick masks can make it difficult to mount SMD components.

REFERENCES

[1]   Coombs, C.F., Jr., ed., *Printed Circuits Handbook*, McGraw-Hill, New York, 1979.

[2]   *Electronic Designer's Handbook*, McGraw-Hill, New York.

[3]     Grace, Frank E., "Tolerance Study of Printed-Circuit Process Steps for a Microwave Application," *IEEE Transaction on Parts, Materials, and Packaging*, Vol. PMP-1, June 1965, No. 1, pp. S-16–S-27.

[4]     Waite, John M., "Impedance Control Discussion," Unpublished, December 6, 1989.

# Chapter 10

# Cable Dielectrics

The dielectrics used in constructing cables are, in general, different from those used in printed-circuit boards and ICs. Table 10.1 presents constants that are useful when calculating the performance of coaxial cables, twisted pairs, and other cables. It is also helpful when a custom cable is to be designed. These constants can be used with the equations of Chapter 3 to make the required calculations.

| Material | $\varepsilon_r$ | tan $\delta$ |
|---|---|---|
| Air | 1.0006 | |
| Cellular FEP | 1.50 | 0.0007 |
| Cellular TFE | 1.40 | 0.0002 |
| Cellular Polyethylene | 1.40–2.10 | 0.0003 |
| Ethylene Propylene | 2.24 | 0.00046 |
| Expanded PTFE[1] (Teflon) | 1.93 | |
| FEP | 2.10 | 0.00007 |
| KEL-F[2] | 2.37 | 0.027–0.0053 |
| Kynar | 6.4 | |
| Nylon | 4.60–3.5 | 0.04–0.03 |
| Perforated TFE | 1.50 | 0.0002 |
| Polyimide | 3–3.5 | 0.0001 |
| PFA | 2.06 | |
| PVC | 3–8 | |
| Si Rubber | 2.08 | 0.007–0.016 |
| SiO$_2$ | 3.78 | 0.0001 |
| Solid Polyethylene | 2.35 | 0.0003 |
| Solid PTFE | 2.10 | 0.0002 |
| Spline PTFE | 1.35 | 0.0002 |
| Tefzel | 2.6 | 0.005 |

**Table 10.1:     Cable Dielectrics**

[1]Polytetrafluoroethylene

[2]Polymonochlorotrifluoroethylene

453

# *Chapter 11*

# *Wire Gauges*

It is convenient to use standard wire sizes when building slab lines, twisted pair, and other transmission lines. Wire can be used as high-impedance lines (substituting for small PC traces, which would be difficult or impossible to fabricate), inductors, and as conductors in various multiple-line configurations. Bond wires are used to interconnect substrates and to connect substrates to their package leads.

In addition to designing for the proper impedance or inductance it is important to make use of sizes that are standard and use a gauge and insulation sufficient for the expected power dissipation. Power dissipation numbers are approximate—they depend heavily on the physical situation (*i.e.*, length, airflow, presence of other heated conductors). The following tables provide a starting point for making these design trade-offs.

## 11.1  FUSING CURRENT

Fusing currents of wire are tabulated in Table 11.2.1. For CAE use the fusing current (long wire in 25°C still air) can be approximated for copper wire with [1]:

$$I_{fusing} = 80.0 \, d^{1.5}$$

and for aluminum wire

$$I_{fusing} = 59.2 \, d^{1.5}$$

REFERENCE

[1]  Giacoletto, L.J., ed., *Electronics Designers' Handbook,* 2nd Edition, McGraw-Hill, New York.

## 11.2  WIRE  GAUGE

Wires are commonly referenced by their gauge.  Table 11.2.1 tabulates the gauges and corresponding wire diameters.  For CAE, the wire gauge is related to the wire diameter by the following approximation [3]:

$$d \cong 0.324883 \, e^{-0.115947 \, g}$$

or

$$g \cong \frac{-\ln \left( \dfrac{d}{0.324883} \right)}{0.115947}$$

where $g$ is the wire gauge and $d$ is the wire diameter in inches.  For zero and negative $g$ the gauges follow the pattern:  $0 = 0$ gauge, $-1 = 00$ gauge, $-2 = 000$ gauge, etc.

| AWG B & S Gauge | Diameter (inches) | Aluminum Wire Fusing Current (A) | Copper Wire Fusing Current (A) |
|---|---|---|---|
| 6 | 0.1620 | 495 | 668 |
| 7 | 0.1443 | 416 | 561 |
| 8 | 0.1285 | 349 | 472 |
| 9 | 0.1144 | 293 | 396 |
| 10 | 0.1019 | 247 | 333 |
| 11 | 0.09074 | 207 | 280 |
| 12 | 0.08081 | 174 | 235 |
| 13 | 0.07196 | 146 | 197 |
| 14 | 0.06408 | 123 | 166 |
| 15 | 0.05707 | 103 | 140 |
| 16 | 0.0582 | 86.8 | 117 |
| 17 | 0.04526 | 72.9 | 98.4 |
| 18 | 0.04030 | 61.4 | 82.9 |
| 19 | 0.03589 | 51.6 | 69.7 |
| 20 | 0.03196 | 43.2 | 58.4 |
| 21 | 0.02846 | | |
| 22 | 0.02535 | 30.5 | 41.2 |
| 23 | 0.02257 | | |
| 24 | 0.02010 | 21.6 | 29.2 |
| 25 | 0.01790 | | |
| 26 | 0.01594 | 15.2 | 20.5 |
| 27 | 0.01420 | | |
| 28 | 0.01264 | 10.7 | 14.4 |
| 29 | 0.01126 | | |
| 30 | 0.0103 | 7.58 | 10.2 |
| 31 | 0.008928 | | |
| 32 | 0.007950 | 5.32 | 7.19 |
| 33 | 0.007080 | | |
| 34 | 0.006305 | 3.79 | 5.12 |
| 35 | 0.005615 | | 4.28 |
| 36 | 0.005000 | 2.68 | 3.62 |
| 37 | 0.004453 | | |
| 38 | 0.003965 | 1.85 | 2.50 |
| 39 | 0.003531 | | |
| 40 | 0.003145 | 1.31 | 1.77 |
| 41 | 0.0028 | | 1.52 |
| 42 | 0.0025 | | 1.28 |
| 43 | 0.0022 | | 1.06 |
| 44 | 0.0020 | | 0.916 |

Table 11.2.1: Wire Gauges and Fusing Currents

REFERENCES

[1]   Giacoletto, L.J., ed., *Electronics Designers' Handbook,* 2nd Edition, McGraw-Hill, New York.

[2]   *Reference Data For Radio Engineers,* Howard W. Sams, Indianapolis, Fifth Edition, 1982.

[3]   Williamsen, Mark, "Calculating Wire Design Factors," *Machine Design,* October 12, 1989, p. 99.

## 11.3 BOND WIRES

| Bond Wire | Material | Fusing Current[1] | Usage | Typical Resisitivity |
|-----------|----------|-------------------|-------|----------------------|
| 1 mil | 99.99% Gold | 0.74 A | thermocompression or thermosonic bonding | 1.17 mΩ/mil |
| 1.3 mil | 99.99% Gold | 0.95 A | thermocompression or thermosonic bonding | .54 mΩ/mil |
| .7 mil | 99.99% Gold | | thermocompression or thermosonic bonding | |
| 1 mil | Al with 1% Si | 0.47 A | ultrasonic bonding | 1.50 mΩ/mil |
| 1.3 mil | Al with 1% Si | 0.66 A | ultrasonic bonding | 1.00 mΩ/mil |
| 5 × 1 mil | 99.99% Gold | 3.50 A | ultrasonic bonding | 0.1 mΩ/mil |

Round bond wires are available in diameters of 0.5 , 0.7, 1.0, 1.25, 1.5, 2, and 3 mils. Ribbon is available in (thickness × width) 0.8 × 1.6, 1.0 × 2.0, 1.5 × 3.0, 2.0 × 4.0, 2.0 × 10.0, 2.0 × 20.0, 2.0 × 30.0, and as small as 0.25 × 1.5 mils.

REFERENCES

[1]    Boulanger, Denis, "Wirebonding to Soft Substrates," *Microwave Journal*, February 1990, pp. 159–164.

[2]    Carlson, Jerry, "From Bonded Wire to Bumpless Ribbon/TAB Fine Pitch Interconnects," *Hybrid Circuit Technology*, April 1990, pp. 26–28.

[3]    Ham, R.E., "Prediction of Bond Wire Temperature Using an Electronic Circuit Analogy," *Hybrid Circuit Technology*, April 1990, pp. 53–54. (Describes an accurate thermal model suitable for predicting failure of bond wires subject to current pulses.)

[4]    Tektronix, "Thick-Film Hybrid Design Guidelines", pp. 3–6.

---

[1]Fusing current is dependent on the wire's length and any heatsink or airflow. These numbers are guidelines.

# Chapter 12

# Special Functions

The functions in this chapter are common in transmission line problems. Often a problem that seems formidable at first look becomes quite easy to solve when a library of the desired routines is available. These functions have been successfully implemented as spreadsheet functions (Microsoft Excel™) as well as C subroutines. Many computers will have some or all of these available as libraries.

There are many ways to implement a given function, each of which has different speed–accuracy trade-offs. Depending on the task, optimization for speed may be required. A useful technique is to curve-fit a simple equation to the more complicated equation over a limited range.

The final topic covered here is the representation of circuits with chain matrices.

## 12.1 STANDARD FUNCTIONS

To build the special functions, standard functions are used that are available on most computers. The C math library includes all of the functions found in the file <math.h>. A typical list includes:

| | | | |
|---|---|---|---|
| acos() | sinh() | log() | frexp() |
| atan2() | tanh() | pow() | log10() |
| cosh() | asinh() | sqrt() | sin() |
| floor() | ceil() | atan() | tan() |
| ldexp() | exp() | cos() | |
| modf() | fmod() | fabs() | |

The meaning of most of the above is obvious. Parameters to trigonometric functions and return values of the inverse trigonometric functions are in radians. The absolute value of doubles is found with fabs(); use abs() for integers. Similarly, fmod() returns the remainder of its operands; use the binary operator % for integers.

The function `ldexp()` returns the mantissa and power of 2 required to represent the passed parameter; `frexp()` is its inverse function. The natural logarithm, ln() is the function `log()`.

## 12.2 SPECIAL FUNCTIONS

### 12.2.1 Hyperbolic Functions

The hyperbolic functions used in transmission line problems are simply related to the exponential function.

$$e^{j\theta} = \cos\theta + j\sin\theta$$

$$e^x = \cosh x + \sinh x$$

$$\sinh x = \frac{e^x - e^{-x}}{2} = \frac{\sin jx}{j}$$

$$\cosh x = \frac{e^x + e^{-x}}{2} = \cos jx$$

$$\tanh x = \frac{e^x - e^{-x}}{e^x + e^{-x}} = \frac{\tan jx}{j}$$

$$\sinh^{-1} x = \ln\left(x + \sqrt{x^2 + 1.0}\right)$$

$$\cosh^{-1} x = \ln\left(x + \sqrt{x^2 - 1.0}\right), \quad x \geq 1.0$$

$$\tanh^{-1} x = \frac{1}{2}\ln\left(\frac{1.0 + x}{1.0 - x}\right), \quad x^2 < 1.0$$

### 12.2.2 Elliptic Integrals

Elliptic integrals occur when evaluating several types of transmission lines, among them helical inductors and coplanar waveguide. We have encountered the elliptic integrals of the first and second kind and the complete integrals of the first and second kind.

The elliptic integral of the first kind is of the form

$$F(k, \phi) = \int_0^\phi \frac{d\Phi}{\sqrt{1.0 - k^2 \sin^2 \Phi}} = \int_0^x \frac{d\xi}{\sqrt{(1.0 - \xi^2)(1.0 - k^2 \xi^2)}}.$$

where

$$x = \sin \phi$$

$$k^2 < 1.0$$

and the elliptic integral of the second kind is of the form

$$E(k, \phi) = \int_0^\phi \sqrt{1.0 - k^2 \sin^2 \Phi} \; d\Phi = \int_0^x \frac{\sqrt{1.0 - k^2 \xi^2} \; d\xi}{\sqrt{1.0 - \xi^2}}$$

where

$$x = \sin \phi$$

$$k^2 < 1.0$$

When the parameter $\phi$ is set to $\pi/2$ we define the complete elliptic integrals of the first [$K(k)$] and second kind [$E(k)$]. The parameter $\phi$ is implicit in these cases.

### 12.2.2.1    Complete Elliptic Integral of the First Kind

The complete elliptic integral of the first kind is encountered both unprimed and primed. The prime is a shorthand form for the parameter $k$.

$$K'(k) = K(k')$$

$$k' = \sqrt{1 - k^2}$$

Miller [6] gives the following recursive equation for use in calculating inductance of helical coils, which is equally applicable for use with other structures using $K(k)$.

$$K(k) = \frac{\pi}{2 \, a_N}$$

where

$$a_n = \frac{a_{n-1} + b_{n-1}}{2}$$

$$b_n = \sqrt{a_{n-1} b_{n-1}}$$

$$c_n = \frac{a_{n-1} - b_{n-1}}{2}$$

and

$$a_0 = 1.0$$

$$b_0 = \sqrt{1.0 - k^2}$$

$$c_0 = k$$

To calculate the complete elliptic integral of the first kind, start the calculation with the initial values $a_0$, $b_0$, and $c_0$, and iterate until $c_N = 0$ to within the desired accuracy.

### 12.2.2.2   Complete Elliptic Integral of the Second Kind

The complete elliptic integral of the second kind, $E(k)$, is also found both primed and unprimed. Again, this is just a shorthand form for the parameters.

$$E'(k) = E(k')$$

$$k' = \sqrt{1 - k^2}$$

As before, a simple recursion is given by Miller [6]

$$E(k) = F(k) \left( 1.0 - 0.5 \sum_{n=0}^{N} 2^N c_n^2 \right)$$

where

$$F(k) = \frac{\pi}{2 a_N}$$

$$a_n = \frac{a_{n-1} + b_{n-1}}{2}$$

$$b_n = \sqrt{a_{n-1} b_{n-1}}$$

$$c_n = \frac{a_{n-1} - b_{n-1}}{2}$$

and

$$a_0 = 1.0$$

$$b_0 = \sqrt{1.0 - k^2}$$

$$c_0 = k$$

Start the calculation with the initial values $a_0$, $b_0$, and $c_0$, and iterate until $c_N = 0$ to within the desired accuracy.

### 12.2.2.3    Ratios of Complete Elliptic Integrals of the First Kind

A form that commonly occurs in transmission line problems is the ratio $K(k)$ / $K(k')$. This ratio has been approximated by Hilberg [4] to very good accuracy using simple functions. For $1.0 \le K(k) / K(k') \le \infty$ and $1.0 / \sqrt{2.0} \le k \le 1.0$:

$$\frac{K(k)}{K(k')} \cong \frac{1.0}{2.0\ \pi}\ \ln\left(2.0\ \frac{\sqrt{1.0 + k}\ + \sqrt[4]{4.0\ k}}{\sqrt{1.0 + k}\ - \sqrt[4]{4.0\ k}}\right)$$

For $0.0 \le K(k) / K(k') \le 1.0$ and $0.0 \le k \le 1.0 / \sqrt{2.0}$

$$\frac{K(k)}{K(k')} \cong \frac{2.0\ \pi}{\ln\left(2.0\ \dfrac{\sqrt{1.0 + k\ '}\ + \sqrt[4]{4.0\ k\ '}}{\sqrt{1.0 + k\ '}\ - \sqrt[4]{4.0\ k'}}\right)}$$

These equations are accurate to within $4 \times 10^{-12}$.

### REFERENCES

[1]    Abramowitz, Milton, and Irene A. Stegun, *Handbook of Mathematical Functions*, Dover Publications, New York, 1965. (If you are going to be using special functions a lot, this is a great all-in-one reference. Includes definitions, graphs, tables, approximation formulas, and more.)

[2]    Fettis, Henry E., and James C. Casling, *Tables of Elliptic Integrals of the First, Second and Third Kind*, Aerospace Research Laboratories, ARL 64-232, December 1964.

[3]    Hart, John F., *et al.*, *Computer Approximations*, John Wiley & Sons, New York, 1968.

[4]    Hilberg, Wolfgang, "From Approximations to Exact Relations for Characteristic Impedances," *IEEE Transactions on Microwave Theory and Techniques*, Vol. MTT-17, No. 5, May 1969, pp. 259–265.

[5]     Luke, Yudell L., *The Special Functions and their Approximations,* Vol. II, Academic Press, NY, 1969. (Has a large number of tables of coefficients of Chebyschev polynomials for approximating functions ranging from $1/x$ to the special functions, and their integrals.)

[6]     Miller, H. Craig, "Inductance Formula for a Single-Layer Circular Coil," *Proceedings of the IEEE*, Vol. 75, No. 2, February 1987, pp. 256–257.

[7]     Press, William H., *et al.*, *Numerical Recipes in C: The Art of Scientific Computing*, Cambridge University Press, New York, 1988. (This is the reference book for software which must solve any type of numerical problem. It is also available in BASIC, and FORTRAN versions.)

### 12.2.3 Bessel Functions

Bessel functions are solutions of the equation

$$x^2 \frac{d^2y}{dx^2} + x \frac{dy}{dx} + (x^2 - v^2)y = 0 \tag{12.2.3.1}$$

There are several types of solutions:

$$J_v(x) = \text{Bessel function of the first kind of order } v \tag{12.2.3.2}$$

$$Y_v(x) = \text{Bessel function of the second kind of order } v \tag{12.2.3.3}$$

$$H_v(x) = \text{Bessel function of the third kind of order } v \tag{12.2.3.4}$$

The modified Bessel functions are solutions of

$$x^2 \frac{d^2y}{dx^2} + x \frac{dy}{dx} - (x^2 + v^2)y = 0 \tag{12.2.3.5}$$

Again there are several solutions:

$$I_v(x) = \text{modified Bessel function of the first kind of order } v \tag{12.2.3.6}$$

$$K_v(x) = \text{modified Bessel function of the second kind of order } v \tag{12.2.3.7}$$

Three types of problems occur in transmission line use of Bessel functions:
- value of Bessel function is required
- zeroes of the Bessel function are required
- zeroes of combinations of Bessel functions

Some useful relations among the Bessel functions are:

$$J_{v+1}(x)\, Y_v(x) - J_v(x)\, Y_{v+1}(x) = \frac{2.0}{\pi x} \qquad (12.2.3.8)$$

$$J_{v+1}(x) = \frac{2.0\, n}{x}\, J_v(x) - J_{v-1}(x) \qquad (12.2.3.9)$$

$$Y_{v+1}(x) = \frac{2.0\, n}{x}\, Y_v(x) - Y_{v-1}(x) \qquad (12.2.3.10)$$

### 12.2.3.1 Values of Bessel Functions

A number of numerical techniques are found in [1, 5] for these functions. Vaughan [6] has given a simple technique of calculating four of the Bessel functions for integer orders. C subroutine listings are given below (the original source is BASIC):

```
#define CONST1    0.57215665
#define CONST2    0.63661978
...
double
bessj(n, x)
int        n;
double     x;
{
        double     J1 = 1.0;
        double     J0 = 1.0;
        double     Q0 = 0.0;
        double     Q1 = 1.0;
        double     Q2 = 1.0;
        double     Q3 = 1.0;
        double     Q6 = 0.0;
        double     Q7;
        double     array[6];
        int        j;

        for(j = 0; j < 6; j++)
                array[j] = 0.0;

        Q7 = x * x / 4.0;

        /* Calculate J0 and J1 */
        do{
                Q6++;
                Q1 = -Q1;
                Q2 *= Q7 / (Q6 * Q6);
                Q3 *= Q7 / (Q6 * (Q6 + 1));

                J0 += Q1 * Q2;
                J1 += Q1 * Q3;
        }while(Q2 > 1.0e-05);
```

```
        J1 *= x/2.0;

        /* Do the rest of the array */
        array[0] = J0;
        array[1] = J1;
        for(j = 2; j <= n; j++){
                array[j] = ((2.0 * (j - 1)) / x) * array[j - 1]
                        - array[j - 2];
        }
        return array[n];
}

double
bessy(n, x)
int        n;
double     x;
{
        double     Y0 = 0.0;
        double     Y1 = 1.0;
        double     J1 = 1.0;
        double     J0 = 1.0;
        double     Q0 = 0.0;
        double     Q1 = 1.0;
        double     Q2 = 1.0;
        double     Q3 = 1.0;
        double     Q4, Q5;
        double     Q6 = 0.0;
        double     Q7;
        double     array[6];
        int        j;

        for(j = 0; j < 6; j++)
                array[j] = 0.0;

        Q7 = x * x / 4.0;

        /* Calculate Y0 and Y1 */
        do{
                Q6++;
                Q0 += 1.0 / Q6;
                Q1 = -Q1;
                Q2 = (Q2 * Q7) / (Q6 * Q6);
                Q3 = (Q3 * Q7) / (Q6 * (Q6 + 1.0));
                Q4 = Q3 * ((2.0 * Q0 + 1.0) / (Q6 + 1.0));

                Y0 += Q1 * Q2 * Q0;
                Y1 += Q1 * Q4;
                J0 += Q1 * Q2;
                J1 += Q1 * Q3;

        }while(Q2 > 1.0e-05);

        J1 *= x/2.0;
        Q5 = 0.57215665 + log(x/2.0);
        Y0 = 0.63661978 * (J0 * Q5 - Y0);
        Y1 = 0.63661978 * (J1 * Q5 - (1.0 / x) - (Y1 * x / 4.0));

        /* Do the rest of the array */
```

```
        array[0] = Y0;
        array[1] = Y1;
        for(j = 2; j <= n; j++){

                array[j] = ((2.0 * (j - 1)) / x) * array[j - 1] -
array[j - 2];

        }
        return array[n];
}


double
bessi(n, x)
int        n;
double     x;
{
        double     I1 = 1.0;
        double     I0 = 1.0;
        double     Q0 = 0.0;
        double     Q1 = 1.0;
        double     Q2 = 1.0;
        double     Q3 = 1.0;
        double     Q6 = 0.0;
        double     Q7;
        double     array[6];
        int        j;

        for(j = 0; j < 6; j++)
                array[j] = 0.0;

        Q7 = x * x / 4.0;

        /* Calculate I0 and I1 */
        do{
                Q6++;
                Q2 *= Q7 / (Q6 * Q6);
                Q3 *= Q7 / (Q6 * (Q6 + 1));

                I0 += Q2;
                I1 += Q3;
        }while(Q2 > 1.0e-05);

        I1 *= x/2.0;

        /* Do the rest of the array */
        array[0] = I0;
        array[1] = I1;
        for(j = 2; j <= n; j++){

                array[j] = ((-2.0 * (j - 1)) / x) * array[j - 1]
                        + array[j - 2];

        }
        return array[n];
}

double
bessk(n, x)
```

```
int             n;
double          x;
{
        double      K0 = 0.0;
        double      K1 = 1.0;
        double      I1 = 1.0;
        double      I0 = 1.0;
        double      Q0 = 0.0;
        double      Q1 = 1.0;
        double      Q2 = 1.0;
        double      Q3 = 1.0;
        double      Q4, Q5;
        double      Q6 = 0.0;
        double      Q7;
        int         j;
        double      array[6];

        for(j = 0; j < 6; j++)
                array[j] = 0.0;

        Q7 = x * x / 4.0;

        /* Calculate Y0 and Y1 */
        do{
                Q6++;
                Q0 += 1 / Q6;
                Q2 *= Q7 / (Q6 * Q6);
                Q3 *= Q7 / (Q6 * (Q6 + 1));
                Q4 = Q3 * ((2 * Q0 + 1) / (Q6 + 1));

                K0 += Q2 * Q0;
                K1 += Q4;
                I0 += Q2;
                I1 += Q3;

        }while(Q2 > 1.0e-05);

        I1 *= x/2.0;
        Q5 = CONST1 + log(x/2.0);
        K0 -= I0 * Q5;
        K1 = I1 * Q5 + (1.0 / x) - (K1 * x / 4.0);

        /* Do the rest of the array */
        array[0] = K0;
        array[1] = K1;
        for(j = 2; j <= n; j++){

                array[j] = ((2.0 * (j - 1)) / x) * array[j - 1]
                        + array[j - 2];

        }
        return array[n];
}
```

**Figure 12.2.3.1.1:**        **Bessel Function Source Code**

The equations are valid for

$0.0 < x \le 10.0$

$0 \le v \le 5$ (integers)

Accuracy is to five places or five significant digits except for small $x$ and large $v$ (4, 5).

### 12.2.3.2 Zeroes of Bessel Functions

The $s$th real zero of Bessel functions can be found by a root-finding program in conjunction with the relations of the previous section or by polynomial expansions that are available. The first zeroes of the Bessel functions of the first and second kind are given approximately by [1]:

$$j_{v,1} \cong v + 1.8557571\ v^{1/3} + 1.033150\ v^{-1/3} - 0.00397\ v^{-1} - 0.0908\ v^{-5/3}$$

$$+ 0.043\ v^{-7/3} \tag{12.2.3.2.1}$$

$$y_{v,1} \cong v + 0.9315768\ v^{1/3} + 0.260351\ v^{-1/3} - 0.01198\ v^{-1} - 0.0060\ v^{-5/3}$$

$$+ 0.001\ v^{-7/3} \tag{12.2.3.2.2}$$

### 12.2.3.3 Zeroes of Combination of Bessel Functions

Another case that occurs in waveguide problems is the solution of

$$J_v(x)\ Y_v(k\,x) - J_v(k\,x)\ Y_v(x) = 0 \tag{12.2.3.3.1}$$

which is also found rearranged as

$$\frac{J_v(x)}{J_v(k\,x)} = \frac{Y_v(x)}{Y_v(k\,x)}$$

Gunston [4] and Cochran [3] have found simple relations for the first zeroes of this equation.

$$z_{v,s} \cong \sqrt{\frac{(s\,\pi)^2}{(k-1.0)^2} + \frac{4.0\ v^2 - 1.0}{(k+1.0)^2}} \tag{12.2.3.3.2}$$

A similar problem is

$$J_v'(x)\ Y_v'(k\,x) - J_v'(k\,x)\ Y_v'(x) = 0 \tag{12.2.3.3.3}$$

again, this can be rearranged to the equivalent form

$$\frac{J_v'(x)}{J_v'(k\,x)} = \frac{Y_v'(x)}{Y_v'(k\,x)}$$

The primed quantities are derivatives with respect to $x$. The solution of the equation is found with

$$z'_{v,s} \cong \sqrt{\frac{(s\,\pi)^2}{(k-1.0)^2} + \frac{4.0\,v^2 + 3.0}{(k+1.0)^2}} \qquad (12.2.3.3.4)$$

$$z'_{v,0} \cong \frac{2.0\,v}{(k+1.0)}\left[\frac{1.0 + (k-1.0)^2}{6.0\,(k+1.0)^2}\right] \qquad (12.2.3.3.5)$$

where the $z$s are the $s$th zeroes of the corresponding combination of Bessel function equations. Accuracy of the above is dependent on the combination of parameters. See [3, 4] for more information. In general accuracy is better than a few percent and improves with increasing $s$ and $v$ and decreasing $k$. See [1, p. 374] for another solution of this problem.

### 12.2.3.4 Kelvin Functions

The Kelvin functions are the real and imaginary parts of Bessel functions with imaginary arguments.

$$\text{ber}_v x + j\,\text{bei}_v x = J_v(x\,e^{3\pi j/4}) \qquad (12.2.3.4.1)$$

$$\text{ker}_v x + j\,\text{kei}_v x = e^{-v\,\pi j/2}\,K_v(x\,e^{\pi\,j/4}) \qquad (12.2.3.4.2)$$

which are defined for

$v = $ real

$x = $ real, $\geq 0$

$n = $ integer, $\geq 0$

See the references for more information on calculating these functions.

REFERENCES

[1]   Abramowitz, Milton, and Irene A. Stegun, *Handbook of Mathematical Functions,* Dover Publications, New York, 1965. (If you are going to be using special functions a lot, this is a great all-in-one reference. Includes definitions, graphs, and tables, approximation formulas, and more.)

[2] Balanis, Constantine A., *Advanced Engineering Electromagnetics,* John Wiley & Sons, New York, 1989.

[3] Cochran, J.A., "Further Formulas for Calculating Approximate Values of the Zeros of Certain Combinations of Bessel Functions," *IEEE Transactions on Microwave Theory and Techniques,* Vol. MTT-11, No. 11, November 1963, pp. 546–547. (Extends Gunston, and corrects errors in accuracy.)

[4] Gunston, M.A.R., "A Simple Formula for Calculating Approximate Values of the First Zeros of a Combination Bessel Function Equation," *IEEE Transactions on Microwave Theory and Techniques,* Vol. MTT-11, No. 1, January 1963, pp. 93–94.

[5] Press, William H., *et al.*, *Numerical Recipes in C: The Art of Scientific Computing,* Cambridge University Press, New York, 1988. (This is the reference book for software that must solve any type of numerical problem. It is also available in BASIC, and FORTRAN versions.)

[6] Vaughan, J. Rodney M., "Routine to Supply Bessel Functions as Required in BASIC Programs," *IEEE Transactions on Microwave Theory and Techniques,* Vol. MTT-18, No. 8, August 1970, pp. 511–512.

### 12.2.4 Sine-Integral Function

The sine integral function was encountered in coaxial discontinuity relations.

$$Si(x) = \int_0^x \frac{\sin y}{y} \, dy \tag{12.2.4.1}$$

An approximation formula [1] for $0.0 \le x \le 1.0$ is

$$Si(x) \cong x - \frac{x^3}{18.0} + \frac{x^5}{600.0} - \frac{x^7}{35,280.0} + \frac{x^9}{3,265,920.0} \tag{12.2.4.2}$$

and for $1.0 \le x \le \infty$

$$Si(x) \cong \frac{\pi}{2.0} - f(x)\cos(x) - g(x)\sin(x) \tag{12.2.4.3}$$

where

$$f(x) = \frac{1.0}{x} \frac{x^8 + 38.027264\,x^6 + 265.187033\,x^4 + 335.677320\,x^2 + 38.102495}{x^8 + 40.021433\,x^6 + 322.624911\,x^4 + 570.236280\,x^2 + 157.105423} \tag{12.2.4.4}$$

$$g(x) = \frac{1.0}{x^2} \frac{x^8 + 42.2242855\, x^6 + 302.757865\, x^4 + 352.018498\, x^2 + 21.821899}{x^8 + 48.196927\, x^6 + 482.485984\, x^4 + 1114.978885\, x^2 + 449.690326}$$

$$(12.2.4.5)$$

These equations give results that are accurate to better than 1 ppm.

REFERENCES:

[1]   Abramowitz, Milton, and Irene A. Stegun, *Handbook of Mathematical Functions,* Dover Publications, New York, 1965. (If you are going to be using special functions a lot, this is a great all-in-one reference. Includes definitions, graphs, tables, approximation formulas, and more.)

### 12.2.5   Gamma Function

The gamma function is used alone and as part of other functions.  The definition is

$$\Gamma(z) = \int_0^\infty t^{z - 1.0}\, e^{-t}\, dt \quad (12.2.5.1)$$

For integer $z$, $\Gamma(z)$ equals $(z - 1)!$.  An approximation formula is

$$\log \Gamma(x) = (x - 1) \log x - x + \frac{1}{2} \log (2\pi) + \frac{1}{2} \log \left(x + \frac{1}{6}\right) + E(x) \qquad (12.2.5.2)$$

where for large $x$

$$E(x) = \left(\frac{1}{12\,x + \frac{46}{15}}\right)^{\!2} \qquad (12.2.5.3)$$

This approximation to $\Gamma(x)$ is accurate to within approximately 2% for all $x$.

REFERENCES

[1]   Lewin, L., "An Accurate Formula for the Gamma Function," *IEEE Transactions on Microwave Theory and Techniques*, Vol. MTT-22, No. 10, October 1974, p. 910. (Corrects the usual approximation formula to improve accuracy near $x = 0$.)

### 12.2.6   Hencken Impedance Function

A function used in calculating optimal impedance tapers is the Hencken impedance function defined as

$$G(B,\xi) = \frac{B}{\sinh B} \int_0^\xi I_0\left(B\sqrt{1.0 - \xi^2}\right) d\xi \qquad (12.2.6.1)$$

where $I_0\{z\}$ is the zeroth-order modified Bessel function of the first kind whose calculation is described earlier. This function can be calculated using the recursion of [1]:

$$G(B,\xi) = \frac{B}{\sinh B} \sum_{k=0}^\infty a_k b_k \qquad (12.2.6.2)$$

where

$$a_0 = 1.0 \qquad (12.2.6.3)$$

$$a_k = \left(\frac{B^2}{4.0\,k^2}\right) a_{k-1}, \quad k = 1, 2, \ldots \qquad (12.2.6.4)$$

$$b_0 = \xi \qquad (12.2.6.5)$$

$$b_k = \frac{\xi(1.0 - \xi^2)^k + 2\,k\,b_{k-1}}{2.0\,k + 1.0} \qquad (12.2.6.6)$$

REFERENCES

[1]    Cloete, J.H., "Computation of the Hecken Impedance Function," *IEEE Transactions on Microwave Theory and Techniques*, Vol. MTT-25, No. 5, May 1977, p. 440.

## 12.3 TWO-PORT REPRESENTATIONS

RF and microwave circuits can be represented by a number of different matrices that describe the two-port network. The most commonly used in measurements is the *s*-parameter or scattering matrix. For CAE calculations it is convenient to have a matrix where the product of the component matrices produces the response of a circuit. These matrices are the ABCD and the T matrix. Matrix operations also allow us to calculate the result of two parallelled two-ports. Conversion among these parameters and low frequency parameters are also possible.

Some circuits have three or more ports. For example, the directional coupler is a four-port circuit.

### 12.3.1 Scattering or *s*-Parameters

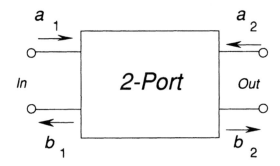

**Figure 12.3.1.1:** *s*-Parameter Definition

The *s*-parameters are defined using a two-port model as shown in Figure 12.3.1.1. The equations are

$$b_1 = s_{11} a_1 + s_{12} a_2 \qquad (12.3.1.1)$$

$$b_2 = s_{21} a_1 + s_{22} a_2 \qquad (12.3.1.2)$$

where[1]

$$s_{11} = \frac{b_1}{a_1} \bigg|_{a_2 = 0} \qquad (12.3.1.3)$$

---

[1]Note that $a_n = 0$ is equivalent to matching port $n$.

$$s_{21} = \left. \frac{b_2}{a_1} \right|_{a_2 = 0} \tag{12.3.1.4}$$

$$s_{22} = \left. \frac{b_2}{a_2} \right|_{a_1 = 0} \tag{12.3.1.5}$$

$$s_{12} = \left. \frac{b_1}{a_2} \right|_{a_1 = 0} \tag{12.3.1.6}$$

and

$a_n$ = incident wave

$b_n$ = reflected wave

Conversions to $T$-parameters and $s$-parameters are given below.

## 12.3.2 Transmission or $T$-Parameters

The $T$-parameters are defined with the same figure as the $s$-parameters.

$$a_1 = T_{11} b_2 + T_{12} a_2 \tag{12.3.2.1}$$

$$b_1 = T_{21} b_2 + T_{22} a_2 \tag{12.3.2.2}$$

These parameters can be derived in terms of the $s$-parameters and have the added advantage that two-port networks described by $T$-parameters can be chained by simply multiplying the matrices.

To transfer to and from $s$-parameters, the relations are

$$\begin{bmatrix} T_{11} & T_{12} \\ T_{21} & T_{22} \end{bmatrix} = \begin{bmatrix} \dfrac{1.0}{s_{21}} & \dfrac{-s_{22}}{s_{21}} \\[2ex] \dfrac{s_{11}}{s_{21}} & s_{12} - \dfrac{s_{11} s_{22}}{s_{21}} \end{bmatrix} \tag{12.3.2.3}$$

$$\begin{bmatrix} s_{11} & s_{12} \\ s_{21} & s_{22} \end{bmatrix} = \begin{bmatrix} \dfrac{T_{21}}{T_{11}} & T_{22} - \dfrac{T_{21} T_{12}}{T_{11}} \\[2ex] \dfrac{1.0}{T_{11}} & \dfrac{-T_{12}}{T_{11}} \end{bmatrix} \tag{12.3.2.4}$$

### 12.3.3 ABCD Matrices

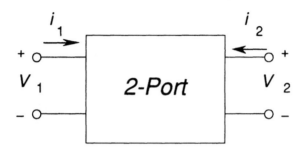

**Figure 12.3.3.1:   ABCD Parameter Definitions**

The equations describing the *ABCD* matrix are:

$$v_1 = A\, v_2 - B\, i_2 \qquad\qquad (12.3.3.1)$$

$$i_1 = C\, v_2 - D\, i_2 \qquad\qquad (12.3.3.2)$$

The relationship to the *s*-parameters is not as straightforward as for *T*-parameters, but the *ABCD* matrix shares the ability to chain matrices to build up a circuit. To convert between *s*-parameters and *ABCD* parameters

$$\begin{bmatrix} A & B \\ C & D \end{bmatrix} = \begin{bmatrix} \dfrac{(1 + s_{11})(1 - s_{22}) + s_{12}\,s_{21}}{2 s_{21}} & \dfrac{(1 + s_{11})(1 + s_{22}) + s_{12}\,s_{21}}{2\,s_{21}} \\[3ex] \dfrac{(1 - s_{11})(1 - s_{22}) + s_{12}\,s_{21}}{2\,s_{21}} & \dfrac{(1 - s_{11})(1 + s_{22}) + s_{12}\,s_{21}}{2\,s_{21}} \end{bmatrix}$$

$$(12.3.3.3)$$

$$\begin{bmatrix} s_{11} & s_{12} \\ s_{21} & s_{22} \end{bmatrix} = \begin{bmatrix} \dfrac{A + B - C - D}{A + B + C + D} & \dfrac{2.0\,(A\,D - B\,C)}{A + B + C + D} \\[3ex] \dfrac{2.0}{A + B + C + D} & \dfrac{-A + B - C + D}{A + B + C + D} \end{bmatrix}$$

$$(12.3.3.4)$$

REFERENCES

[1]   Choma, Jr., J., "Signal Flow Analysis of Feedback Networks," *IEEE Transactions on Circuits and Systems*, Vol. CS-37, No. 4, April 1990, pp. 455–463.

[2]     Gonzalez, Guillermo, *Microwave Transistor Amplifers: Analysis and Design,* Prentice-Hall, NJ, 1984. (A very succinct and thorough book with good coverage of *s*-parameter design applied to amplifiers and oscillators.)

[3]     Liao, Samuel Y., *Microwave Circuit Analysis and Amplifier Design,* Prentice-Hall, NJ, 1987.

[4]     Teyssier, Edward M., "Two-Port to Three-Port S-Parameter Conversion," *MSN&CT,* September 1987, pp. 84–91.

[5]     Young, Leo, and H. Sobol, eds., *Advances in Microwaves,* Volume 8, "Computer Aided Design, Simulation and Optimization," by B.S. Perlman and V.G. Gelnovatch, Academic Press, New York, 1974.

### 12.3.4 Series Combinations of Two-Port Networks

To analyze a circuit made up of transmission lines and lumped and active elements, we multiply the individual *ABCD* or *T*-parameter matrices. The product matrix then represents the matrix of the combination. Any arbitrary set of *s*-parameters can be converted with the above equations into the required chain matrix form. The multiplication of two $2 \times 2$ matrices is

$$[\mathbf{A}]\,[\mathbf{B}] = \begin{bmatrix} a_{00} & a_{01} \\ a_{10} & a_{11} \end{bmatrix} \begin{bmatrix} b_{00} & b_{01} \\ b_{10} & b_{11} \end{bmatrix}$$

$$= \begin{bmatrix} a_{00}\,b_{00} + a_{01}\,b_{10} & a_{00}\,b_{01} + a_{01}\,b_{11} \\ a_{10}\,b_{00} + a_{11}\,b_{10} & a_{10}\,b_{01} + a_{11}\,b_{11} \end{bmatrix} \qquad (12.3.4.1)$$

The chain matrix can then be converted to *s*-parameters, impedance, gain, etc., as required.

### 12.3.5 Parallel Combinations of Two-Port Networks

In many cases it is desireable to connect two-port networks in other than simple cascades. To do this we can transform the desired two-ports using the following *s*-parameter matrix relations [1, 2].

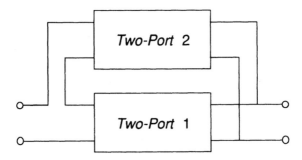

**Figure 12.3.5.1:** Series-Parallel Connection of Two-Port Networks

$$S_{sp} = \begin{bmatrix} -1.0 & 0.0 \\ 0.0 & 1.0 \end{bmatrix} \times f\!\left( \begin{bmatrix} -1.0 & 0.0 \\ 0.0 & 1.0 \end{bmatrix} \times S_1, \begin{bmatrix} -1.0 & 0.0 \\ 0.0 & 1.0 \end{bmatrix} \times S_2 \right)$$

$$(12.3.5.1)$$

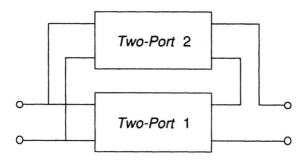

**Figure 15.3.5.2:** **Parallel-Series Connection of Two-Port Networks**

$$S_{ps} = \begin{bmatrix} 1.0 & 0.0 \\ 0.0 & -1.0 \end{bmatrix} \times f\left( \begin{bmatrix} 1.0 & 0.0 \\ 0.0 & -1.0 \end{bmatrix} \times S_1, \begin{bmatrix} 1.0 & 0.0 \\ 0.0 & -1.0 \end{bmatrix} \times S_2 \right)$$

(12.3.5.2)

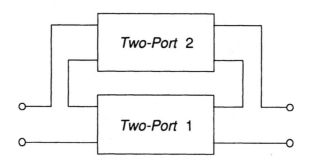

**Figure 15.3.5.3:** **Series-Series Connection of Two-Ports**

$$S_{ss} = \begin{bmatrix} -1.0 & 0.0 \\ 0.0 & -1.0 \end{bmatrix} \times f\left( \begin{bmatrix} -1.0 & 0.0 \\ 0.0 & -1.0 \end{bmatrix} \times S_1, \begin{bmatrix} -1.0 & 0.0 \\ 0.0 & -1.0 \end{bmatrix} \times S_2 \right)$$

(12.3.5.3)

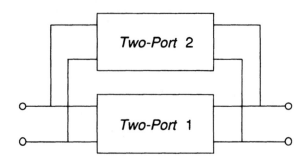

**Figure   15.3.5.4.:**   **Parallel-Parallel   Connection   of   Two-Port**
**Networks**

$$S_{pp} = \begin{bmatrix} 1.0 & 0.0 \\ 0.0 & 1.0 \end{bmatrix} \times f \left( \begin{bmatrix} 1.0 & 0.0 \\ 0.0 & 1.0 \end{bmatrix} \times S_1, \begin{bmatrix} 1.0 & 0.0 \\ 0.0 & 1.0 \end{bmatrix} \times S_2 \right)$$

$$(12.3.5.4)$$

In the above relations

$$A = 3 \times I \qquad\qquad (12.3.5.5)$$

$$B = S_1 - I \qquad\qquad (12.3.5.6)$$

$$C = S_1 + I \qquad\qquad (12.3.5.7)$$

$$f(S_1, S_2) = A^{-1} \times \left[ B + 4.0 \times C \times S_2 \times (A - B \times S_2)^{-1} \times C \right]$$

$$(12.3.5.8)$$

$$I = \begin{bmatrix} 1.0 & 0.0 \\ 0.0 & 1.0 \end{bmatrix}, \text{ the identity matrix} \qquad (12.3.5.9)$$

$$S_1 = \text{the } s\text{-parameter matrix of two-port 1} \qquad (12.3.5.10)$$

$$S_2 = \text{the } s\text{-parameter matrix of two-port 2} \qquad (12.3.5.11)$$

REFERENCES

[1]     Besser, Les, "A Fast Computer Routine to Design High Frequency Circuits,"
        *IEEE International Conference on Communications,* San Francisco, June 8–10,
        1970, pp. 24-13–24-16.

[2]     Besser, Les, and R.W. Newcomb, "A Scattering Matrix Program for High
        Frequency Circuit Analysis," *Proceedings of the IEEE Conference on Systems,
        Networks, and Computers,* Mexico, January 1971, pp. 512–515.

[3]     Bodharamik, Phichani, *et al.,* "Two Scattering Matrix Programs for Active
        Circuit Analysis," *IEEE Transactions on Circuit Theory,* Vol. CT-18, No. 6,
        November 1971, pp. 610–619.

[4]     Vendelin, George D., *et al., Microwave Circuit Design Using Linear and
        Nonlinear Techniques,* John Wiley & Sons, New York, 1990.

# *Glossary*

**ANISOTROPIC.** A material having parameters ($\sigma$, $\varepsilon_r$, etc.) that vary with direction in the material. For example, woven materials have one $\varepsilon_r$ with the grain and another perpendicular to it.

**B STAGE.** Partially cured dielectric, used to bond the layers together

**COMMENSURATE.** Equal length.

**CONFORMAL MAPPING.** Conformal mapping is a technique used for analytically solving EM field problems that are otherwise too complex. The technique maps the physical structure into another domain via a transforming equation that preserves the structure's capacitance and $Z_0$. The transformed equations are then easily solvable. This is analogous to a Mercatur projection, which transforms the surface of the earth (essentially a sphere) onto a map (two-dimensional).

**C STAGE.** Fully cured epoxy; C stage dielectric usually has the foil already attached to the dielectric.

$\varepsilon_{effective}$. The effective relative dielectric constant taking into account that a portion of the wave is propagating in air ($\varepsilon_r = 1$) and a portion in the dielectric ($\varepsilon_r$).

**FEA OR FINITE ELEMENT ANALYSIS.** A technique used to solve Maxwell's equations that divides the transmission line space into a grid of small elements. A simplified version of Maxwell's equations is then applied iteratively to these elements until the solution is achieved.

**FRINGING CAPACITANCE.** The capacitance of a line's edges; the increased capacitance beyond the ideal parallel-plate capacitance; capacitance due to edge fields that do not reach from one edge to the other.

**GENERALIZED SCATTERING MATRIX TECHNIQUE.** A technique of calculating discontinuities by combining the $s$-parameters of the simple discontinuities making up the discontinuity. The simple discontinuity *s*-parameters are calculated with mode-matching or other techniques.

**GREEN'S FUNCTIONS.**   A technique used to solve EM problems. In this technique, the driving function is a Dirac delta. The response to the delta is the Green's function. The actual driving function is then written as the sum or integral of offset Dirac deltas. The Green's function can then be used to solve for the response to the actual driving function.

**MAGNETIC.** A material is considered magnetic when $\mu_r$ is greater than one.

**MAXWELL'S EQUATIONS.**
The physical laws that describe the relationships between electric and magnetic fields. They can be expressed in a set of differential or integral equations.

**NEWTON'S METHOD.**   A method of finding roots of an equation which evaluates the local slope of the function and uses this to estimate the next guess at the root. Successive iterations quickly home in on the root's value.

**NEWTON-RAPHSON METHOD.**
See *Newton's Method* above.

**NONHOMOGENEOUS**   A dielectric or other material is non-homogeneous when its physical parameters vary with position in the material. For example, microstrip line has dielectric below the center conductor and air above—this is non-homogenous.

**SPECTRAL DOMAIN METHOD.**   A technique used for EM analysis of structures with thin conductors. The Fourier transform is applied converting the space equations to the spectral domain.

**SUPERCONDUCTOR.**   A material is a superconductor when $\sigma = \infty$.

**TELEGRAPHER'S EQUATIONS.**   A set of differential equations that describes a one-dimensional transmission line.

**TLM OR TRANSMISSION LINE MATRIX METHOD.**
The space of a transmission line structure is divided into a mesh of transmission lines excited with a Dirac pulse. An equivalence between this and the solution to Maxwell's equations is established.

**TRANSVERSE RESONANCE TECHNIQUE.**   The line or discontinuity is modeled as a resonant waveguide structure. The resonance is used to calculate

the propagation constants of the
structure.

**QUASI-STATIC.**    Quasi-static
analyses are carried out with
simplifications that are only
strictly valid at dc; however, the
ac (time-varying) solution is
found to be similar.

**WAVEGUIDE MODEL.**    The actual
structure is mapped into an
equivalent waveguide structure.
This waveguide is then analyzed
with existing techniques.  Finally
this is converted back to the
original structure, which is the
desired solution.

$W_{EFFECTIVE}$.  The effective trace width
taking into account nonzero trace
thickness fringing fields;  the
width of a parallel-plate
waveguide having the same
thickness dielectric, impedance,
and propagation velocity.

$Z_0$.  The characteristic impedance.

# Appendix A
# List of Symbols, Acronyms, and Abbreviations

| | | | |
|---|---|---|---|
| $\alpha$ | attenuation constant | $\delta$ | skin depth |
| $\alpha_c$ | attenuation due to conductor losses | $e$ | base of natural logarithm, 2.718281828... |
| $\alpha_d$ | attenuation due to dielectric losses | $E(k)$ | complete elliptic integral of the second |
| $\alpha_r$ | attenuation due to radiative losses | $E(k,\phi)$ | elliptic integral of the second kind |
| $B$ | susceptance | $\varepsilon$ | $\varepsilon = \varepsilon_0\,\varepsilon_r$, permittivity |
| $\beta$ | phase constant (rad / m) | $\varepsilon_0$ | permittivity of free space, $8.854183 \times 10^{-12}$ F / m |
| $b$ | dielectric thickness, used with stripline | $\varepsilon_{eff}$ | relative effective permittivity |
| $\mathrm{bei}_\nu x$ | Kelvin function | $\varepsilon_r$ | relative permittivity, dielectric constant |
| $\mathrm{ber}_\nu x$ | Kelvin function | | |
| $C$ | capacitance | F | farad |
| $C'$ | capacitance per length, incremental capacitance | $f$ | frequency |
| | | $F(k,\phi)$ | elliptic integral of the first kind |
| $c, c_0$ | speed of light in a vacuum, $2.99792456 \times 10^8$ m / s | $g$ | gap width |
| | | $G$ | conductance, S |
| CPW | coplanar waveguide | $G'$ | incremental conductance, conductance per unit length |
| CPWG | coplanar waveguide with ground plane | | |
| | | $G(B,\xi)$ | Hencken function |
| $\mathrm{Ct}(x, y)$ | large radial cotangent function | $\Gamma$ | reflection coefficient |
| $D$ | outer (or larger) diameter | $\Gamma(z)$ | gamma function |
| $d$ | inner (or smaller) diameter | $\gamma$ | propagation constant |

| | | | |
|---|---|---|---|
| H | henry | log | logarithm base 10, also known as common logarithm |
| $H_v(x)$ | Bessel function of the third kind of order $v$, Hankel function | m | meter |
| $h$ | height, the dielectric thickness | $\theta$ | electrical length |
| $\eta$ | wave impedance, $\Omega$ | $\mu$ | permeability |
| $\eta_0$ | characteristic impedance of free space, $120\,\pi\,\Omega$ | $\mu_0$ | permeability of free space, $4\,\pi \times 10^{-7}\,H/m$ |
| $j$ | $\sqrt{-1}$ | $\mu_r$ | relative permeability |
| $j_{v,s}$ | the $s$th zero of the Bessel function of the first kind of order $v$ | $\pi$ | the ratio of a circle's circumference to its diameter, $3.14159265\ldots$ |
| $j'_{v,s}$ | the $s$th zero of the derivative of the Bessel function of the first kind of order $v$ | $q$ | filling factor |
| | | $\theta$ | electrical length |
| $J_v(x)$ | Bessel function of the first kind of order $v$ | $\rho$ | resistivity |
| $k$ | coupling | $\rho_{Cu}$ | resistivity of copper, $1.72 \times 10^{-8}\,\Omega\text{-m}$ |
| $k$ | modulus of elliptic functions | $R$ | resistance |
| $k'$ | complementary modulus of elliptic functions | $R'$ | incremental resistance, resistance per unit length |
| $K(k)$ | complete elliptic integral of the first kind | $R_s$ | surface resistivity |
| $K(k')$ | complete elliptic integral of the first kind and complementary modulus | $s$ | seconds |
| | | $s$ | spacing |
| $\text{kei}_v x$ | Kelvin function | $Si(x)$ | sine integral function |
| $\text{ker}_v x$ | Kelvin function | $\sigma$ | conductivity |
| $L$ | inductance | $t$ | thickness |
| $L'$ | inductance per length, incremental inductance | $\tan\delta$ | tan delta, loss tan, loss tangent |
| | | $v_p$ | propagation velocity |
| $l$ | length | $\omega$ | angular frequency, $\omega = 2\,\pi f$ |
| $\lambda$ | wavelength | W | watts |
| $\lambda_g$ | wavelength in the guide | $w$ | width |
| $\lambda_0$ | wavelength in free space | $w_{eff}$ | effective width taking into acount thickness or frequency |
| ln | logarithm base e, also known as natural or Napierian log | $X$ | reactance |
| | | $y_{v,s}$ | the $s$th zero of the Bessel function of the second kind of order $v$ |

| | |
|---|---|
| $y'_{v,s}$ | the $s$th zero of the derivative of the Bessel function of the second kind of order $v$ |
| $Y$ | admittance |
| $Y_v(x)$ | Bessel function of the second kind of order $v$, Weber's function |
| $Y'_v(x)$ | derivative of the Bessel function of the second kind of order $v$, Weber's function |
| $z_{v,s}$ | the $s$th zero of the combined Bessel function equation of order $v$ |
| $z'_{v,s}$ | the $s$th zero of combined Bessel function derivatives of the first kind of order $v$ |
| $Z$ | impedance |
| $Z_0$ | characteristic impedance |
| $Z_{0,e}$ | even mode characteristic impedance |
| $Z_{0,o}$ | odd mode characteristic impedance |
| $\Omega$ | ohm |

# Appendix B

# Common Questions and

# Answers


*Does surface roughness matter?*

At frequencies where the skin depth is of the same order as the surface roughness, the roughness will result in increased losses. Rolled copper (smoother) can be used instead of electroplated copper when this is important.

*How close can a shield come to trace top before affecting $Z_0$?*

Shielding lowers $\varepsilon_{eff}$ and $Z_0$ and creates resonances. Neglect the effects of shield for $W_{wall} > 5 \times w$ and $h_2 > 5 \times h$. Equations for enclosure resonances agree pretty well with experience. When resonances occur, the box must be partitioned or the cavity loaded with a dissipative material.

*What is the coupling between two traces? How far apart should they be spaced for no coupling?*

Unless one conductor is completely shielded, it is impossible to have no ($-\infty$ dB) coupling. To calculate the coupling, use the equations of Chapter 4 for the correct situation. These will give the signal level in the second line relative to the first line (sometimes called the "backward coupling").

*What happens when microstrip line is embedded? ($v_p$, capacitance, $Z_0$)*

The capacitance increases, lowering $v_p$ and $Z_0$. To calculate this effect, see Chapter 3.

*When should one use microstrip, when should one use stripline?*

The advantage of stripline is that it almost completely confines the fields between two ground planes. This reduces coupling to adjacent traces and makes it less sensitive to packaging. On the negative side, the cost is increased and vias must be provided to connect components to the center conductor.

*When does one use coplanar waveguide?*

Coplanar waveguide is useful when components need to be mounted from the transmission path to ground. Plated-through holes to the other side of the PCB are not required. Coplanar waveguide also allows the trace to be narrowed to the width of a device's leads while keeping the impedance constant.

*How does one choose a dielectric thickness for multilayer boards?*

Determine the number of layers from the density. Programs are available to assist with this calculation. Determine the number of signal, ground, and power planes needed. Determine the overall board thickness required for the card cage and connectors.

Now the goal is to create a symmetric layup to minimize warpage. This means that the planes and signal layers are placed symmetrically on either side of the PCB center line. Corresponding planes on opposite sides of the center line should have similar amounts of metallization after etching. The dielectric thicknesses should also be symmetric about the center line. To minimize the effects of tolerances, controlled impedance signal layers should have C stage dielectric layers as thick as possible in a standard size. The other layers can be thinned to accomplish this down to the limitations tabulated in Chapter 9.

*How does one choose a dielectric thickness for two-sided boards?*

The main consideration here will often be the physical requirements of the card cage and connectors. The thicker the dielectric, the less sensitive you will be to tolerances in the manufacturing process. Thinner dielectrics reduce the trace dimensions for a given impedance and may make it easier avoid stray couplings.

*How much effect does the trace thickness (inner layer vs. outer layer) have on the conductor width? ($Z_0$?)*

To answer this, calculations were made using the microstrip line equations of Chapter 3 for 0.0625 in thick G-10 at 1.6 GHz. As shown in the table below, the line width for 50 $\Omega$ changes about 1% when the thickness is changed from 1 oz. to 2 oz. The trace length for an electrical 90° changed about 0.04% due to the change in $\varepsilon_{eff}$. This translates into about 0.03° or 0.1 ps length difference at 1.6 GHz over 1 in of length.

| Trace Thickness (mil) | 50 Ω Width (in) | 90° length (in) |
|---|---|---|
| 1.4 | 0.11984 | 1.01375 |
| 2.8 | 0.11878 | 1.01411 |
| 4.2 | 0.11785 | 1.01445 |
| 5.6 | 0.11699 | 1.01476 |

*What happens at a T or crossing of lines?*

The model is a series inductance in each arm and a shunt capacitance at their junction.

*Why use a dielectric other than G-10 or FR-4?*

The main reasons are loss characteristics, impedance tolerance, frequency range, and feature size. At high frequencies the loss (tan $\delta$) of G-10 will cause the transmission paths to be lossy, which increases insertion losses and lowers the $Q$ achievable in the circuit. PTFE dielectrics tend to have a more uniform dielectric constant and etch characteristics (at a greater cost) as well as lower loss. Dielectrics with $\varepsilon_r$'s greater than that of G-10 will have narrower lines for a given impedance resulting in less radiation, better match to device leads, and more dense circuitry.

*What is the difference between G-10 and FR-4?*

FR-4 is a fire-resistant version of G-10.

*What effect does solder mask have on microstripline impedance?*

Solder mask increases the capacitance and lowers $v_p$. See Chapters 3 and 9 for more information.

*How far from a transmission line does the dielectric need to extend to guarantee $Z_0$?*

Referring to the figure for microstripline with truncated dielectric (Figure 3.5.2.1): $T / w > 1.0$ will result in insignificant changes in the propagation characteristics of the line ($Z_0$, $v_p$, $\alpha$, etc.).

For striplines, ground extending beyond the trace $> 0.8\ h$ (for $w /$ $h = 0.6$) will cause $< 0.2\%$ error in trace capacitance[1]. Figure 7 of the reference indicates that extending the dielectric beyond the trace $> 3.0$ $h$ (for $w / h = 0.6$) will give $< 0.2\%$ error in trace capacitance.

This effect is also frequency dependent. As frequency increases the fields concentrate more and more into the dielectric directly below the signal trace.

### What are even and odd mode impedances?

There is only the characteristic impedance of a line whenever there are only two conductors involved; however, when there are three the signal can propagate in two different ways. The even mode corresponds to two of the conductors being at the same potential relative to the third ("common mode"). The odd mode corresponds to the two conductors having opposite potentials ("differential mode"). Usually one or the other is desired and is driven by the source. Source or load imperfections or discontinuities can cause the other mode to propagate.

### What is the best way to make a constant impedance bend in a trace?

A mitered bend will take up less room than an equivalent VSWR rounded corner bend. Use Chapter 4 to calculate the required miter and the required minimum bend radius.

### What should I do to achieve a very precise $Z_0$? What accuracy can be expected?

For tightest control over the characteristic impedance of a line:

- Use as thin a trace as possible. This minimizes etching tolerances.
- Keep controlled $Z$ traces in the center of the PCB where etching is most uniform.
- Keep controlled $Z$ traces on PCB inner layers. These layers are not plated and this eliminates tolerances caused by the plating and other processes.
- Use as thick a dielectric (and as wide a trace) as possible.
- Specify that you want controlled $Z$ on the PCB fabrication documentation.

---

[1]Figure 6 of Ahmadouche, A., and J. Chilo, "Optimum Computation of Capacitance Coefficients of Multilevel Interconnecting Lines for Advanced Package," *IEEE Transactions on Components, Hybrids, and Manufacturing Technology*, Vol. 12, No. 1, March 1989, pp. 124–129.

- Use a standard C stage dielectric thickness. Calculate line widths using the dielectric constant of the actual thickness dielectric used (applies to woven dielectrics).
- Include the effect of any overlaid dielectric. This includes solder mask and *skim coats*.
- Keep traces separated from each other and any adjacent ground plane by several line widths.
- Follow controlled impedance bend guidelines.
- Use as low a dielectric constant as possible (do not forget air!).
- Add thieving to keep trace density constant over the surface. This allows etching to be even over the board's surface.

*Why 50 Ω? Couldn't I use 60 Ω?*

The value 50 Ω comes from a trade-off in coaxial lines that was made between attenuation (loss) and power-handling capability. The maximum power point is at 30 Ω and the minimum attenuation occurs at 77 Ω. From these came the 75 Ω used for TV cabling and the compromise value of 50 Ω.[1] For maximum transmission of power from a source to a load, the source, the load, and the line should have the same impedance.

*Some references give 4.3 and others give 4.8 as the dielectric constant of FR-4. Which is correct?*

The amount of epoxy in the dielectric affects the dielectric constant. Thinner dielectrics tend to have a greater percentage of epoxy than thicker dielectrics. For thin dielectrics (less than ~30 mils), 4.3 is a good approximation to the permittivity. For thicker dielectrics, 4.8 is a better number. See Chapter 9 for more detailed information.

*How accurate are the MECL handbook equations relative to the Wheeler-Schneider ones?*

As the graph below shows for 0.0625 in G-10 and 0.0014 in thick traces, the MECL Handbook equations are most accurate for impedances near and above 50 Ω:

[1]Laverghetta, Thomas S., *Practical Microwaves*, Howard W. Sams, Indianapolis, IN, 1984.

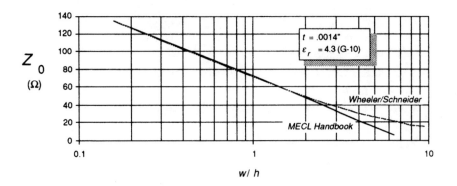

**MECL Handbook Compared to Wheeler-Schneider Equations**

# Appendix C

# VSWR, Reflection

# Coefficient, and Impedance

There are a number of ways to express the impedance of a circuit. The terms *VSWR*, reflection coefficient, and impedance are all interrelated and convey roughly the same information.

The *voltage standing wave ratio* is physically the ratio of the maximum voltage along a line to the minimum. It can actually be measured this way—*e.g.,* when we use a detector probe in the field of a slotted line. These maximum and minimum voltages along the line can also be written in terms of the waves incident on and reflected from the load:

$$VSWR \equiv \frac{E_{max}}{E_{min}} = \frac{E_{inc} + E_{refl}}{E_{inc} - E_{refl}} \tag{C.1}$$

The *reflection coefficient*, $\rho$, is also related to the incident and reflected waves:

$$\Gamma = \rho \equiv \frac{E_-}{E_+} = \frac{Z_L - Z_0}{Z_L + Z_0} \quad \text{unitless} \tag{C.2}$$

Similarly, we define the *transmission coefficient*:

$$\tau \equiv \frac{E_L}{E_+} = \frac{2 Z_0}{Z_L + Z_0} \tag{C.3}$$

The reflection and transmission coefficients have both magnitude and phase. Note that

$$\tau - \rho = 1 \tag{C.4}$$

We can now write the *VSWR* in terms of the reflection coefficient:

$$VSWR = \frac{1.0 + |\rho|}{1.0 - |\rho|} \quad \text{(unitless)} \tag{C.5}$$

The return loss is the ratio of the reflected power to the incident power:

$$return\ loss = -20 \log_{10}|\rho| \quad \text{(dB)} \tag{C.6a}$$

$$= -20 \log_{10}\frac{VSWR - 1}{VSWR + 1} \tag{C.6b}$$

In the above equations $Z_0$ is the characteristic impedance of the system.

## EXAMPLE 1:    PERFECT MATCH

For a perfect match, $Z_L = Z_0$ and we have no reflected power, so using (C1), (C2), and (C6):

$\rho = 0$

$VSWR = 1.0{:}1$

$return\ loss = \infty$

## EXAMPLE 2:    SHORT CIRCUIT

For a short circuit, $Z_L = 0$ and all of the power is reflected, so

$\rho = -1$

$VSWR = \infty{:}1$

$return\ loss = 0$ dB

## EXAMPLE 3:    OPEN CIRCUIT

For an open circuit, $Z_L = \infty$ and all of the power is again reflected, so

$\rho = +1$

$VSWR = \infty{:}1$

$return\ loss = 0$ dB

Table C.1 summarizes the relationships for a range of cases

| $Z_L$ | $\rho$ | Return Loss | VSWR |
|---|---|---|---|
| 0 | $-1.00$ | 0 dB | $\infty{:}1$ |
| $.5\,Z_0$ | $-.33$ | 9.6 | 2.0:1 |
| $1.0\,Z_0$ | 0.00 | $\infty$ | 1.0:1 |
| $1.1\,Z_0$ | $+.05$ | 26.4 | 1.1:1 |
| $2\,Z_0$ | $+.33$ | 9.6 | 2.0:1 |
| $10\,Z_0$ | $+.82$ | 1.7 | 10.0:1 |
| $100\,Z_0$ | $+.98$ | .2 | 100:1 |
| $\infty$ | $+1.00$ | 0 | $\infty{:}1$ |

**Table C.1:** **Impedance, $\rho$, Return Loss, and VSWR**

Some other useful relations are:

$$|\rho| = 10^{-\text{return loss}/20} \tag{C.7}$$

$$VSWR = \frac{1 + 10^{-\text{return loss}/20}}{1 - 10^{-\text{return loss}/20}} = \begin{cases} \frac{Z_L}{Z_0} & \text{iff } Z_0 < Z_L \\ \frac{Z_0}{Z_L} & \text{iff } Z_0 \geq Z_L \end{cases} \tag{C.8}$$

# *Index*

*RF Design Guide: Systems, Circuits, and Equations,* Peter Vizmuller

*The RF and Microwave Circuit Design Handbook,* Stephen A. Maas

*RF and Microwave Coupled-Line Circuits,* Rajesh Mongia, Inder Bahl, and Prakash Bhartia

*RF Power Amplifiers for Wireless Communications,* Steve C. Cripps

*RF Systems, Components, and Circuits Handbook,* Ferril Losee

*TRAVIS Pro: Transmission Line Visualization Software and User's Manual, Professional Version,* Robert G. Kaires and Barton T. Hickman

*Understanding Microwave Heating Cavities,* Tse V. Chow Ting Chan and Howard C. Reader

For further information on these and other Artech House titles, including previously considered out-of-print books now available through our In-Print-Forever® (IPF®) program, contact:

| Artech House | Artech House |
|---|---|
| 685 Canton Street | 46 Gillingham Street |
| Norwood, MA 02062 | London SW1V 1AH UK |
| Phone: 781-769-9750 | Phone: +44 (0)20 7596-8750 |
| Fax: 781-769-6334 | Fax: +44 (0)20 7630 0166 |
| e-mail: artech@artechhouse.com | e-mail: artech-uk@artechhouse.com |

Find us on the World Wide Web at:
www.artechhouse.com

1.  Correct (4.5.1.47) p. 203 to read the following:

$$\varepsilon_{eff}(0) = 0.5\,(\varepsilon_r + 1.0) + 0.5\,(\varepsilon_r - 1.0)\left(1.0 + \frac{10.0}{v}\right)^{-a_c(v)\,b_c(\varepsilon_r)} \tag{4.5.1.42}$$

The equation given as (4.5.1.42) should be deleted and (3.5.1.2) and (3.5.1.3) should be used instead.

Thanks to Marvin Green.

2.  The equations for $Q_0$ were omitted in section 4.5.1 beginning on p. 199. These are:

$$Q_0 = R_7\left[1.0 - \frac{1.1241\,R_{12}}{R_{16}}\,e^{\left(-0.026\,f_n^{1.15656} - R_{15}\right)}\right] \tag{4.5.1.3a}$$

$$R_1 = 0.03891\,\varepsilon_r^{1.4}$$

$$R_2 = 0.267\,u^{7.0}$$

$$R_7 = 1.206 - 0.3144\,e^{-R_1}\left[1.0 - e^{-R_2}\right]$$

$$R_{10} = 0.00044\,\varepsilon_r^{2.136} + 0.0184$$

$$R_{11} = \frac{\left(\dfrac{f_n}{19.47}\right)^{6.0}}{\left[1.0 + 0.0962\left(\dfrac{f_n}{19.47}\right)^{6.0}\right]}$$

$$R_{12} = \frac{1.0}{1.0 + 0.00245\,u^2}$$

$$R_{15} = 0.707\,R_{10}\left(\frac{f_n}{12.3}\right)^{1.097}$$

$$R_{16} = 1.0 + 0.0503\,\varepsilon_r^2\,R_{11}\left\{1.0 - \exp\left[-(u/15)^6\right]\right\}$$

Thanks to Marvin Green.

3.  Equation (4.5.1.13) p. 200 is missing a subscript. It should read:

$$Z_{0,e}(f_n) = \text{Error!)} \tag{4.5.1.13}$$

Thanks to Marvin Green.

4.    Several clarifications should be added to section 4.5.1 and an omission must be corrected. $Z_0(f)$ i
the frequency-dependent single-strip characteristic impedance. The permittivity $\varepsilon_{eff}(f_n)$ is the single-strip
frequency-dependent permittivities. The dc values are $Z_0(0)$ and $\varepsilon_{eff}(0)$.

$\varepsilon_{eff}(f_n) = $ (3.5.1.7)

$\varepsilon_{eff}(0) = $ (3.5.1.2) and (3.5.1.3)

$Z_0(0) = $ (3.5.1.1)

$Z_0(f_n) = Z_0(0) \, (R_{13} / R_{14})^{R_{17}}$

$R_{17} = Q_0$ (4.5.1.3a)

$R_{13} = 0.9408 \, \varepsilon_{eff}(f_n)^{R_8} - 0.9603$

$R_{14} = (0.9408 - R_9) \, \varepsilon_{eff}(0)^{R_8} - 0.9603$

$R_8 = 1.0 + 1.275 \left\{ 1.0 - \exp\left[ -0.004625 \, R_3 \, \varepsilon_r^{1.674} \, (f_n / 18.365)^{2.745} \right] \right\}$

$R_9 = 5.086 \, R_4 \dfrac{R_5}{0.3838 + 0.386 \, R_4} \dfrac{e^{-R_6}}{1.0 + 1.2992 \, R_5} \dfrac{(\varepsilon_r - 1.0)^6}{1.0 + 10.0 \, (\varepsilon_r - 1.0)^6}$

$R_3 = 4.766 \, \exp(-3.228 \, u^{0.641})$

$R_4 = 0.016 + (0.0514 \, \varepsilon_r)^{4.524}$

$R_5 = (f_n / 28.843)^{12.0}$

$R_6 = 22.20 \, u^{1.92}$

Thanks to Marvin Green.

5.    Equation (4.5.1.21) on page 201 is typeset poorly. For clarity, it is:

$$Q_{12} = 2.121 \frac{(f_n / 20)^{4.91}}{\left[ 1.0 + Q_{11} \, (f_n / 20)^{4.91} \right]} \exp(-2.87 \, g) \, g^{0.902} \quad (4.5.1.21)$$

Thanks to Marvin Green.

6. Page 207 is missing a reference for section 4.5.1:

[7a] Jansen, Rolf, and Martin Kirschning "Arguments and an Accurate Model for the Power-Current Formula of Microstrip Characteristic Impedance," *Archiv Für Elektronik und Übertragungstechnik*, Band 37, Heft 3/4, 1983, pp. 108–112.

Thanks to Marvin Green.

7. In section 12.3.5, pp. 482–485 a secondary source was referenced for the techniques of combining two-ports. The original development of this technique was carried out by Les Besser which is described in his original papers (add to References page 485):

[0] Besser, Les, "A Fast Computer Routine to Design High Frequency Circuits," *IEEE International Conference on Communications,* San Francisco, June 8–10, 1970, pp. 24-13–24-16.

[0a] Besser, Les, and R.W. Newcomb, "A Scattering Matrix Program for High Frequency Circuit Analysis," *Proceedings of the IEEE Conference on Systems, Networks, and Computers,* Mexico, January 1971, pp. 512–515.

Thanks to Les Besser.

8. In equation (4.5.3.43), page 213 the function should be sin() rather than csc(). It should read:

$$E = 0.725 \sin\left(\frac{\pi\, y}{2.0}\right) \tag{4.5.3.43}$$

Thanks to Doug White.

9. In equation (3.5.1.7) on page 95 the superscript is missing. It should read:

$$P(f) = P_1\, P_2 \left[ (0.1844 + P_3\, P_4)\, (10.0\; f h) \right]^{1.5763}$$

The second line of 3.5.1.1 should read:

with ([28], see also [35]):

Thanks to A.J. DeKremer.

10. In equation (3.5.1.3) page 94 "12.o" should read "12.0".

11. On page 386, in equation (6.3.1.2) $d$ is not defined. It is

$$d = \sqrt{4.0\, a^2 + b^2} \tag{6.3.1.3a}$$

Thanks to Bob Stengel.

12. Equation (3.4.1.5), p. 74 should read

$$k_t' = \sqrt{1.0 - k_t^2}$$

Thanks to Paul Levin.

13.　　On page 133 equations (3.6.3.21)–(3.6.3.30)

$$h_2 > h_1$$

Figures 3.6.3.2 and 3.6.3.3 appear to have $h_1 > h_2$ which is not correct.
Thanks to Donald Ingram.

14.　　On page 126 and 127, the variable $b$ in (3.6.2.1), (3.6.2.4), (3.6.2.5), (3.6.2.6), and (3.6.2.7) should be replaced with $(b - t)$ to improve accuracy for
$t \varnothing b$. Equations (3.6.2.1–6) are due to [9]. Equation (3.6.2.2) becomes:

$$b = 2.0 \, h + t \tag{3.6.2.2}$$

Thanks to Steve Hauptman.

15.　　The last line on page 23 should read "and we calculate the guide wavelengths of the fundamental and harmonics as (2.3.12):"

16.　　On page 479 the $s$-parameter suffixes are scrambled. Equations (12.3.1.4–6) should read:

$$s_{21} = \frac{b_2}{a_1} \bigg|_{a_2 = 0} \tag{12.3.1.4}$$

$$s_{22} = \frac{b_2}{a_2} \bigg|_{a_1 = 0} \tag{12.3.1.5}$$

$$s_{12} = \frac{b_1}{a_2} \bigg|_{a_1 = 0} \tag{12.3.1.6}$$

Thanks to Les Besser.

17.　　In the software example on page 470 constants CONST1 and CONST2 are undefined. They are defined in the program header with
```
#define CONST1 0.57215665
#define CONST2 0.63661978
```

Thanks to Robert Lafevre.

18.　　Figure 5.9.1.3 on page 371 is missing captions. The upper arrow should be labeled microstrip lin and the lower is pointing to the ground plane.

19.　　A new relation for via inductance was presented recently in the literature. Numerical simulation and measurements show good agreement with a relatively simple closed-form equation. The interested reader should refer to:

[4a]　　Goldfarb, Marc E., and Robert A. Pucel, "Modeling Via Hole Grounds in Microstrip," *IEEE Microwave and Guided Wave Letters,* Vol.1, No. 6, June 1991, pp. 135–137.

20.    A number of people have requested information on calculation of complete elliptic integrals. A detailed example is attached as the last page of this document.

21.    In the solution to the example on pp. 75–76 the answer should read:

$$b = 0.0221 \text{ in.} = 0.0561 \text{ cm}$$
$$\varepsilon_{eff} = 4.45$$

22.    On page 310, Figure 5.5.10.2 should be

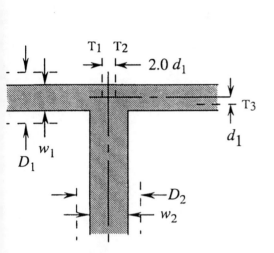

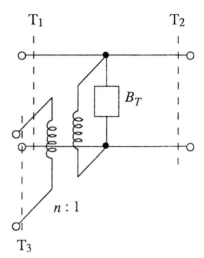

Thanks to Robert Lafevre.

23.    On page 126, Equation (3.6.1.1) should read:

$$Z_0 = \frac{h_0}{4.0\sqrt{\varepsilon_r}} \frac{K(k)}{K(k')}$$

(3.6.1.1)

Thanks to Juan Milton Garduño.

24.    On page 475, Equation (12.2.4.2a) has a variable $z$. Replace $z$ with $x$ everywhere.

25.    On page 79, Equations (3.4.2.1) and (3.4.2.2) do not define $\varepsilon_1$, and $\varepsilon_2$. The replacement figure and equations below define them:

$$Z_0 = \frac{\eta_0}{\left(\sqrt{\varepsilon_{r1}} + \sqrt{\varepsilon_{r2}}\right)} \frac{K(k)}{K'(k)} \quad (\bullet)$$

(3.4.2.1)

$$\varepsilon_{eff} = \frac{\left(\sqrt{\varepsilon_{r1}} + \sqrt{\varepsilon_{r2}}\right)^2}{4.0}$$

(3.4.2.2)

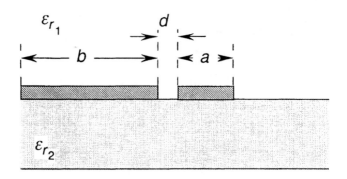

Figure 3.4.2.1:   Micro-Coplanar Stripline

26.     In section 4.4.3 $\beta_1$ is undefined. The necessary relations are:

$$\beta_1 = \sqrt{\frac{1.0 - y^2}{1.0 - k_1^2 \, y^2}}$$
(4.4.3.13)

$$y = \frac{s}{s + 2.0 \, w}$$
(4.4.3.14)

$$k_1 = \frac{s + 2.0 \, w}{s + 2.0 \, w + 2.0 \, d}$$
(4.4.3.15)

27.     On page 79, the equation is correct but does not follow the format used elsewhere in the book. Replace Equation (3.4.3.1) with

$$Z_0 = \frac{h_0}{2.0 \sqrt{\varepsilon_{eff}}} \frac{1.0}{\dfrac{K(k)}{K(k')} + \dfrac{K(k_1)}{K(k_1')}}$$
(3.4.3.1)

28.     On page 233, the first line references Cohn [3]. The correct reference is Cohn [2].

29.     The equations for three coplanar strips (Section 3.4.8) have errors present in the original reference. The corrections are:

$$Z_0 = \frac{\eta_0}{4.0 \sqrt{\varepsilon_{eff}}} \frac{1.0}{\dfrac{K(k_1')}{K(k_1)} + \dfrac{t}{b - a}} \qquad (\bullet)$$
(3.4.8.1)

$$\varepsilon_{eff} = 1.0 + \frac{(\varepsilon_r - 1.0)}{2.0} \frac{\dfrac{K(k_2')}{K(k_2)}}{\dfrac{K(k_1')}{K(k_1)} + \dfrac{t}{b - a}}$$
(3.4.8.2)

Thanks to Paul Levin, and Stuart M. Wentworth.

30.     On page 204, (4.5.1.59) should read:

$$P_4 = 1.0 + 2.751 \left[ 1.0 - e^{-(\varepsilon_r / 15.916)^8} \right] \tag{4.5.1.59}$$

1.      On page 71, in (3.3.5.1) the variable $\varepsilon$ is $e$, the Napierian constant (2.71828...).

2.      On page 246, the equation for $k_2$, (4.8.1.8), should read (sign change in denominator):

$$k_2 = \sqrt{\frac{1.0 + \tan^2 \vartheta_2'}{1.0 + \tan^2 \vartheta_1'}} \tag{4.8.1.8}$$

Note, that in (4.8.1.6), (4.8.1.8), (4.8.1.9), and (4.8.1.10) the function tg is the tangent and th is the hyperbolic tangent.

Thanks to Santana Roy, Derek Tong, and Yunyi Wang.

3.      On page 158, (3.9.6.7) the first line should read (change to first coefficient in numerator of third term):

$$= \left\{ 1.045 - 0.365 \ln \varepsilon_r + \frac{6.3 \, (w/h) \, \varepsilon_r^{0.945}}{238.64 + 100.0 \, w/h} \right.$$

Equations (3.9.6.1-4) were incorrectly converted from $\log_{10}$ to ln and should read:

$$Z_0 = 72.62 - 15.283 \ln \varepsilon_r + \frac{50.0 \, (w/h - 0.02)(w/h - 0.1)}{w/h}$$

$$+ \, 0.434 \ln \left( w/h \infty 10^2 \right) \left( 44.28 - 8.503 \ln \varepsilon_r \right)$$

$$- \left[ 0.139 \ln \varepsilon_r - 0.11 + w/h \, (0.465 \ln \varepsilon_r + 1.44) \right]$$

$$\infty \left( 11.4 - 2.636 \ln \varepsilon_r - h/\lambda \infty 10^2 \right)^2 \tag{3.9.6.1}$$

$$\varepsilon_{eff} = \left( \frac{\lambda_g}{\lambda_0} \right)^{-2}$$

$$= \left[ 0.923 - 0.195 \ln \varepsilon_r + 0.2 \, w/h \right.$$

$$\left. - \, (0.126 \, w/h + 0.020) \ln (h/\dot{\lambda}_0 \infty 10^2) \right]^{-2} \tag{3.9.6.2}$$

and for $0.2 \bullet w/h \bullet 1.0$

$$Z_0 = 113.19 - 23.257 \ln \varepsilon_r + 1.25 \, w/h \, (114.59 - 22.531 \ln \varepsilon_r)$$

$$+ \, 20.0 \, (w/h - 0.2) \, (1.0 - w/h)$$

$$- \left[ 0.15 + 0.0999 \ln \varepsilon_r + w/h \, (-0.79 + 0.899 \ln \varepsilon_r) \right]$$

$$\infty \left\{ \left[ 10.25 - 2.171 \ln \varepsilon_r + w/h \, (2.1 - 0.617 \ln \varepsilon_r) - h/\lambda \, \infty \, 10^2 \right]^2 \right\} \tag{3.9.6.3}$$

$$\varepsilon_{eff} = \left( \frac{\lambda_g}{\lambda_0} \right)^{-2}$$

$$= \left[ 0.987 - 0.210 \ln \varepsilon_r + w/h \, (0.111 - 0.0022 \, \varepsilon_r) \right.$$

$$\left. - (0.0525 + 0.0408 \, w/h - 0.00139 \, \varepsilon_r) \ln(h/\lambda \, \infty \, 10^2) \right]^{-2} \tag{3.9.6.4}$$

Thanks to Ján Zehentner.

33.    On page 250, (4.8.2.4) the variables $H$ and $P$ should be $H_e$ and $P_e$. Equation (4.8.2.10) should be the equation for $T_e$ rather than $P_e$.

34.    On page 347, the $\lambda$ in equation (5.6.2.3) should be $\lambda_g$, the guide wavelength.

Thanks to Dave Mesaros.

35.    On page 130, equation (3.6.3.7) should read:

$$C_f = \frac{\varepsilon_r \varepsilon_0}{p} \left\{ 2.0 \ln \left[ \frac{1.0}{g(b-g)} \right] + \frac{1.0}{g(b-g)} \left[ F \left( \frac{t}{2.0 \, b} \right) - F(c_f/ \, b) \right] \right\}$$

36.    On page 495, *Appendix B Common Questions and Answers* the first sentence of the third question's answer should be corrected to read:

Unless one conductor is completely shielded, it is impossible to have no (–• dB) coupling.

Thanks to Larry Tichauer.

37.    Equation (2.7.1.5) should be:

$$d = \sqrt{\frac{2.0}{\omega \mu \sigma}} = \frac{1.0}{\sqrt{\pi f \mu \sigma}} = \text{skin depth} \tag{2.7.1.5}$$

38.    Correct (4.5.4.1) to read:

$$h = \left( \frac{b_{sl} - s_{sl}}{4.0} \right) \tag{4.5.4.1}$$

Thanks to Dennis Hand.

39.    The second equation in section 12.2.1 Hyperbolic Functions on page 465 should read:

$$e^x = \cosh x + \sinh x$$

Thanks to Anson Whealler.

40. Equations (3.9.6.7-8) have typos, the factors 6.4 and 1.8 should be changed to 6.3 and 1.0:

$$\varepsilon_{eff} = \left(\frac{\lambda_g}{\lambda_0}\right)^{-2}$$

$$= \left\{1.045 - 0.365 \ln \varepsilon_r + \frac{6.3\ (w/h)\ \varepsilon_r^{0.945}}{238.64 + 100.0\ w/h}\right.$$

$$\left. - \left[0.148 - \frac{8.81\ (\varepsilon_r + 0.95)}{100.0\ \varepsilon_r}\right] \ln \frac{h}{\lambda_0}\right\}^{-2} \tag{3.9.6.7}$$

$$Z_0 = 133.0 + 10.34\ (\varepsilon_r - 1.8)^2 + 2.87 \left[2.96 + (\varepsilon_r - 1.582)^2\right]$$

$$\left\{(w/h + 2.32\varepsilon_r - 0.56)\left[(32.5 - 6.67\ \varepsilon_r)(100.0\ h/\lambda_0)^2 - 1.0\right]\right\}^{1/2}$$

$$- \left(684.45\ h/\lambda_0\right)\left(\varepsilon_r + 1.35\right)^2 + 13.23 \left[(\varepsilon_r - 1.722)\ w/\lambda_0\right]^2 \tag{3.9.6.8}$$

Thanks to Stephen Maas.

41. In 5.5.13.6 the numerator of the second term should be "+". In 5.5.13.7 the sign for the last term should be "+".

$$A = \left(\frac{1.0 + a}{1.0 - a}\right)^{2.0\ a} \left[\frac{1.0 + \sqrt{1.0 - \left(\frac{2.0\ w_{eff,1}}{\lambda_1}\right)^2}}{1.0 - \sqrt{1.0 - \left(\frac{2.0\ w_{eff,1}}{\lambda_1}\right)^2}}\right] - \frac{1.0 + 3.0\ a^2}{1.0 - a^2} \tag{5.5.13.6}$$

$$B = \left(\frac{1.0 + a}{1.0 - a}\right)^{a/2.0} \left[\frac{1.0 + \sqrt{1.0 - \left(\frac{2.0\ w_{eff,2}}{\lambda_2}\right)^2}}{1.0 - \sqrt{1.0 - \left(\frac{2.0\ w_{eff,2}}{\lambda_2}\right)^2}}\right] + \frac{3.0 + a^2}{1.0 - a^2} \tag{5.5.13.7}$$

Thanks to Stephen Maas.

42. In (3.2.3.7) the "−" sign should be "+."

$$\alpha_c = \alpha_{c,centered} \left[1.0 + \frac{2.0\ e^2}{k} + \frac{e^2\ k^2}{(k^2 - 1.0)\log k}\right] \quad (\text{dB}/\text{m}) \tag{3.2.3.7}$$

Thanks to Edward Wollack.

43. In (5.5.13.11) swap the words "symmetrical" and "asymmetrical."

$$\Delta = \begin{cases} 1 & \text{for asymmetrical steps} \\ 2 & \text{for symmetrical steps} \end{cases} \qquad (5.5.13.11$$

Thanks to Stephen Maas.

44.     In Table 9.2.4.1 the tan $\delta$ for Kapton should read "0.0025." The deckmal point is missing.

45.     In (5.5.1.1) the equation is for the total inductance (2 $L$) so it should be divided by 2.

$$L/h = \frac{100.0}{2.0} \left( 4.0 \sqrt{w/h} - 4.21 \right) \ (\text{nH}/\text{m}) \qquad (5.5.1.1$$

Also (5.5.1.5) is an equation for $L/h$ not $L$ as indicated.

$$L = 0.22\, h \left[ 1.0 - 1.35\, e^{-0.18(w/h)^{1.39}} \right] \ (\text{nH}) \qquad (5.5.1.5$$

Thanks to C. James.

46.     Equation (5.5.1.4) has an incorrect constant. The constant 5.44 should be 5.64.

$$C = 0.001\, h \left[ (10.35\, \varepsilon_r + 2.5\,)\,(w/h)^2 + (2.6\, \varepsilon_r + 5.64)\,(w/h) \right] \ (\text{pF}) \qquad (5.5.1.4$$

47.     In section 4.3.2 there are two undefined variables, $C$ and $D$. They are:

$$C = s + d \qquad (4.3.2.1.7b$$

$$D = \frac{w - C}{2.0} \qquad (4.3.2.1.7c$$

Thanks to John Beeley.

48.     In (1.9) the $X$ should be subscripted

$$C = \frac{-1.0}{X_C \omega} \qquad (1.9$$

Thanks to Dennis Han.

49.     In (3.4.1.12) the units should read $\cdot/\square$.

50.     In reference [9] in section 6.4.2 the word *Microwave* is misspelled.

51.     Figure 6.4.5.1 is missing notations for $d_o$ and $d_i$ and $a$, $c$, $n$ in the associated equation 6.4.5.1 are undefined. The missing information is:

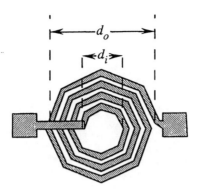

$$a = \frac{d_o + d_i}{4.0}$$

$$c = \frac{d_o - d_i}{2.0}$$

and $n$ is the number of turns.

52. In (6.5.1.1) the units of $h$ are cm. The reference should be [3]. The reference of table 6.5.1 should be [2].

Thanks to Walid Ali-Ahmad.

53. Equation (6.5.2.1) gives $\mu$H when $a$ and $R$ are in cm.

54. Equation (6.2.1.2) yields a unitless number for $x$ in cm.

55. Equations (6.2.2.1), (6.2.2.2), and (6.2.2.3) are in $\mu$H.

56. Equation (6.2.3.1) a parenthesis is in the wrong place. It should read:

$$L = 0.002 \, l \left[ \ln \left( \frac{2.0 \, l}{\rho_1} \right) + \ln \zeta - 1.0 \right] \quad (\mu H) \tag{6.2.3.1}$$

# Example of calculating complete elliptic integrals of the first, K(k) and second kind E(k)

B.C. Wadell    REV 911228    911228

For example, let's calculate K(0.3420)    (complete elliptic integral of the first kind)

using equations of 12.2.2.1, p. 465:

| iteration n | an | bn | cn | K(k) at iteration N = n |
|---|---|---|---|---|
| 0 | 1 | 0.939693 | 0.34202 | 1.57079 63268 |
| 1 | 0.969846 | 0.969377 | 0.030154 | 1.61963 42758 |
| 2 | 0.969612 | 0.969612 | 0.000234 | 1.62002 58752 |
| 3 | 0.969612 | 0.969612 | 1.42E-08 | 1.62002 58989 |
| 4 | 0.969612 | 0.969612 | 0 | 1.62002 58989 |
| 5 | 0.969612 | 0.969612 | 0 | 1.62002 58989 |
| 6 | 0.969612 | 0.969612 | 0 | 1.62002 58989 |

To check with the CRC values we need to calculate
arcsin(k) = 0.349066   radians =    20  degrees

As you can see, this agrees to 4 places with the values given in the CRC p. 433 after only a few iterations.
Similarly, to calculate E(0.4226):    (complete elliptic integral of the second kind)

| iteration n | an | bn | cn | 2**n | F(k) at iteration N = n | E(k) at iteration N = n |
|---|---|---|---|---|---|---|
| 0 | 1 | 0.906308 | 0.422618 | 1 | 1.57079 63268 | 1.43051 93524 |
| 1 | 0.953154 | 0.952002 | 0.046846 | 2 | 1.64799 86449 | 1.49721 06366 |
| 2 | 0.952578 | 0.952578 | 0.000576 | 4 | 1.64899 50658 | 1.49811 47932 |
| 3 | 0.952578 | 0.952578 | 8.71E-08 | 8 | 1.64899 52165 | 1.49811 49302 |
| 4 | 0.952578 | 0.952578 | 2E-15 | 16 | 1.64899 52165 | 1.49811 49302 |
| 5 | 0.952578 | 0.952578 | 0 | 32 | 1.64899 52165 | 1.49811 49302 |
| 6 | 0.952578 | 0.952578 | 0 | 64 | 1.64899 52165 | 1.49811 49302 |

1.49811 49302

Again, to check against the CRC we need a different parameter, so we calculate
arcsin(k) = 0.436332   radians =    25  degrees

and again, we see that this agrees to 4 places with the CRC p. 435 after only a few iterations.
Compare Abromowitz and Stegun, Tables 17.2, pp. 610-611. Agreement is in 8-9 places.

Printed in the United States
61121LVS00003B/2